Physics
and
Combinatorics
2000

Proceedings of the Nagoya 2000 International Workshop

Physics and Combinatorics 2000

Graduate School of Mathematics, Nagoya University
21–26 August, 2000

Editors

Anatol N. Kirillov
Professor of Graduate School of Mathematics,
Nagoya University

Nadejda Liskova
PhD in Physics and Mathematics

World Scientific
Singapore • New Jersey • London • Hong Kong

Published by

World Scientific Publishing Co. Pte. Ltd.
P O Box 128, Farrer Road, Singapore 912805
USA office: Suite 1B, 1060 Main Street, River Edge, NJ 07661
UK office: 57 Shelton Street, Covent Garden, London WC2H 9HE

British Library Cataloguing-in-Publication Data
A catalogue record for this book is available from the British Library.

Cover: Illustration by Nadya Kirillova

PHYSICS AND COMBINATORICS
Proceedings of the Nagoya 2000 International Workshop

ISBN 981-02-4642-0

Printed in Singapore by World Scientific Printers

PREFACE

This volume contains the Proceedings of the Workshop "Physics and Combinatorics" held at the Graduate School of Mathematics, Nagoya University, Japan, during August 21–26, 2000. The workshop organizing committee consisted of Kazuhiko Aomoto, Fumiyasu Hirashita, Anatol Kirillov, Ryoichi Kobayashi, Akihiro Tsuchiya, and Hiroshi Umemura.

This is the second Workshop in a series of workshops with common title "Physics and Combinatorics" which have been held at the Graduate School of Mathematics, Nagoya University. The first one was held in 1999 and happened to be successful. In the preface to the Proceedings of the Workshop "Physics and Combinatorics, Nagoya 1999" (World Scientific Publisher, 2000) we had explained the purpose and ideas behind the Workshop. Here we would like to repeat:

"The purpose of the Workshop in Nagoya was to get together a group of scientists actively working in Combinatorics, Representation Theory, Special Functions, Number Theory, and Mathematical Physics to acquaint the participants with some basic results in their fields and discuss existing and possible interactions between the mentioned subjects."

The present volume contains contributions on

- algebra–geometric approach to the representation theory;
- algebraic and tropical combinatorics;
- birational representations of affine symmetric group and integrable systems;
- Grothendieck polynomials;
- (t, q)–analogue of characters of finite dimensional representations of quantum affine algebras;
- quantum Teichmüller theory;
- complex reflections groups;
- Integrable systems: quantum Cologero–Moser models;
- Statistical Physics: Bethe ansatz, exclusion statistics, ...

The Workshop "Physics and Combinatorics, Nagoya 2000" appeared to be successful, and we hope both respective researchers and graduate students can find many interesting and useful facts and results in this volume of Proceedings.

Organizers would like to take an opportunity and to thank all participants of the Workshop, all contributors to this volume, and anonymous referees for their speedy work.

Anatol Kirillov
Nadejda Liskova

CONTENTS

VANISHING THEOREMS AND CHARACTER FORMULAS FOR THE HILBERT SCHEME OF POINTS IN THE PLANE (ABBREVIATED VERSION)

MARK HAIMAN
Dept. of Mathematics, U.C. San Diego
La Jolla, CA, 92093-0112
E-mail: mhaiman@macaulay.ucsd.edu

In an earlier paper,[13] we showed that the Hilbert scheme of points in the plane $H_n = \mathrm{Hilb}^n(\mathbb{C}^2)$ can be identified with the Hilbert scheme of regular orbits $\mathbb{C}^{2n}//S_n$. Using our earlier result and a recent result of Bridgeland, King and Reid,[4] we prove vanishing theorems for tautological bundles on the Hilbert scheme. We apply the vanishing theorems to establish the conjectured character formula for diagonal harmonics of Garsia and the author.[8] In particular we prove that the dimension of the space of diagonal harmonics is

$$(n+1)^{n-1}.$$

This is a preliminary report. We state the main results and outline the proofs. Detailed proofs, a more systematic study of the applications, and a fuller exposition will be given in a future publication.

1 Introduction

In an earlier paper,[13] we showed that the Hilbert scheme of points in the plane $H_n = \mathrm{Hilb}^n(\mathbb{C}^2)$ can be identified with the Hilbert scheme of regular orbits $\mathbb{C}^{2n}//S_n$ for the action of S_n permuting the factors in the Cartesian product $\mathbb{C}^{2n} = (\mathbb{C}^2)^n$. This identification gives rise to two different *tautological vector bundles* on the Hilbert scheme. From the universal family

$$\begin{array}{c} F \subseteq H_n \times \mathbb{C}^2 \\ {\scriptstyle\pi}\big\downarrow \\ H_n \end{array} \tag{1}$$

we get the usual tautological bundle B of rank n, whose sheaf of sections is $\pi_* \mathcal{O}_F$, the sheaf of regular functions on F, pushed down to H_n. The Hilbert scheme of S_n-orbits also has a universal family

$$\begin{array}{l} X_n \subseteq (\mathbb{C}^{2n}//S_n) \times \mathbb{C}^{2n} \\ {\scriptstyle\rho}\big\downarrow \\ H_n = \mathbb{C}^{2n}//S_n \end{array} \tag{2}$$

giving rise to a bundle P of rank $n!$ with sections $\rho_* \mathcal{O}_{X_n}$. This bundle P carries an S_n action affording the regular representation on every fiber.

Here we give a vanishing theorem for the higher sheaf cohomology of all bundles of the form $P \otimes B^{\otimes l}$ on H_n. We also identify the space of global sections of $P \otimes B^{\otimes l}$ with the coordinate ring $R(n, l)$ of a subspace arrangement $Z(n, l)$ in $\mathbb{C}^{2n+2l}$, called a *polygraph*. The polygraph was introduced in ([13]), where we used a freeness property of its coordinate ring as the main technical tool to derive other results. Here we shall see more clearly the nature of the link between the polygraph and the Hilbert scheme, and the reason why the former carries geometric information about the latter.

The trivial bundle $\mathcal{O}_{H_n}$ occurs as a direct summand of P, and the natural ample line bundle $\mathcal{O}_{H_n}(1)$ is the highest exterior power of B. As special cases of our vanishing theorem we therefore recover the previously known vanishing theorems of Danila[9] for the tautological bundle B and of Kumar and Thomsen[14] for the ample line bundles $\mathcal{O}_{H_n}(k)$, $k > 0$.

From our first vanishing theorem we derive a second, giving higher cohomology vanishing for the bundles $P \otimes B^{\otimes l}$ over the zero fiber H_n^0 in H_n. Again we explicitly identify the space of global sections. For $l = 0$ in particular, we find that the space of global sections $H^0(H_n^0, P)$ is isomorphic to the ring R_n of coinvariants for the "diagonal" action of S_n on $\mathbb{C}^{2n}$, or equivalently to the space of harmonics for that action.

The coinvariant ring R_n was the subject of a series of conjectures, presented in ([10]), relating its character as a doubly graded S_n-module to enumerations of various well-known combinatorial objects, such as trees and parking functions. In the spring of 1992, Garsia and the author discussed these conjectures with C. Procesi. Procesi suggested that it might be possible to determine the dimension and character of R_n by identifying it as the space of global sections of a vector bundle on H_n^0. Following this suggestion, the author obtained a formula expressing the doubly graded character of R_n in terms of Macdonald polynomials, assuming the validity of suitable geometric hypotheses. Subsequently, Garsia and the author[8] showed that the combinatorial conjectures would follow from this master formula. Using the results in ([13]) together with the second vanishing theorem here, we can now prove the geometric hypotheses needed to justify the formula. As a particular consequence, we obtain the dimension formula

$$\dim R_n = (n+1)^{n-1}. \tag{3}$$

Another consequence is that the q, t-Catalan polynomial $C_n(q, t)$ studied in ([8]) and ([11]) is the Hilbert series of the S_n-alternating part of R_n, and therefore has non-negative coefficients. This has also been proven very recently by Garsia and Haglund,[7] who gave a remarkably simple combinatorial interpretation of $C_n(q,t)$.

Our methods can be applied to show that other expressions related to the character formula for R_n are character formulas for suitable doubly-graded S_n-modules, and hence have non-negative coefficients. This establishes some positivity conjectures made in ([2]), as will be explained in the expanded version of this report.

Now we state our primary vanishing theorem, whose proof is discussed in Section 4.

Theorem 1 *For all l we have*

$$H^i(H_n, P \otimes B^{\otimes l}) = 0 \quad \textit{for } i > 0, \tag{4}$$

and

$$H^0(H_n, P \otimes B^{\otimes l}) = R(n,l), \tag{5}$$

where $R(n,l)$ is the coordinate ring of the polygraph $Z(n,l) \subseteq \mathbb{C}^{2n+2l}$.

To properly explain the meaning of the identity in (5) we must review the definition of the polygraph. It was defined in ([13]) as a certain union of linear subspaces in $\mathbb{C}^{2n+2l}$, but it is better here first to describe it from a Hilbert scheme point of view. Let

$$Z = X_n \times F^l / H_n \tag{6}$$

be the fiber product over H_n of X_n and l copies of F. Since X_n is a closed subscheme of $H_n \times \mathbb{C}^{2n}$ and F is a closed subscheme of $H_n \times \mathbb{C}^2$, we have that Z is a closed subscheme of $H_n \times \mathbb{C}^{2n} \times (\mathbb{C}^2)^l = H_n \times \mathbb{C}^{2n+2l}$. The polygraph $Z(n,l)$ is the image of the projection of Z on the factor $\mathbb{C}^{2n+2l}$.

Next we describe $Z(n,l)$ in elementary terms. To each point $I \in H_n$ of the Hilbert scheme there corresponds a subscheme $V(I) \subseteq \mathbb{C}^2$ of the plane. Counting the points of $V(I)$ with multiplicities we get a 0-cycle $\sigma(I) = \sum m_i P_i$ of total weight $\sum m_i = n$. We may identify $\sigma(I)$ with a multiset, or unordered n-tuple with possible repetitions, of points in the plane $\sigma(I) = [\![P_1, \ldots, P_n]\!]$. This defines a projective morphism

$$\sigma\colon H_n \to S^n\mathbb{C}^2 = \mathbb{C}^{2n}/S_n, \tag{7}$$

called the *Chow morphism.* A point of X_n is just a tuple $(I, P_1, \ldots, P_n)$, such that $\sigma(I)$ is equal to the unordered n-tuple $[\![P_1, \ldots, P_n]\!]$. A point of F is a pair (I, Q) such that $Q \in V(I)$. Hence a point of Z is a tuple

$$(I, P_1, \ldots, P_n, Q_1, \ldots, Q_l) \tag{8}$$

such that $\sigma(I) = [\![P_1, \dots, P_n]\!]$ and

$$Q_i \in \{P_1, \dots, P_n\} \quad \text{for all } 1 \leq i \leq l. \tag{9}$$

Projecting on $\mathbb{C}^{2n+2l}$, we see that $Z(n,l)$ is the set of points $(P_1, \dots, P_n, Q_1, \dots, Q_l) \in \mathbb{C}^{2n+2l}$ satisfying (9).

Given a global regular function on $Z(n,l)$, we may compose it with the projection $Z \to Z(n,l)$ to get a global regular function on Z, which is the same thing as a global section of $P \otimes B^{\otimes l}$. Hence we have a natural injective ring homomorphism

$$R(n,l) \hookrightarrow H^0(H_n, P \otimes B^{\otimes l}) = \mathcal{O}(Z), \tag{10}$$

where $R(n,l) = \mathcal{O}(Z(n,l))$ is the coordinate ring of the polygraph. The meaning of the equal sign in (5) is that this homomorphism is an isomorphism.

For the character formula application we will need a vanishing theorem along the lines of Theorem 1, but for the restriction of the tautological bundles to the *zero fiber* $H_n^0 = \sigma^{-1}(0)$ in the Hilbert scheme H_n. In ([11]) we showed that the set-theoretic zero fiber, with the induced reduced subscheme structure, is the same as scheme-theoretic zero fiber $\sigma^{-1}(0)$, so there is no ambiguity in the definition of H_n^0. We also proved there that H_n^0 is Cohen-Macaulay, and constructed an explicit locally free resolution of its structure sheaf as a sheaf of $\mathcal{O}_{H_n}$ modules. As will be explained in Section 4, the terms of the resolution are essentially exterior powers of B. Combined with Theorem 1, this yields the following result.

Theorem 2 *For all l we have*

$$H^i(H_n^0, P \otimes B^{\otimes l}) = 0 \quad \textit{for } i > 0, \tag{11}$$

and

$$H^0(H_n^0, P \otimes B^{\otimes l}) = R(n,l)/I, \tag{12}$$

where $R(n,l)$ is the polygraph coordinate ring and $I = \mathfrak{m}R(n,l)$ is the ideal generated by the homogeneous maximal ideal $\mathfrak{m}$ in the subring $\mathbb{C}[\mathbf{x},\mathbf{y}]^{S_n} \subseteq R(n,l)$. Here $\mathbb{C}[\mathbf{x},\mathbf{y}] = \mathbb{C}[x_1, y_1, \dots, x_n, y_n]$ is the coordinate ring of $\mathbb{C}^{2n}$.

Again let us make a few clarifying remarks. As explained above, there is a geometrically natural homomorphism $R(n,l) \to H^0(H_n, P \otimes B^{\otimes l})$. Restricting global sections of $P \otimes B^{\otimes l}$ from H_n to the zero fiber, we get a homomorphism

$$R(n,l) \to H^0(H_n^0, P \otimes B^{\otimes l}), \tag{13}$$

which *a priori* might not be surjective. The content of (12) is that this homomorphism is surjective and its kernel is the ideal I. We remark that it is obvious that the ideal I is contained in the kernel of the homomorphism in

(13), since it is the ideal of the scheme-theoretic fiber over 0 of the projection $Z(n,l) \to S^n\mathbb{C}^2$. What is not obvious, but is so according to the theorem, is that the kernel is not larger than this.

2 The Bridgeland–King–Reid theorem

We derive our main results using our earlier results on Hilbert schemes and polygraphs, combined with a recent general theorem of Bridgeland, King and Reid concerning Hilbert schemes of orbits $M//G$ such that $M//G$ is what is known as a crepant resolution of singularities of M/G. It is well-known that H_n is a crepant resolution of $\mathbb{C}^{2n}/S_n$ via the Chow morphism. Therefore, as a consequence of our identification of $\mathbb{C}^{2n}//S_n$ with H_n, the Bridgeland–King–Reid theorem applies in the case $M = \mathbb{C}^{2n}$, $G = S_n$.

Let us describe the relevant set-up and state their theorem. In general, M is a non-singular complex projective variety and G is a finite group of automorphisms of M such that the induced action on the canonical sheaf ω_M is locally trivial. This implies that M/G has Gorenstein singularities, with canonical sheaf ω descended from ω_M. In our situation G will act linearly on a complex vector space M, so the condition means that G is a subgroup of $\mathrm{SL}(M)$. More specifically, we are interested in the action of S_n on $\mathbb{C}^{2n}$. This is a subgroup of $\mathrm{SL}(2n)$, since every element $w \in S_n$ acts with determinant $\epsilon(w) = \pm 1$ on $\mathbb{C}^n$, and therefore with determinant $\epsilon(w)^2 = 1$ on $\mathbb{C}^{2n} = \mathbb{C}^n \oplus \mathbb{C}^n$.

For a generically chosen point $x \in M$, the stabilizer of x is trivial, so the orbit Gx has $|G|$ elements. Such an orbit is said to be *regular*. Each regular orbit corresponds to a point of $\mathrm{Hilb}^{|G|}(M)$, and the Hilbert scheme of G-orbits $M//G$ is defined to be the closure of the set of these points in $\mathrm{Hilb}^{|G|}(M)$, with the induced reduced subscheme structure. It is a component of the locus of all points $I \in \mathrm{Hilb}^{|G|}(M)$ having the property that I is G-invariant and $\mathcal{O}(M)/I$ affords the regular representation of G. We ignore other components of this locus, if any.

There is a Chow morphism

$$\sigma\colon M//G \to M/G \tag{14}$$

which restricts to the obvious isomorphism on the open subset of $M//G$ consisting of points corresponding to regular orbits. The Chow morphism is projective, so in the event that $M//G$ is non-singular, it is a resolution of singularities. In that case, if we also have $\omega_{M//G} \cong \sigma^*\omega_{M/G}$, the resolution is said to be *crepant*. If M is a vector space this just means that $\omega_{M//G} \cong \mathcal{O}$ is trivial.

Let $X \subseteq (M//G) \times M$ be the universal family, inherited from the universal family on $\mathrm{Hilb}^{|G|}(M)$. We have a G-equivariant commutative diagram

$$\begin{array}{ccc} X & \xrightarrow{f} & M \\ {\scriptstyle\rho}\downarrow & & \downarrow \\ M//G & \xrightarrow{\sigma} & M/G. \end{array} \tag{15}$$

Bridgeland, King and Reid define a functor

$$\Phi\colon D(M//G) \to D^G(M) \tag{16}$$

from the bounded derived category of coherent sheaves on $M//G$ to the bounded derived category of coherent G-equivariant sheaves on M, by the formula

$$\Phi = Rf_* \circ \rho^*. \tag{17}$$

Note that ρ is flat, so we can write ρ^* instead of $L\rho^*$ here.

The Bridgeland–King–Reid theorem has two parts: a criterion for $M//G$ to be a crepant resolution of M/G, and, more importantly for us, a generalized McKay correspondence, expressed in a strong form as an equivalence of derived categories, whenever their criterion holds.

Theorem 3 ([4]) *Suppose that the Chow morphism $M//G \to M/G$ satisfies the following smallness condition: for every d, the locus of points $x \in M/G$ such that $\dim \sigma^{-1}(x) \geq d$ has codimension at least $2d-1$. Then $M//G$ is a crepant resolution of singularities of M/G, and the functor Φ is an equivalence of categories.*

Let us apply this to our case, $M = \mathbb{C}^{2n}$ and $G = S_n$. Identifying $\mathbb{C}^{2n}//S_n$ with H_n, the diagram in (15) becomes

$$\begin{array}{ccc} X_n & \xrightarrow{f} & \mathbb{C}^{2n} \\ {\scriptstyle\rho}\downarrow & & \downarrow \\ H_n & \xrightarrow{\sigma} & S^n\mathbb{C}^2. \end{array} \tag{18}$$

The Hilbert scheme H_n is non-singular by a theorem of Fogarty,[6] and it is known[13,15] that $\omega_{H_n} = \mathcal{O}_{H_n}$ (more generally, $\mathrm{Hilb}^n(M)$ possesses a non-degenerate symplectic form whenever M does). This shows directly that $\mathbb{C}^{2n}//S_n$ is a crepant resolution of $\mathbb{C}^{2n}/S_n$; we don't need the Bridgeland–King–Reid criterion for this. The Bridgeland–King–Reid criterion does hold, however. This follows either from the description of the fibers of the Chow morphism due to Briançon,[3] or from the observation in ([4]) that, conversely to Theorem 3, the criterion holds whenever G preserves a symplectic form on M and $M//G$ is a crepant resolution. Hence we have the following result.

Corollary 2.1 *The Bridgeland–King–Reid functor $\Phi = Rf_* \circ \rho^*$ for the diagram* (18) *is an equivalence of categories*

$$\Phi\colon D(H_n) \to D^{S_n}(\mathbb{C}^{2n}). \tag{19}$$

Let $\mathbf{x}, \mathbf{y} = x_1, y_1, \ldots, x_n, y_n$ be the coordinates on $\mathbb{C}^{2n}$. Since $\mathbb{C}^{2n}$ is affine, we may identify $D^{S_n}(\mathbb{C}^{2n})$ with the bounded derived category of S_n-equivariant finitely generated $\mathbb{C}[\mathbf{x}, \mathbf{y}]$-modules. Then the functor Rf_* is identified with $R\Gamma_{X_n}$. Since ρ is finite and therefore affine, $R\Gamma_{X_n} \circ \rho^*$ is naturally isomorphic to $R\Gamma_{H_n}(P \otimes -)$. This gives us an alternate description of the Bridgeland–King–Reid functor and a corresponding reformulation of Corollary 2.1.

Corollary 2.2 *The functor $\Phi = R\Gamma(P \otimes -)$ is an equivalence of categories $\Phi\colon D(H_n) \to D^{S_n}(\mathbb{C}^{2n})$.*

Using this we can also reformulate our main theorem.

Proposition 2.3 *Theorem 1 is equivalent to the identity in $D^{S_n}(\mathbb{C}^{2n})$*

$$\Phi B^{\otimes l} \cong R(n, l), \tag{20}$$

where the isomorphism is given by the map $R(n,l) \to \Phi B^{\otimes l}$ obtained by composing the canonical natural transformation $\Gamma \to R\Gamma$ with the homomorphism $R(n,l) \to \Gamma(P \otimes B^{\otimes l})$ in (10).

We prove identity (20), and thus Theorem 1, by using the *inverse* Bridgeland–King–Reid functor $\Psi\colon D^{S_n}(\mathbb{C}^{2n}) \to D(H_n)$, which also has a simple description in our case. In general, as observed in ([4]), the inverse functor Ψ can be calculated using Grothendieck duality as the right adjoint of Φ, given by the formula

$$\Psi = (\rho_*(\omega_X \overset{L}{\otimes} Lf^* -))^G. \tag{21}$$

We can simplify this using the following result from ([13]).

Proposition 2.4 *The line bundle $\mathcal{O}(1) = \wedge^n B$ is the twisting sheaf induced by a natural embedding of H_n as a scheme projective over $S^n\mathbb{C}^2$. Writing $\mathcal{O}(1)$ also for its pullback to X_n, we have that X_n is Gorenstein with canonical sheaf $\omega_{X_n} \cong \mathcal{O}(-1)$.*

We need an extra bit of information not contained in the proposition. There are two possible equivariant S_n actions on $\mathcal{O}_{X_n}(1)$. One is the trivial action coming from the definition of $\mathcal{O}_{X_n}(1)$ as $\rho^*\mathcal{O}_{H_n}(1)$. The other is the twist of the trivial action by the sign character of S_n. The latter action is the correct one, in the sense that identification $\omega_{X_n} \cong \mathcal{O}(-1)$ is an S_n-equivariant isomorphism for this action. This can be seen by a careful examination of the proof of Proposition 2.4 given in ([13]). Taking this into account, and using the fact that $\mathcal{O}_{X_n}(-1)$ is pulled back from H_n, we have the following description of the inverse functor.

Proposition 2.5 *The inverse of the functor Φ in Corollary 2.2 is given by*

$$\Psi = \mathcal{O}(-1) \otimes (\rho_* L f^* -)^{\epsilon}. \tag{22}$$

Here $-^{\epsilon}$ denotes the functor of S_n-alternants, i.e., $A^{\epsilon} = \mathrm{Hom}_{S_n}(\varepsilon, A)$, where ε is the sign representation.

3 Prior results on Hilbert schemes and polygraphs

To derive our vanishing theorem from the Bridgeland–King–Reid isomorphism, we need some results from ([13]). First, of course, we need the identification of H_n with $\mathbb{C}^{2n}//S_n$, in order to have Theorem 3 apply at all. More importantly, we need the theorem on polygraphs that was the main technical tool in ([13]), in order to calculate the inverse isomorphism Ψ applied to the polygraph coordinate ring $R(n,l)$. For this calculation we also need a proposition describing X_n locally where $V(I)$ is not concentrated at a point. Finally, for the application to character formulas, we need the characters of the fibers of the tautological bundle at certain distinguished points of H_n. We now briefly review all these results.

In ([13]), we defined the *isospectral Hilbert scheme* to be the set-theoretic fiber product X_n in the diagram (18), with its induced reduced structure as a subscheme of $H_n \times \mathbb{C}^{2n}$. The identification of H_n with $\mathbb{C}^{2n}//S_n$ and of X_n with the universal family over $\mathbb{C}^{2n}//S_n$ then follows once it is shown that the morphism $\rho\colon X_n \to H_n$ in (18) is flat. Since the Hilbert scheme H_n is nonsingular, this is equivalent to X_n being Cohen-Macaulay. What we actually proved in ([13]) is the stronger result cited above as Proposition 2.4.

Via the projection $f\colon X_n \to \mathbb{C}^{2n}$, the coordinates $\mathbf{x}, \mathbf{y}$ on $\mathbb{C}^{2n}$ can be regarded as global functions on X_n. For the geometric argument in ([13]), we needed to know that the sheaf of regular functions $\mathcal{O}_{X_n}$ is flat as a sheaf of $\mathbb{C}[\mathbf{y}]$-modules, a fact which we obtained using a theorem on the coordinate rings of polygraphs. We will restate this theorem for use again here.

The polygraph $Z(n,l)$ and its coordinate ring $R(n,l)$ have already been defined in the introduction. Let us rephrase the definition in the form given in ([13]). We write

$$\mathbf{x}, \mathbf{y}, \mathbf{a}, \mathbf{b} = x_1, y_1, \ldots, x_n, y_n, a_1, b_1, \ldots, a_l, b_l \tag{23}$$

for the coordinates on $\mathbb{C}^{2n+2l}$. To each function $f\colon \{1,\dots,l\} \to \{1,\dots,n\}$ there corresponds a linear subspace

$$W_f = V(I_f) \subseteq \mathbb{C}^{2n+2l}, \quad \text{where} \quad I_f = (a_i - x_{f(i)}, b_i - y_{f(i)} : 1 \le i \le l). \tag{24}$$

The polygraph $Z(n,l)$ is the union of these subspaces. Hence its coordinate ring is

$$R(n,l) = \mathbb{C}[\mathbf{x},\mathbf{y},\mathbf{a},\mathbf{b}]/I(n,l), \quad \text{where} \quad I(n,l) = \bigcap_f I_f. \tag{25}$$

The main technical theorem on polygraphs is as follows.

Proposition 3.1 *The polygraph coordinate ring $R(n,l)$ is a free $\mathbb{C}[\mathbf{y}]$-module.*

For present purposes, we need to strengthen this slightly in two ways. Any automorphism of $\mathbb{C}^2$ induces an automorphism of $\mathbb{C}^{2n+2l}$, and the corresponding automorphism of $\mathbb{C}[\mathbf{x},\mathbf{y},\mathbf{a},\mathbf{b}]$ leaves invariant the defining ideal $I(n,l)$ of $Z(n,l)$. In particular, this is the case for translations in the x-direction, which also leave invariant the ideal $(\mathbf{y})$ and hence $I(n,l)+(\mathbf{y})$. This implies that any of the coordinates x_i, a_i is a non-zero-divisor in $R(n,l)/(\mathbf{y})$, yielding the following two corollaries.

Corollary 3.2 *Thc coordinatc ring $R(n,l)$ is a free $\mathbb{C}[x_1,\mathbf{y}]$-module.*

Corollary 3.3 *The coordinate ring $R(n,l)$ has a free resolution of length $n-1$ as a $\mathbb{C}[\mathbf{x},\mathbf{y}]$-module.*

The polygraph ring $R(n,l)$ can be defined with any ground ring S in place of $\mathbb{C}$, following the same recipe as in (24)–(25). In ([13]) we showed that Proposition 3.1 is valid in this more general setting. Here we will only need the case when S is a polynomial ring over $\mathbb{C}$. Any automorphism of $S[x,y]$ as an S-algebra extends to an automorphism of $R(n,l)$, giving the next corollary.

Corollary 3.4 *Let S be a $\mathbb{C}$-algebra and let y' denote the image of y under some automorphism of $S[x,y]$ as an S-algebra. Then $S\otimes_{\mathbb{C}} R(n,l)$ is a free $S[y_1',\dots,y_n']$-module.*

In addition to the results on polygraphs we need a local structure theorem for X_n, which enables us to assume by induction on n that desired geometric results hold locally over the open locus in H_n consisting of points I such that $V(I)$ is not concentrated at a single point.

Proposition 3.5 *Let $U_k \subseteq X_n$ be the open set consisting of points $(I,P_1,\dots,P_n)$ for which $\{P_1,\dots,P_k\}$ and $\{P_{k+1},\dots,P_n\}$ are disjoint. Then U_k is isomorphic to an open set in $X_k \times X_{n-k}$, in a manner compatible with the projection on $\mathbb{C}^{2n}$. The pullback $F' = F \times X_n/H_n$ of the universal family to X_n decomposes over U_k as the disjoint union of the pullbacks of the universal families from H_k and H_{n-k}.*

In Section 5, we will make reference to the fixed points of a natural torus action on H_n. The two-dimensional torus group

$$\mathbb{T}^2 = (\mathbb{C}^*)^2 \tag{26}$$

acts on $\mathbb{C}^2$ as the group of 2×2 invertible diagonal matrices. This action induces an equivariant action on all schemes and bundles under discussion. The $\mathbb{T}^2$-action on the Hilbert scheme H_n has finitely many fixed points, namely, the points corresponding to monomial ideals $I \subseteq \mathbb{C}[x, y]$. For such an I, the exponents (r, s) of monomials $x^r y^s$ not in I form the diagram of a partition μ of n. We denote the corresponding fixed point I by I_μ. The motivating application for the geometric results in ([13]) was the proof of the positivity conjecture for Macdonald polynomials via the identification of the Macdonald polynomial $\tilde{H}_\mu(z; q, t)$ with the character of the fiber of the tautological bundle P at the distinguished point I_μ.

Proposition 3.6 *The character as an $S_n \times \mathbb{T}^2$-module of the fiber $P(I_\mu)$ of P at I_μ is given by*

$$\mathcal{F}P(I_\mu) = \tilde{H}_\mu(z; q, t), \tag{27}$$

in the notation of Section 5 (specifically, $\mathcal{F}$ denotes the Frobenius series as defined by (44)*, and $\tilde{H}_\mu(z; q, t)$ the transformed Macdonald polynomial in* (46)*).*

4 Main results

In this section we outline the proofs of Theorems 1 and 2. We begin with Theorem 1. We have a map

$$R(n, l) \to \Phi B^{\otimes l} \tag{28}$$

in the derived category $D^{S_n}(\mathbb{C}^{2n})$, and by Proposition 2.3, it suffices to show that this map is an isomorphism. Applying the inverse functor Ψ yields a map

$$\Psi R(n, l) \to B^{\otimes l} \tag{29}$$

in $D(H_n)$, and we can equally well show that this is an isomorphism. Let C be the third vertex of a distinguished triangle

$$C[-1] \to \Psi R(n, l) \to B^{\otimes l} \to C. \tag{30}$$

We are to show that $C = 0$.

We may compute $\Psi R(n,l)$ as follows. By Corollary 3.3, the $\mathbb{C}[\mathbf{x},\mathbf{y}]$-algebra $R(n,l)$ has a free resolution of length $n-1$. In derived category terminology this means that there is complex of free $\mathbb{C}[\mathbf{x},\mathbf{y}]$-modules

$$A_{\bullet} = \cdots \to 0 \to A_{n-1} \to \cdots \to A_1 \to A_0 \to 0 \to \cdots \qquad (31)$$

quasi-isomorphic to $R(n,l)$. Using the formula for Ψ from Proposition 2.5, we have $\Psi R(n,l) = \mathcal{O}(-1) \otimes (\rho_* f^* A_{\bullet})^{\epsilon}$. Moreover, since ρ is flat, and since the functor $-^{\epsilon}$ is a natural direct summand of the identity functor, this formulation provides us with a locally free resolution of $\Psi R(n,l)$. As $B^{\otimes l}$ is a sheaf, the map $\Psi R(n,l) \to B^{\otimes l}$ in (29) is represented by an honest homomorphism of complexes. The object C is represented by the mapping cone of this homomorphism, namely, the complex of locally free sheaves

$$0 \to C_n \to \cdots \to C_2 \to C_1 \to B^{\otimes l} \to 0, \qquad (32)$$

where $C_i = \mathcal{O}(-1) \otimes (\rho_* f^* A_{i-1})^{\epsilon}$. We are to prove that this complex is exact. For this we will make use of the following fundamental result of homological algebra.

Proposition 4.1 (Intersection Theorem[16,17,18]) *Let $0 \to C_n \to \cdots \to C_1 \to C_0 \to 0$ be a bounded complex of locally free coherent sheaves on a Noetherian scheme X. Denote by* Supp$(C_{\bullet})$ *the union of the supports of the homology sheaves $H_i(C_{\bullet})$. Then every component of* Supp$(C_{\bullet})$ *has codimension at most n in X, where n is the length of the complex. In particular, if $C_{\bullet}$ is exact on an open set $U \subseteq X$ whose complement has codimension exceeding n, then $C_{\bullet}$ is exact.*

Let $U \subseteq H_n$ be the open set of points I such that $V(I)$ contains at least two distinct points, that is, such that $\sigma(I)$ is not a single point with multiplicity n. Using Proposition 3.5, we can show by induction on n that the map $R(n,l) \to \Phi B^{\otimes l}$ in (28) restricts to an isomorphism on the open set corresponding to U in $\mathbb{C}^{2n}$. Although the derived category is not a local object, the sheaf operations used to define the functors Φ and Ψ do localize, so we can conclude that the map $\Psi R(n,l) \to B^{\otimes l}$ in (29) is an isomorphism on U, and hence the mapping cone complex in (32) is exact on U. In other words, the support Supp(C) of the object C is disjoint from U.

The complement of U in H_n is isomorphic to $\mathbb{C}^2 \times H_n^0$, so it has dimension $n+1$ and codimension $n-1$. We are not yet ready to apply Proposition 4.1; we first need to enlarge U to an open set whose complement has codimension $n+1$.

Let $U_x \subseteq H_n$ be the open set consisting of ideals I such that x generates the tautological fiber $B(I) = \mathbb{C}[x,y]/I$ as a $\mathbb{C}$-algebra, or equivalently, such that $\{1, x, \ldots, x^{n-1}\}$ is a basis of $B(I)$. Let U_y be defined similarly, with y

in place of x. Our desired open set will be $U \cup U_x \cup U_y$. Its complement is isomorphic to $\mathbb{C}^2 \times (H_n^0 \setminus (U_x \cup U_y))$, which has codimension $n+1$ by the following lemma.

Lemma 4.2 *The complement $H_n^0 \setminus (U_x \cup U_y)$ of of $U_x \cup U_y$ in the zero fiber has dimension $n-3$.*

Proof. Let V denote the complement. We interpret the statement that a locus has negative dimension to mean that it is empty. Then the lemma holds trivially for $n = 1$, so we can assume $n \geq 2$. We consider the decomposition of H_n^0 into affine cells as in ([3,5]), and show that each cell intersects V in a locus of dimension at most $n-3$. There is one open cell, of dimension $n-1$. This cell is actually $U_x \cap H_n^0$, so it is disjoint from V. There is also one cell of dimension $n-2$. It has non-empty intersection with U_y, so its intersection with V also has dimension at most $n-3$. In fact this intersection has dimension exactly $n-3$, since the complement of U_y is the zero locus of a section of the line bundle $\wedge^n B = \mathcal{O}(1)$. All remaining cells have dimension less than or equal to $n-3$. □

We digress briefly to point out the geometric meaning of this lemma. For I in the zero fiber, the fiber $B(I)$ is an Artin local $\mathbb{C}$-algebra with maximal ideal (x, y). The point I belongs to $U_x \cup U_y$ if and only if the maximal ideal is principal, that is, $B(I)$ has embedding dimension one, or equivalently, the corresponding closed subscheme $V(I)$ is a subscheme of some smooth curve through the origin in $\mathbb{C}^2$. In this case I is said to be *curvilinear*. Lemma 4.2 says that the non-curvilinear locus has codimension two in the zero fiber.

All that we now require to complete the proof of Theorem 1 is the following lemma, together with a corresponding version with U_y in place of U_x that clearly follows from it by symmetry.

Lemma 4.3 *The map $\Psi R(n,l) \to B^{\otimes l}$ restricts to an isomorphism on U_x.*

Outline of proof. The lemma is proven by a calculation in local coordinates on the open set U_x and its preimage U_x' in X_n. The calculation has two ingredients. First, using Corollary 3.4, we can show that $Lf^*R(n,l)$ reduces to the sheaf $f^*R(n,l)$ on U_x'. In local coordinates, this sheaf is associated to the algebra $\mathcal{O}(U_x') \otimes_{\mathbb{C}[\mathbf{x},\mathbf{y}]} R(n,l)$. The desired result takes the form of an isomorphism between the S_n-alternating part of this algebra and another algebra representing the sheaf $B^{\otimes l}$ on U_x. The isomorphism in question and its inverse can be written down explicitly. □

Next we turn to the proof of Theorem 2. As we will see, it follows as a corollary to Theorem 1, once we have an appropriate resolution of the structure sheaf of the zero fiber. Such a resolution was found in ([11]), and the demonstration that Theorem 1 implies Theorem 2 in the case $l = 0$ was given in ([12]). Here we treat the case where l is arbitrary, and give a somewhat simpler proof using the functorial interpretation.

To begin with, we describe the relevant resolution. The tautological sheaf B is a sheaf of $\mathcal{O}_{H_n}$-algebras, the quotient of $\mathcal{O} \otimes \mathbb{C}[x, y]$ by the ideal sheaf of the universal family. As such it comes with a canonical homomorphism $i\colon \mathcal{O} \to B$. Since B is locally free, there is a trace homomorphism $\tau\colon B \to \mathcal{O}$ sending a section f of B to the trace of multiplication by f. More explicitly, for a section represented by a polynomial $f(x, y)$, we have

$$\tau f = \sum_{i=1}^{n} f(x_i, y_i). \tag{33}$$

The right hand side here is a symmetric polynomial, an element of $\mathbb{C}[\mathbf{x}, \mathbf{y}]^{S_n}$, and thus makes sense as a regular function on H_n, via the Chow morphism. Since $\frac{1}{n}\tau(1) = 1$, we see that $\frac{1}{n}\tau$ splits the canonical map i, giving a direct sum decomposition

$$B = \mathcal{O} \oplus B', \tag{34}$$

where B' is the kernel of $\frac{1}{n}\tau$.

Let Q be the rank-2 free $\mathcal{O}_{H_n}$-module sheaf $Q = \mathcal{O} \otimes_{\mathbb{C}} (\mathbb{C}^2)^*$, where $(\mathbb{C}^2)^*$ is the dual vector space of $\mathbb{C}^2$. We may identify $(\mathbb{C}^2)^*$ with the homogeneous component of degree 1 in the polynomial ring $\mathbb{C}[x, y]$. Then the realization of B as a quotient algebra of $\mathcal{O} \otimes \mathbb{C}[x, y]$ yields a homomorphism of sheaves of $\mathcal{O}$-modules $s\colon Q \to B$. We have the following characterization of the structure sheaf of the zero fiber.

Proposition 4.4 ([11,12]) *Let J be the sheaf of ideals in B generated by the subsheaf B' and the image of the homomorphism $s\colon Q \to B$. Then B/J is isomorphic to $\mathcal{O}_{H_n^0}$ as a sheaf of $\mathcal{O}_{H_n}$-algebras.*

The content of this can be rephrased as follows. Let $j\colon B' \hookrightarrow B$ be the inclusion homomorphism. We can compose the homomorphism $(j \oplus s) \otimes 1_B :$ $(B' \oplus Q) \otimes B \to B \otimes B$ with multiplication in B to get a homomorphism $\mu_B \circ ((j \oplus s) \otimes 1_B)\colon (B' \oplus Q) \otimes B \to B$ whose image is the sheaf of ideals J in the proposition. Then we have an exact sequence of sheaves of $\mathcal{O}_{H_n}$ modules

$$(B' \oplus Q) \otimes B \to B \to \mathcal{O}_{H_n^0} \to 0. \tag{35}$$

Now $\operatorname{Spec} B$ is the universal family F, which is Cohen-Macaulay and $2n$-dimensional, since it is flat and finite over the smooth variety H_n. The zero fiber H_n^0 has dimension $n - 1$. The sheaf $B' \otimes Q$ is locally free of rank $n + 1$, equal to the codimension of the zero fiber in F. Hence the ideal sheaf J is locally a complete intersection ideal, minimally generated by any local basis of $B' \oplus Q$. It follows that the sequence in (35) extends to a Koszul complex

$$\begin{aligned} 0 \to \wedge^{n+1}(B' \oplus Q) \otimes B \to \\ \cdots \to \wedge^2(B' \oplus Q) \otimes B \to (B' \oplus Q) \otimes B \to B \to \mathcal{O}_{H_n^0} \to 0, \end{aligned} \tag{36}$$

which is a locally free resolution of $\mathcal{O}_{H_n^0}$.

Let $V_\bullet$ be the deleted resolution, that is, the complex in (36), but with the final term $\mathcal{O}_{H_n^0}$ omitted. For every l, the complex of locally free sheaves $V_\bullet \otimes B^{\otimes l}$ is isomorphic in the derived category to $\mathcal{O}_{H_n^0} \otimes B^{\otimes l}$. Note that every term in $V_\bullet \otimes B^{\otimes l}$ is a direct summand of a sum of tensor powers of B.

The functor Φ is the right derived functor of $\Gamma_{X_n} \circ \rho^*$, and Theorem 1 implies that the terms in the complex $V_\bullet \otimes B^{\otimes l}$ are acyclic for the latter functor. Hence we can compute $\Phi(\mathcal{O}_{H_n^0} \otimes B^{\otimes l})$ as $\Gamma_{X_n}\rho^*(V_\bullet \otimes B^{\otimes l})$. This last complex has non-zero terms only in degrees $i \leq 0$, while the homology modules $H^i\Phi(\mathcal{O}_{H_n^0} \otimes B^{\otimes l})$ are non-zero only in degrees $i \geq 0$. Together these facts imply that $\Phi(\mathcal{O}_{H_n^0} \otimes B^{\otimes l})$ reduces to a complex concentrated in degree zero, or in other words, to a module, and that the complex $\Gamma_{X_n}\rho^*(V_\bullet \otimes B^{\otimes l})$ is a resolution of that module. Theorem 1 also allows us to write down the terms of the resolution. Its tail looks like this:

$$\Phi(B' \otimes B^{\otimes l+1}) \oplus ((\mathbb{C}^2)^* \otimes R(n, l+1)) \xrightarrow{\phi} R(n, l+1) \to \Phi(\mathcal{O}_{H_n^0} \otimes B^{\otimes l}) \to 0. \tag{37}$$

To complete the calculation of $\Phi(\mathcal{O}_{H_n^0} \otimes B^{\otimes l})$, we need to make the map ϕ explicit. This can be done, and the result is that the image of ϕ is the the ideal $J \subseteq R(n, l+1)$ generated by x, y and all expressions of the form

$$f(x, y) - \frac{1}{n}\sum_{i=1}^{n} f(x_i, y_i), \tag{38}$$

for $f \in \mathbb{C}[x, y]$. Here we have written x, y instead of a, b for the coordinates in $R(n, l+1) = \Phi(B \otimes B^{\otimes l})$ corresponding to the generators of the first tensor factor B. We write $\mathbf{a}, \mathbf{b} = a_1, b_1, \ldots, a_l, b_l$ as usual for the coordinates corresponding to generators of $B^{\otimes l}$.

Let $I \subseteq R(n, l)$ be the ideal generated by the homogeneous maximal ideal $\mathfrak{m}$ in the subring $\mathbb{C}[\mathbf{x}, \mathbf{y}]^{S_n}$, as in the statement of Theorem 2. To complete the proof, we show that the inclusion of $R(n, l)$ as the subalgebra of $R(n, l+1)$ generated by the variables $\mathbf{x}, \mathbf{y}, \mathbf{a}, \mathbf{b}$ induces an isomorphism

$$\xi\colon R(n, l)/I \to R(n, l+1)/J. \tag{39}$$

This suffices, as Theorem 2 amounts to the identity

$$\Phi(\mathcal{O}_{H_n^\circ} \otimes B^{\otimes l}) \cong R(n,l)/I. \tag{40}$$

By a theorem of Weyl,[19] the ideal I is generated by the power-sums $p_{h,k} = \sum_{i=1}^n x_i^h y_i^k$, for $h+k>0$. If $f(x,y)$ is any polynomial, then the expression in (38) is congruent modulo (x,y) to $f(0,0) - \frac{1}{n}\sum_{i=1}^n f(x_i,y_i)$. This last quantity is symmetric and vanishes at $\mathbf{x}=\mathbf{y}=0$, so it belongs to I. By Weyl's theorem, I is generated by such expressions, so we have $J = IR(n,l+1)+(x,y)$. In particular, the inclusion $R(n,l) \hookrightarrow R(n,l+1)$ makes I a subset of J, so the homomorphism ξ in (39) is well-defined. The variables x,y generate $R(n,l+1)$ as an $R(n,l)$-algebra, and since they vanish modulo J, we see that ξ is surjective.

To show that ξ is injective, we construct its left inverse. The inclusion $R(n,l) \hookrightarrow R(n,l+1)$ is split by a trace map $\frac{1}{n}\tau\colon R(n,l+1) \to R(n,l)$, which is a homomorphism of $R(n,l)$-modules satisfying $\frac{1}{n}\tau f(x,y) = \frac{1}{n}\sum_{i=1}^n f(x_i,y_i)$. Since $\frac{1}{n}\tau$ is an $R(n,l)$-module homomorphism, it carries $IR(n,l+1)$ into I. The ideal $(x,y) \subseteq R(n,l+1)$ is generated as an $R(n,l)$-module by the monomials $x^h y^k$ with $h+k>0$. We have $\frac{1}{n}\tau(x^h y^k) = p_{h,k}$, so the trace map carries (x,y) and hence also J into I. Therefore it induces a well-defined homomorphism $\frac{1}{n}\tau\colon R(n,l+1)/J \to R(n,l)/I$, and the fact that it is a homomorphism of $R(n,l)$-modules implies that $\frac{1}{n}\tau \circ \xi = 1$.

5 Application to character formulas

In ([10]), we presented a series of combinatorial conjectures by the author and others concerning the ring of coinvariants

$$R_n = \mathbb{C}[\mathbf{x},\mathbf{y}]/I \tag{41}$$

for the action of S_n on $\mathbb{C}^{2n}$. Here $I = \mathfrak{m}\mathbb{C}[\mathbf{x},\mathbf{y}]$ is the ideal generated by the homogeneous maximal ideal $\mathfrak{m}$ of the subring of invariants $\mathbb{C}[\mathbf{x},\mathbf{y}]^{S_n}$. As a doubly-graded S_n-module, R_n is isomorphic to the space DH_n of harmonics for the S_n action on $\mathbb{C}^{2n}$, that is, the space of polynomials $f(\mathbf{x},\mathbf{y})$ annihilated by all S_n-invariant differential operators $p(\partial\mathbf{x},\partial\mathbf{y})$ without constant term. We call DH_n the space of *diagonal harmonics*, because the S_n action on $\mathbb{C}^{2n}$ is the diagonal action on the direct sum of two copies of the natural permutation representation on $\mathbb{C}^n$.

It is convenient to keep track of the character of R_n or any doubly graded S_n-module using its *Frobenius series*, a notion which combines the notions of Hilbert series and Frobenius characteristic. Let Λ be the algebra of symmetric functions in variables $z = z_1, z_2, \ldots$, and let Λ_n be its homogeneous component of degree n. Recall that the Frobenius characteristic is the linear map

$$F\colon (S_n \text{ characters}) \to \Lambda_n \tag{42}$$

sending the irreducible character χ^λ to the Schur function $s_\lambda(z)$. It has the explicit formula

$$F\chi = \frac{1}{n!} \sum_{w \in S_n} \chi(w) p_{\tau(w)}(z), \tag{43}$$

where $p_{\tau(w)}(z)$ is the power-sum symmetric function indexed by the partition of n corresponding to the decomposition of w into disjoint cycles. Using the Frobenius characteristic, one can extract the characters and other information about representations of the symmetric groups S_n from calculations in the algebra of symmetric functions.

We define the Frobenius series of a doubly-graded S_n-module $A = \bigoplus_{r,s} A_{r,s}$ to be the symmetric function with coefficients in the ring of formal Laurent series $\mathbb{Z}[[q,t]][q^{-1},t^{-1}]$

$$\mathcal{F}A(z;q,t) = \sum_{r,s} t^r q^s F \operatorname{char}(A_{r,s}). \tag{44}$$

By construction, the coefficients of $\mathcal{F}A(z;q,t)$ are actually in $\mathbb{N}[[q,t]][q^{-1},t^{-1}]$, and if the grading of A is positive, they are in $\mathbb{N}[[q,t]]$. If A is finite-dimensional, they are (Laurent) polynomials. The univariate Frobenius series $\mathcal{F}A(z;q)$ of a singly-graded S_n-module A is defined analogously.

The main conjectures on diagonal harmonics in ([10]) can be expressed as formulas for the specializations at $t = 1$ and $t = q^{-1}$ of the Frobenius series $\mathcal{F}R_n(z;q,t)$. The first specialization is most conveniently expressed combinatorially in terms of parking functions, whose definition we pause to review.

A *parking function* is a function $f\colon \{1,\dots,n\} \to \{1,\dots,n\}$ with the property that $|f^{-1}(\{1,\dots,k\})| \geq k$, for all $1 \leq k \leq n$. The idea behind the name is this: picture a one-way street with n parking spaces numbered 1 through n. Suppose that n cars arrive in succession, each with a preferred parking space given by $f(i)$ for the i-th car. Each driver proceeds directly to his or her preferred space and parks there, or in the next available space, if the desired space is already taken. The necessary and sufficient condition for everyone to park without being forced to the end of the street is that f is a parking function. The *weight* of f is the quantity $w(f) = \sum_{i=1}^{n} f(i) - i$. It measures the total frustration experienced by the drivers in having to pass up occupied parking spaces.

The symmetric group acts on the set P of parking functions by permuting the cars (that is, the domain of f) and this action preserves the weight. Let A be the graded permutation representation $A = \bigoplus_d \mathbb{C}P_d$, where $P_d = \{f \in P : w(f) = d\}$, and let ε be the sign representation. We can now state the two conjectured character specializations for the diagonal harmonics.

Conjecture 5.1 *The Frobenius series of the diagonal harmonics DH_n, or equivalently of the ring of coinvariants R_n, satisfies*

(i) $\mathcal{F}R_n(z; q, 1) = \mathcal{F}(\varepsilon \otimes A)(z; q)$, *where A is the parking function module, as above, and*

(ii) $q^{\binom{n}{2}} \mathcal{F}R_n(z; q, q^{-1}) = \sum_{|\lambda|=n} \frac{s_\lambda(1,q,\dots,q^n)}{1+q+\cdots q^n} s_\lambda(z)$.

Part (i) of the conjecture describes the singly graded character of R_n with respect to degree in the $\mathbf{y}$ variables, ignoring the $\mathbf{x}$-degree. Part (ii) describes the character with respect to the difference between the $\mathbf{x}$ and $\mathbf{y}$-degrees, or what is the same, the character of the SL(2)-action on R_n. Ignoring the grading entirely, the character of R_n as an S_n-module is given by the Frobenius characteristic $\mathcal{F}R_n(z; 1, 1)$. Setting $q = 1$ in part (i) of Conjecture 5.1 we have the following corollary.

Corollary 5.2 *The representation of S_n on DH_n and R_n is equivalent to $\varepsilon \otimes A$, and its dimension is therefore the number of parking functions,*

$$\dim R_n = (n+1)^{n-1}. \tag{45}$$

We remark that the permutation action of S_n on parking functions is known to be equivalent to the action of S_n on the finite Abelian group $Q/(n+1)Q$, where $Q = \mathbb{Z}^n/\mathbb{Z}$, S_n acts on $\mathbb{Z}^n$ by permuting the factors, and the subgroup $\mathbb{Z}$ under the fraction bar is the diagonal subgroup generated by $(1, 1, \dots, 1)$. Since Q is free of rank $n-1$, $Q/(n+1)Q$ has order $(n+1)^{n-1}$. One can show directly that this description agrees with the $q = 1$ specialization of part (ii) of the conjecture. It is also interesting to note that Q is the root lattice of type A_{n-1} and S_n is the Weyl group. The reader may consult ([10]) for discussion of the extension (or lack of it) to other Weyl groups.

In ([8]), Garsia and the author conjectured an exact formula for $\mathcal{F}R_n(z; q, t)$ in terms of Macdonald polynomials and proved that it implies the two specializations in Conjecture 5.1. To state the conjectured formula we use the so-called *transformed integral form* Macdonald polynomials

$$\tilde{H}_\mu(z; q, t), \tag{46}$$

as defined in ([8]), ([12]), or ([13]). We work in the algebra $\Lambda_{\mathbb{Q}(q,t)}$ of symmetric functions with coefficients in $\mathbb{Q}(q, t)$, and define as in ([2,8]) a linear operator

∇ on $\Lambda_{\mathbb{Q}(q,t)}$ whose eigenfunctions are the Macdonald polynomials, by the formula

$$\nabla \tilde{H}_\mu = t^{n(\mu)} q^{n(\mu')} \tilde{H}_\mu. \tag{47}$$

Here μ' denotes the partition conjugate to μ, and $n(\mu)$ is the statistic

$$n(\mu) = \sum_i (i-1)\mu_i. \tag{48}$$

Our conjectured character formula for the diagonal harmonics is as follows.

Conjecture 5.3 *We have*

$$\mathcal{F}R_n(z; q, t) = \nabla e_n(z), \tag{49}$$

where $e_n(z)$ is the n-th elementary symmetric function.

The motivation for this conjecture was exactly the geometric picture by means of which it will be proven here. At the time, we had to take the geometric facts as unproved assumptions, although the ability of these assumptions to explain Conjecture 5.1 struck us as strong evidence that they must be true. The results of this paper and ([13]) confirm their validity.

Theorem 4 *Conjecture 5.3 is true, and consequently Conjecture 5.1 and Corollary 5.2 are true also.*

Using our preceding results, we can reduce the proof of this theorem to a calculation. The Bridgeland–King–Reid isomorphism is equivariant with respect to the $\mathbb{T}^2$ action, so it induces an isomorphism $K^{\mathbb{T}^2}(H_n) \to K^{S_n \times \mathbb{T}^2}(\mathbb{C}^{2n})$ from the Grothendieck group of torus-equivariant coherent sheaves on H_n to the Grothendieck group of finitely-generated doubly-graded S_n-equivariant $\mathbb{C}[\mathbf{x}, \mathbf{y}]$-modules. (Note that $\mathbb{T}^2$-equivariance is the same thing as double grading for $\mathbb{C}[\mathbf{x}, \mathbf{y}]$-modules.) By Theorem 2 in the case $l = 0$, this isomorphism carries $\mathcal{O}_{H_n^0}$ to R_n.

If A is any $\mathbb{T}^2$-equivariant coherent sheaf on H_n, then its cohomology modules $M_i = H^i(H_n, A)$ are finitely-generated doubly-graded $\mathbb{C}[\mathbf{x}, \mathbf{y}]^{S_n}$-modules. As such, each has a Hilbert series $\mathcal{H}M_i$, and we write

$$\chi A(q, t) = \sum_i (-1)^i \mathcal{H}M_i(q, t) \tag{50}$$

for the Hilbert series Euler characteristic. The Euler characteristic is additive on exact sequences, so it is well-defined as a functional on objects A in the Grothendieck group $K^{\mathbb{T}^2}(H_n)$. To carry out our calculation, we need the following fundamental result, known as the Atiyah–Bott Lefschetz formula.[1] For simplicity we state it in the notation of our particular case.

Proposition 5.4 *Let $H_n^{\mathbb{T}^2} = \{I_\mu : |\mu| = n\}$ be the set of $\mathbb{T}^2$-fixed points of H_n. For every $\mathbb{T}^2$-equivariant algebraic vector bundle A on H_n, we have*

$$\chi A = \sum_{I_\mu \in H_n^{\mathbb{T}^2}} \frac{\mathcal{H}A(I_\mu)}{\sum_i (-1)^i \mathcal{H}\Omega^i(I_\mu)}, \tag{51}$$

where $\Omega^i = \wedge^i T^* H_n$ is the i-th exterior power of the cotangent bundle, that is, the bundle whose sections are the regular differential i-forms.

The Frobenius series $\mathcal{F}M(z; q, t)$ is additive on exact sequences and hence well-defined for $M \in K^{S_n \times \mathbb{T}^2}(\mathbb{C}^{2n})$. Of course $\mathcal{F}M(z; q, t)$ is merely a compact notation for the simultaneous Hilbert series of all the doubly-graded modules $\mathrm{Hom}_{S_n}(V^\lambda, M)$, as V^λ ranges through irreducible representations of S_n. Thus the Atiyah–Bott formula immediately implies the corresponding result with $\mathcal{F}A$ in place of $\mathcal{H}A$ (in which case we write $\chi_{\mathcal{F}}A$ in place of χA).

By Theorem 2, we have $\mathcal{F}R_n = \chi_{\mathcal{F}}(P \otimes \mathcal{O}_{H_n^0})$, and we can use the resolution (36) to replace $P \otimes \mathcal{O}_{H_n^0}$ with the alternating sum of the vector bundles $P \otimes B \otimes \wedge^i(B' \oplus Q)$. The denominator in (51) is the product $\prod(1 - q^r t^s)$ as $q^r t^s$ ranges over characters of the $\mathbb{T}^2$-eigenspaces on the cotangent bundle of H_n at I_μ. In ([11]), we worked out the eigenvalues and found the denominator to be given by

$$\prod_{x \in \mu} (1 - q^{-a(x)} t^{1+l(x)})(1 - q^{1+a(x)} t^{-l(x)}). \tag{52}$$

Here x ranges over the cells in the diagram of the partition μ, and the *arm* $a(x)$ and *leg* $l(x)$ denote the number of cells strictly east and north of x, respectively, with the diagram drawn in the French manner with the origin at the southwest corner.

At I_μ we have $\mathcal{H}B(I_\mu) = B_\mu(q, t)$, where

$$B_\mu(q, t) = \sum_{(r,s) \in \mu} t^r q^s. \tag{53}$$

Here r and s are the row and column coordinates of a cell in μ, indexed from $(0, 0)$ for the cell at the origin. Since $B \cong B' \oplus \mathcal{O}$, we have $\mathcal{H}B'(I_\mu) = B_\mu(q, t) - 1$, and hence

$$\sum_i (-1)^i \mathcal{H}(\wedge^i B'(I_\mu)) = \Pi_\mu(q, t) = \prod_{\substack{(r,s) \in \mu \\ (r,s) \neq (0,0)}} (1 - t^r q^s). \tag{54}$$

We have $\mathcal{H}Q(I_\mu) = q + t$ at every point and hence

$$\sum_i (-1)^i \mathcal{H}(\wedge^i Q(I_\mu)) = (1 - q)(1 - t). \tag{55}$$

Combining these yields

$$\sum_i (-1)^i \mathcal{H}(\wedge^i(B' \oplus Q)(I_\mu)) = (1-q)(1-t)\Pi_\mu(q,t). \tag{56}$$

Putting all this together with Proposition 3.6, we get the explicit formula for $\mathcal{F}R_n$ in terms of Macdonald polynomials

$$\mathcal{F}R_n(z;q,t) = \sum_{|\mu|=n} \frac{(1-q)(1-t)\Pi_\mu(q,t)B_\mu(q,t)\tilde{H}_\mu(z;q,t)}{\prod_{x\in\mu}(1-q^{-a(x)}t^{1+l(x)})(1-q^{1+a(x)}t^{-l(x)})}. \tag{57}$$

In ([8]), we found the expansion of $e_n(z)$ in terms of the Macdonald polynomials $\tilde{H}_\mu(z;q,t)$ and thereby showed that the expression in (57) is just $\nabla e_n(z)$. This proves Theorem 4.

Acknowledgment

This research was supported in part by N.S.F. Mathematical Sciences grants DMS-9701218 and DMS-0070772.

References

1. M. F. Atiyah and R. Bott, *A Lefschetz fixed point formula for elliptic differential operators*, Bull. Amer. Math. Soc. **72** (1966), 245–250.
2. F. Bergeron, A. M. Garsia, M. Haiman, and G. Tesler, *Identities and positivity conjectures for some remarkable operators in the theory of symmetric functions*, Methods and applications of analysis **6** (1999), no. 3, 363–420.
3. Joël Briançon, *Description de* $\mathrm{Hilb}^n C\{x,y\}$, Invent. Math. **41** (1977), no. 1, 45–89.
4. Tom Bridgeland, Alastair King, and Miles Reid, *Mukai implies McKay: the McKay correspondence as an equivalence of derived categories*, Electronic preprint, arXiv:math.AG/9908027, 1999.
5. Geir Ellingsrud and Stein Arild Strømme, *On the homology of the Hilbert scheme of points in the plane*, Invent. Math. **87** (1987), no. 2, 343–352.
6. John Fogarty, *Algebraic families on an algebraic surface*, Amer. J. Math **90** (1968), 511–521.
7. A. M. Garsia and J. Haglund, *A proof of the q,t-Catalan positivity conjecture*, Submitted to Discrete Mathematics (September, 2000).
8. A. M. Garsia and M. Haiman, *A remarkable q,t-Catalan sequence and q-Lagrange inversion*, J. Algebraic Combin. **5** (1996), no. 3, 191–244.

9. Danila Gentiana, *Sur la cohomologie d'un fibre tautologique sur le schema de Hilbert d'une surface*, Electronic preprint, arXiv:math.AG/9904004, 1999.
10. Mark Haiman, *Conjectures on the quotient ring by diagonal invariants*, J. Algebraic Combin. **3** (1994), no. 1, 17–76.
11. ———, *t, q-Catalan numbers and the Hilbert scheme*, Discrete Math. **193** (1998), no. 1-3, 201–224, Selected papers in honor of Adriano Garsia (Taormina, 1994).
12. ———, *Macdonald polynomials and geometry*, New perspectives in geometric combinatorics (Billera, Björner, Greene, Simion, and Stanley, eds.), MSRI Publications, vol. 38, Cambridge University Press, 1999, pp. 207–254.
13. ———, *Hilbert schemes, polygraphs, and the Macdonald positivity conjecture.*, Submitted to Journal of the A.M.S. (October, 2000).
14. Shrawan Kumar and Jesper Funch Thomsen, *Frobenius splitting of Hilbert schemes of points on surfaces*, Electronic preprint, arXiv:math.AG/9911181, 1999.
15. Hiraku Nakajima, *Lectures on Hilbert schemes of points on surfaces*, American Mathematical Society, Providence, RI, 1999.
16. Christian Peskine and Lucien Szpiro, *Syzygies et multiplicités*, C. R. Acad. Sci. Paris Sér. A **278** (1974), 1421–1424.
17. Paul Roberts, *Le théorème d'intersection*, C. R. Acad. Sci. Paris Sér. I Math. **304** (1987), no. 7, 177–180.
18. ———, *Intersection theorems*, Commutative algebra (Berkeley, CA, 1987), Springer, New York, 1989, pp. 417–436.
19. Hermann Weyl, *The Classical Groups. Their Invariants and Representations*, Princeton University Press, Princeton, N.J., 1939.

EXCLUSION STATISTICS AND CHIRAL PARTITION FUNCTION

KAZUHIRO HIKAMI
Department of Physics, Graduate School of Science
University of Tokyo
Hongo 7–3–1, Bunkyo, Tokyo 113–0033, Japan
E-mail: hikami@phys.s.u-tokyo.ac.jp

We consider the statistical mechanics of quasi-particles which obey the exclusion statistics. Based on the recursion relation of the restricted partition function, we study the thermodynamic properties. We also apply this method to the $s\ell_{M,N}$ supersymmetric system based on the super Yangian symmetry and the path configuration sum.

1 Introduction

In 1991 Haldane introduced the exclusion statistics as quantum fractional statistics which interpolates the boson and fermion [1]. The exclusion statistics is a generalization of the Pauli exclusion principle, and is defined by

$$\delta G_\alpha = -\sum_\beta g_{\alpha\beta}\, \delta N_\beta, \tag{1.1}$$

where N_β is the number of β-particle, and G_α corresponds to the dimension of the Hilbert space of α-particle. Parameters $g_{\alpha\beta}$ is called the statistical interaction; $g_{\alpha\beta} = 0$ denotes the free bosonic system and $g_{\alpha\beta} = \delta_{\alpha\beta}$ is the free fermionic system. We see that the exclusion statistics differs from a notion of "anyon" which is related to the braid group (see, *e.g.*, Ref. [2]).

Since Haldane applied the exclusion statistics to the quasi-particles of the fractional quantum Hall effect and of the spin chains, thermodynamics of the exclusion gas has been widely studied [3,4,5]. The connection with the Bethe ansatz was also discussed [6,7]. In the case of the one-component exclusion gas, the partition function was shown to coincide with the partition function of the Calogero model confined in the harmonic potential [8]. This result is consistent with the fact that the idea of the exclusion statistics originates from the eigenfunction of the Calogero–Sutherland model, which is mathematically called the Jack polynomial. On the other hand, based on studies of the character of the conformal field theory, the universal chiral partition function

(UCPF) is proposed [9];

$$S_{\mathbf{B}} \begin{bmatrix} Q \\ \underline{A} \end{bmatrix} (\underline{u}, \underline{y}|q) = \sum_{\substack{\underline{m}=0 \\ \text{restrictions } Q}} q^{\frac{1}{2}\underline{m}^t \mathbf{B}\,\underline{m} - \frac{1}{2}\underline{A}^t\,\underline{m}} \prod_{\alpha=1}^{n} {y_\alpha}^{m_\alpha} \begin{bmatrix} ((1-\mathbf{B})\,\underline{m} + \frac{1}{2}\,\underline{u})_\alpha \\ m_\alpha \end{bmatrix}_q, \tag{1.2}$$

where

$$\begin{bmatrix} m \\ n \end{bmatrix}_q = \frac{(q;q)_m}{(q;q)_n\,(q;q)_{m-n}},$$

$$(t;q)_n = (1-t)\,(1-t\,q)\cdots(1-t\,q^{n-1}), \qquad (t;q)_0 = 1.$$

It is conjectured that the UCPF with

$$g_{\alpha\beta} = B_{\alpha\beta}, \tag{1.3}$$

becomes the partition function of the exclusion gas satisfying Eq. (1.1). Our purpose below is to reveal the relationship between the exclusion statistics and the *restricted* partition function, and to study the thermodynamics of the quasi-particles.

This paper is organized as follows. In Section 2 we study the ideal g-on gas in detail. We introduce the restricted partition function, and we shall show that the recursion relation of the restricted partition function gives the thermodynamics of the system without the explicit form of the partition function. In Section 3 we consider a generalization to the multi-component case. We first construct the super Rogers–Szegö polynomial in terms of the path configuration sum. As seen from the correspondence with the character of affine Lie algebra for non-super case [10], we can expect that the SRS polynomial gives the character of the affine super Lie algebra. In fact this SRS polynomial coincides with the restricted partition function of the super Yangian invariant spin chains, and this supports our observation. We then apply the recursion relation method to this (super) Yangian invariant system, and study the thermodynamic properties of the quasi-particles. We shall especially give the distribution function and the central charge. The last section is devoted to the concluding remarks.

2 Ideal g-on Gas and Recursion Relation

2.1 Exclusion Statistics

We consider the one-component exclusion gas satisfying

$$g = -\frac{\delta G}{\delta N}, \tag{2.1}$$

where $g = 0$ and $g = 1$ respectively denotes the free boson and fermion. We call the particle obeying Eq. (2.1) "g-on", and study the partition function for the ideal g-on gas. For a meanwhile, we suppose that g is a positive integer. We set for brevity

$$V_k = \mathrm{e}^{-\beta(\varepsilon_k - \mu)}, \tag{2.2}$$

$$\lambda = \mathrm{e}^{\beta\,\mu_{\text{hole}}}, \tag{2.3}$$

where β is an inverse of the temperature $\beta = \frac{1}{k_{\mathrm{B}}\,T}$, and ε_k is the energy of the k-th level. Parameters μ and μ_{hole} denote the chemical potentials for a particle and a hole, respectively. We further introduce the *restricted* partition function φ_k; "restricted" means that φ_k denotes the partition function when the particle can occupy the energy level only up to the k-th level from the ground state. We then obtain the partition function by a large k limit;

$$\Xi = \lim_{k\to\infty} \varphi_k. \tag{2.4}$$

Due to the definition of the exclusion statistics (2.1), we have a constraint for the restricted partition function. To help an understanding, we introduce a kind of "motif". The restricted partition function φ_n is represented by a summation of n sequences of 0 and 1, where 0 (resp. 1) at k-th position from the left denotes an empty (resp. occupancy) at the k-th energy level. For example in a case of $g = 2$, a consecutive 1's is forbidden, and we have explicit forms as

$$\begin{aligned}
\varphi_1^{(g=2)} &= (0) + (1) \\
&= \lambda + V_0, \\
\varphi_2^{(g=2)} &= (00) + (10) + (01) \\
&= \lambda^2 + \lambda\,(V_0 + V_1), \\
\varphi_3^{(g=2)} &= (000) + (100) + (010) + (001) + (101) \\
&= \lambda^3 + \lambda^2\,(V_0 + V_1 + V_2) + \lambda\,V_0\,V_2.
\end{aligned}$$

One easily finds that positions n_j of '1' in motif satisfy a constraint,

$$n_{j+1} - n_j \geq g, \tag{2.5}$$

which is a translation of the definition of the exclusion statistics (2.1). See also that the number of states of N particles occupying G states is given by [3]

$$W = \frac{(G + (N-1)\,(1-g))!}{N!\,(G - g\,N - (1-g))!}. \tag{2.6}$$

From a condition (2.5), we see that the restricted partition function $\varphi_n^{(g)}$ for the ideal g-on gas (2.1) satisfies the recursion relation [8],

$$\varphi_{n+1}^{(g)} = \lambda\, \varphi_n^{(g)} + \lambda^{g-1}\, V_n\, \varphi_{n-g+1}^{(g)}. \tag{2.7}$$

Initial conditions for these recursion relation are given by

$$\begin{aligned} \varphi_0 &= 1, \\ \varphi_1 &= \lambda + V_0, \\ \varphi_2 &= \lambda\,(\lambda + V_0 + V_1), \\ &\cdots\cdots \\ \varphi_g &= \lambda^{g-1}\,(\lambda + V_0 + \cdots + V_{g-1}). \end{aligned} \tag{2.8}$$

The grand partition function $\Xi^{(g)}$ is thus computed by use of the recursion relation (2.7) in a large k limit (2.4). It should be noted that the recursion relation (2.7) is solved easily for $g = 0$ and $g = 1$ cases. We get

$$\Xi^{(g=0)} = \prod_{n=0}^{\infty} \frac{\lambda}{1 - \lambda^{-1}\, e^{-\beta(\varepsilon_n - \mu)}}, \qquad \Xi^{(g=1)} = \prod_{n=0}^{\infty} \left(\lambda + e^{-\beta(\varepsilon_n - \mu)}\right), \tag{2.9}$$

which respectively denote the well-known grand partition functions for the free (chiral) boson and fermion. We thus see that a parameter g indeed intertwines the quantum statistics of the boson and the fermion. We should note that above computation of the partition function of g-on is essentially same with that of the monomer-polymer problem [11].

2.2 *Character for g-on Gas*

We shall solve the recursion relation (2.7) for arbitrary g. Hereafter we suppose that the particle has a linear dispersion relation, $\varepsilon(k) = k$. Here the momentum k is written as

$$k_j = \frac{2\pi}{L} n_j, \qquad \text{for } n_j \in \mathbb{Z}_{\geq 0},$$

where L denotes the size of the system, and the particle is assumed to be right-moving. By this reason character such as Eq. (1.2) is called *chiral* partition function. With these assumptions we substitute in Eq. (2.7)

$$V_n = q^n\, x, \tag{2.10}$$

where q and the fugacity x are defined by

$$q = \exp(-2\pi\beta/L), \qquad\qquad x = \exp(\beta\,\mu). \tag{2.11}$$

By setting $\mu_{\text{hole}} = 0$, the recursion relation (2.7) reduces to

$$\varphi_{n+1}^{(g)} = \varphi_n^{(g)} + x\, q^n\, \varphi_{n-g+1}^{(g)}. \tag{2.12}$$

To solve this recursion relation with an initial condition (2.8), we introduce the generating function $F^{(g)}(t) \equiv F^{(g)}(t; x, q)$ by

$$F^{(g)}(t) = \sum_{n=0}^{\infty} \varphi_n^{(g)}(x, q)\, t^n. \tag{2.13}$$

We see that the generating function satisfies the q-difference equation as

$$(1-t)\, F^{(g)}(t) = 1 + x \sum_{k=1}^{g-1} q^{k-1} t^k + x\, q^{g-1} t^g\, F^{(g)}(q\, t). \tag{2.14}$$

This difference equation is a linear combination of the simplified q-difference equation,

$$(1-t)\, K(t) = 1 + x\, t^{2\,a} q^b\, K(q\, t),$$

where a and b are arbitrary parameters. We find that a solution of this equation is given by

$$K(t) = \sum_{n=0}^{\infty} x^n t^{2\,a\,n} \frac{q^{a\,n^2+(b-a)\,n}}{(t;q)_{n+1}}.$$

Proof is rather straightforward;

$$\begin{aligned}
K(t) &= \frac{1}{1-t} + \sum_{n=1}^{\infty} x^n t^{2\,a\,n} \frac{q^{a\,n^2+(b-a)\,n}}{(t;q)_{n+1}} \\
&= \frac{1}{1-t} + \sum_{n=0}^{\infty} x^{n+1} t^{2\,a\,(n+1)} \frac{q^{a\,n^2+(b+a)\,n+b}}{(1-t)\,(q\,t;q)_{n+1}} \\
&= \frac{1}{1-t} + \frac{x\,t^{2\,a} q^b}{1-t} \sum_{n=0}^{\infty} x^n\, (q\,t)^{2\,a\,n} \frac{q^{a\,n^2+(b-a)\,n}}{(q\,t;q)_{n+1}} \\
&= \frac{1}{1-t} + \frac{x\,t^{2\,a} q^b}{1-t} K(q\,t).
\end{aligned}$$

By use of this result, we find that the q-difference equation (2.14) is solved by

$$F^{(g)}(t) = \sum_{k=0}^{g-1} \sum_{n=0}^{\infty} x^{n+1-\delta_{k,0}}\, t^{g\,n+k} \frac{q^{\frac{g}{2}n^2+(\frac{g}{2}-1+k)\,n+k-1+\delta_{k,0}}}{(t;q)_{n+1}}. \tag{2.15}$$

As a result we obtain the restricted partition function as [8]

$$\varphi_N^{(g)} = \sum_{k=0}^{g-1} \sum_{0 \le g\,n \le N-k} \begin{bmatrix} N-(g-1)\,n-k \\ n \end{bmatrix}_q x^{n+1-\delta_{k,0}}\, q^{\frac{g}{2}n^2+(\frac{g}{2}-1+k)\,n+k-1+\delta_{k,0}}, \tag{2.16}$$

which gives the grand partition function by Eq. (2.4) as

$$\Xi^{(g)}(x,q) = \sum_{n=0}^{\infty} x^n \frac{q^{\frac{g}{2}(n^2-n)}}{(q;q)_n}. \tag{2.17}$$

Though g may take only positive integer in the recursion relation (2.7) of the restricted partition function, the character (2.17) is naturally extended to arbitrary non-negative g. This formula exactly coincides with the grand partition function for the Calogero model confined in an external harmonic potential [12],

$$\mathcal{H}_{\mathrm{C}} = \sum_{j=1}^{N}\Bigl(-\frac{\partial^2}{\partial {x_j}^2} + \omega^2 {x_j}^2\Bigr) + 2\,g\,(g-1) \sum_{1\le j<k\le N} \frac{1}{(x_j-x_k)^2}. \tag{2.18}$$

See that Eq. (2.17) supports an equality (1.3). We note that character (2.17) is factorized for several g's. As was pointed out in eqs. (2.9), the (chiral) boson and fermion cases are rewritten respectively as

$$\Xi^{(g=0)}(x,q) = \prod_{n=0}^{\infty} \frac{1}{1-x\,q^n}, \qquad \Xi^{(g=1)}(x,q) = \prod_{n=0}^{\infty} (1+x\,q^n).$$

Due to the Rogers–Ramanujan identity ($a = 0, 1$),

$$\begin{aligned} \sum_{n=0}^{\infty} \frac{q^{n\,(n+a)}}{(q;q)_n} &= \prod_{n=1}^{\infty} \frac{1}{(1-q^{5\,n-1-a})\,(1-q^{5\,n-4+a})} \\ &= \frac{1}{(q;q)_\infty} \sum_{n=-\infty}^{\infty} \left(q^{n\,(10\,n+1+2\,a)} - q^{(5\,n+2-a)\,(2\,n+1)}\right), \end{aligned} \tag{2.19}$$

we can interpret that that the $g = 2$-on gas acts like a *fractional* boson (only every 2 per 5 energy levels can be occupied). It should be noted that the recursion relation (2.12) with $g = 2$ was used to prove these Rogers–Ramanujan identities [13].

The character (2.17) as a grand partition function for the Calogero model (2.18) can be constructed based on the Bethe ansatz type equation without using the recursion relation of the restricted partition function. We introduce the Bethe ansatz type equation [14] as

$$n_j = I_j + (g-1)\sum_{\substack{l=1 \\ l\neq j}}^{N} \theta(n_j - n_l), \tag{2.20}$$

where $\theta(x)$ is a step function;

$$\theta(x) = \begin{cases} 1, & x > 0, \\ 0, & x < 0. \end{cases}$$

Quantum number I_j takes a positive integer, and n_j corresponds to the momentum of the chiral particles. The ground state is given by $I_j = j-1$. The total energy is a summation of the momenta,

$$E(\{I\}) = \sum_{j=1}^{N} n_j, \tag{2.21}$$

and the partition function is given by

$$\Xi^{(g)}(x,q) = \sum_{N=0}^{\infty} x^N\, Z_N^{(g)}(x,q), \qquad Z_N^{(g)}(x,q) = \sum_{\substack{\text{config. } I_j \\ |I|=N}} q^{E(\{I\})}. \tag{2.22}$$

It is easy to see that, by fixing $n_1 < n_2 < \cdots < n_N$, the momenta n_j fulfill a condition (2.5) of the exclusion statistics. To compute the partition function from the Bethe ansatz equation (2.20), we rewrite the energy as

$$E(\{I\}) \equiv \sum_j n_j = \sum_j I_j + (g-1)\sum_{j,k} \theta(n_j - n_k) = \sum_j I_j + \frac{g-1}{2} N\,(N-1).$$

As a consequence, we get the partition function for N particles as

$$Z_N^{(g)}(x,q) = q^{\frac{g-1}{2} N\,(N-1)} \sum_{\substack{I_j \geq 0 \\ I_{j+1} > I_j}} q^{\sum_{j=1}^{N} I_j} = \frac{q^{\frac{g}{2} N\,(N-1)}}{(q;q)_N}, \tag{2.23}$$

which coincides with Eq. (2.17).

2.3 Thermodynamics I

Once we have the explicit form of the partition function (2.17) in a q-series, we can study the physical properties in a thermodynamic limit $q \to 1$ (or $L \to \infty$ in Eq. (2.11)) (see, e.g., Refs. 9, 15). This method was originally used to derive the nontrivial identity of the dilogarithm function from the asymptotic behavior of the Andrews–Gordon identity [16].

When we set $x = 1$ in Eq. (2.17), we have in a limit $q \to 1$

$$\Xi^{(g)} \simeq \int \mathrm{d}n \, \exp\left(n \log x + \frac{g}{2} n^2 \log q - \frac{1}{\log q} \int_1^{q^n} \mathrm{d}t \, \frac{\log(1-t)}{t} \right).$$

The saddle point of the integral is fixed by

$$w^g = 1 - w, \tag{2.24}$$

where we have set $w = q^n$. With a solution of Eq. (2.24), we have

$$\Xi^{(g)} \sim \exp\left(- \frac{1}{\log q} \, L(1-w) \right),$$

which gives the effective central charge as

$$c(g) = \frac{6}{\pi^2} \, L(1-w). \tag{2.25}$$

Here the function $L(z)$ is the Rogers dilogarithm function defined by

$$\begin{aligned} L(z) &= \sum_{n=1}^{\infty} \frac{z^n}{n^2} + \frac{1}{2} \log z \log(1-z) \\ &= - \int_0^z \frac{\log(1-t)}{t} \mathrm{d}t + \frac{1}{2} \log z \log(1-z), \end{aligned} \tag{2.26}$$

and satisfies the following identities (see, *e.g.*, Ref. 17 for review);

$$L(x) + L(1-x) = L(1) = \frac{\pi^2}{6}, \tag{2.27a}$$

$$L(x) + L(y) = L(x\,y) + L\left(\frac{x\,(1-y)}{1-x\,y}\right) + L\left(\frac{y\,(1-x)}{1-x\,y}\right). \tag{2.27b}$$

We plot the central charge (2.25) as a function of the statistical interaction g in Figure 1. From the identities of the dilogarithm function, we find a duality,

$$c(g^{-1}) = 1 - c(g), \tag{2.28}$$

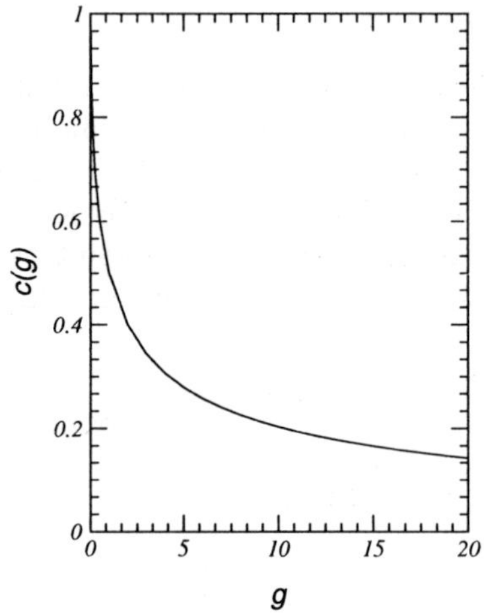

Figure 1. Effective central charge $c(g)$ is plotted for the statistical interaction g.

which represents a particle-hole duality for an ideal g-on gas [5, 18]. We also have

$$c(2) = \frac{2}{5}, \qquad\qquad c(\frac{1}{2}) = \frac{3}{5},$$

which supports a result of the Rogers–Ramanujan identity (2.19).

2.4 *Thermodynamics II*

In the previous section, we have computed the thermodynamics of the ideal g-on gas from the explicit form of the grand partition function $\Xi^{(g)}(x, q)$. In this section we apply another method; based on the recursion relation of the restricted partition function, we shall derive the central charge without using the explicit form of the partition function. We suppose that the restricted partition function φ_k is asymptotically given by

$$\varphi_k \simeq w^{-k}, \tag{2.29}$$

in $k \to \infty$. In a terminology of the inverse scattering method, this function corresponds to the Jost function and the recursion relation can be viewed as the Lax formalism. By substituting this function into the recursion relation (2.7), we see that w satisfies

$$V \lambda^{g-1} w^g = 1 - \lambda w, \tag{2.30}$$

where we have set $V = \mathrm{e}^{-\beta(\varepsilon-\mu)}$ with the energy ε. See that the same equation has already appeared in Eq. (2.24) as a saddle point equation. As the restricted partition function φ_k is approximately regarded as the partition function Ξ in a large k limit (2.4), we see that an average occupation number $\langle n_{\mathrm{av}}\rangle$ is given by [19]

$$\langle n_{\mathrm{av}}\rangle \simeq \frac{1}{k}\cdot\frac{\partial}{\partial(\beta\,\mu)}\log\varphi_k. \tag{2.31}$$

By use of the asymptotic form (2.29) and the algebraic equation (2.30), we get

$$\langle n_{\mathrm{av}}\rangle = \left(g + \frac{\lambda\,w}{1-\lambda\,w}\right)^{-1}. \tag{2.32}$$

Figure 2 denotes an average number of particle $\langle n_{\mathrm{av}}\rangle$ (2.32) for several g. We have $\langle n_{\mathrm{av}}\rangle|_{\varepsilon\to-\infty} = 1/g$. We can check that the well-known results for the boson ($g=0$) and the fermion ($g=1$) are recovered from Eq. (2.32);

$$\langle n_{\mathrm{av}}\rangle_{\mathrm{Bose}} = \frac{1}{\lambda\,V^{-1}-1}, \qquad \langle n_{\mathrm{av}}\rangle_{\mathrm{Fermi}} = \frac{1}{1+\lambda\,V^{-1}}.$$

We also have a particle-hole duality,

$$\langle n_{\mathrm{av}}\rangle + \langle n_{\mathrm{hole}}\rangle = 1, \tag{2.33}$$

where the average number of hole is computed by

$$\langle n_{\mathrm{hole}}\rangle \simeq \frac{1}{k}\cdot\frac{\partial}{\partial(\beta\mu_{\mathrm{hole}})}\log\varphi_k.$$

Based on a distribution function (2.31) we can get the central charge in a simple way. We set $\lambda = 1$ hereafter. As the energy is given by

$$E = \int \mathrm{d}\varepsilon\,\varepsilon\,\langle n_{\mathrm{av}}\rangle,$$

the specific heat is

$$C_V = -2\,k_{\mathrm{B}}{}^2\,T\int_0^1 \mathrm{d}V\,\frac{1}{V}\log w(V). \tag{2.34}$$

Recall that $w(V)$ is a solution of the algebraic equation (2.30) with $\lambda = 1$ which follows from the recursion relation (2.7). As the specific heat C_V is proportional to the central charge, we finally obtain the effective central charge as [20]

$$c(g) = -\frac{6}{\pi^2}\int_0^1 \mathrm{d}V\,\frac{1}{V}\log w(V). \tag{2.35}$$

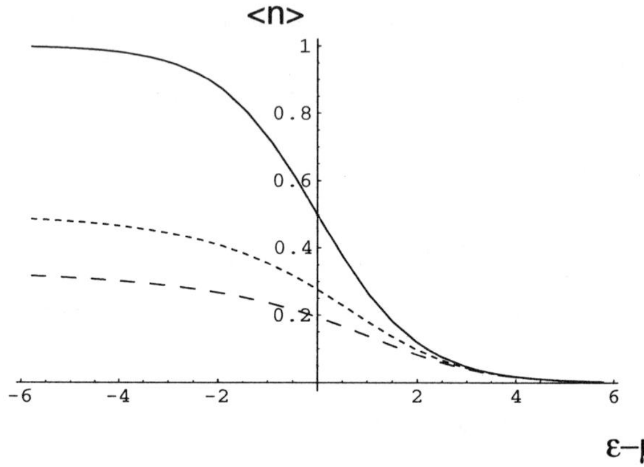

Figure 2. Average number $\langle n_{\rm av}\rangle$ of particle $\langle n_{\rm av}\rangle$ is plotted for energy ε in units β^{-1}. Solid, dotted, and dashed lines respectively denote $g = 1, 2, 3$.

This formula gives the same result with Eq. (2.25). We stress that the central charge can be computed only from the recursion relation of the restricted partition function without an explicit form of the grand partition function.

3 Generalization: Yangian Symmetry

3.1 Background: Physical Spin Chains

In our previous section we have demonstrated in a case of the ideal g-on gas that the thermodynamic properties can be derived only from the recursion relation for the restricted partition function. Hereafter we shall try to generalize this strategy for multicomponent cases. A crucial key in the following is the Yangian symmetry [21] and its intimate relationship with the level-1 WZNW theory [22–24]. This relationship has revealed from studies of the one-dimensional spin chains with inverse-square interactions whose Hamiltonians are given by

$$\mathcal{H}_{\rm HS} = \sum_{1\le j<k\le L} \frac{z_j\, z_k}{(z_j - z_k)\,(z_k - z_j)}\, \mathbf{P}_{jk}, \tag{3.1a}$$

$$\mathcal{H}_{\rm PF} = \sum_{1\le j<k\le L} \frac{1}{(x_j - x_k)^2}\, \mathbf{P}_{jk}. \tag{3.1b}$$

Here L denotes the number of spins, and $\mathbf{P}_{jk}$ is the permutation operator acting on the j-th and k-th spins. The former is called the Haldane–Shastry (HS) model [25, 26], and a position z_j of the j-th spin is on a unit circle; $z_j = \exp(2\pi \mathrm{i}\, j/L)$. The latter is the Polychronakos–Frahm (PF) model [27, 28], and a position x_j of spin is the zero point of the L-th order Hermite polynomial satisfying

$$x_j = \sum_{k=1,k\neq j}^{L} \frac{2}{(x_j - x_k)^3}.$$

Both two Hamiltonians are integrable and Yangian invariant for arbitrary L; explicit form of the Yangian currents are constructed in Refs. 22, 29. Remarkable is that the simultaneous eigenstates of these systems can be constructed combinatorially in terms of "motif", a sequence of '0' and '1' which respectively represents absence or presence of quasi-particles at each energy level [22]. The dispersion relation of the quasi-particle excitations of the HS spin chain (3.1a) is square while that of the PF chain (3.1b) is linear. This fact is consistent with the representation of the Yangian conserved operators in terms of the level-1 current algebra [23, 24]. Therein the first conserved operator of the Yangian invariant systems is identified with the Virasoro generator L_0, and consequently we can conclude that the Hamiltonian of the PF spin chain is proportional to L_0 in a large L limit, $L_0 \propto \lim_{L\to\infty} \mathcal{H}_{\mathrm{PF}}$ [29]. Furthermore thanks to the linear dispersion relation of the PF chain, the energy spectrum is equidistant, and we can exactly compute the partition function $Z_L(q)$ of the PF chain for any L spins. This partition function $Z_L(q)$ is shown to be related to a generalization of the Rogers–Szegö (RS) polynomial, and we have the recursion relation for this *restricted* partition function $Z_L(q)$ [29]. This property holds also in the supersymmetric case, *i.e.*, $\mathbf{P}$ in the Hamiltonians (3.1) is the supersymmetric permutation operator (see below), and the partition function for the supersymmetric $s\ell_{M,N}$ PF chain is given as [30]

$$Z_L^{(M,N)}(q) = \mathrm{Tr}\, q^{E_0 - \mathcal{H}_{\mathrm{PF}}} = H_L^{(M,N)}(x, y; q)\Big|_{x=y=1}, \tag{3.2}$$

where E_0 is a constant and $H_L^{(M,N)}(x, y; q)$ is the super Rogers–Szegö polynomial (see Appendix for definition). As the PF spin chain (3.1b) preserves the Yangian symmetry for arbitrary L spins, it is natural to conclude that the recursion relation of the SRS polynomial is responsible for the thermodynamics of the quasi-particle excitations.

In the following, we shall consider the supersymmetric case based on a result of Ref. 30. We consider the super Lie algebra $g\ell_{M,N}$ (see e.g., Ref. 31).

We suppose that $M \geq 1$. For our convention, we set $\mathbf{B} = \mathbf{B}_+ \sqcup \mathbf{B}_-$ where

$$\mathbf{B}_+ = \{\epsilon_1, \ldots, \epsilon_M\}, \qquad \mathbf{B}_- = \{\epsilon_{M+1}, \epsilon_{M+2}, \ldots, \epsilon_{M+N}\}.$$

We use $\mathbf{E}^{ab}$ as a set of generators of the supersymmetric Lie algebra $g\ell_{M,N}$;

$$\big[\mathbf{E}^{ab}\ ,\ \mathbf{E}^{cd}\big\} = \delta^{bc}\,\mathbf{E}^{ad} - (-1)^{(p(a)+p(b))\,(p(c)+p(d))}\,\delta^{ad}\,\mathbf{E}^{cb},$$

where we assign the parity function $p(a)$ as

$$p(a) = \begin{cases} 0, & \text{for } \epsilon_a \in \mathbf{B}_+, \\ 1, & \text{for } \epsilon_a \in \mathbf{B}_-. \end{cases}$$

With these generators, the permutation operator $\mathbf{P}$ is written as

$$\mathbf{P} = \sum_{a,b=1}^{M+N} (-1)^{p(b)}\,\mathbf{E}^{ab} \otimes \mathbf{E}^{ba}.$$

The generators $\mathbf{E}^{ab}$ and the permutation operator $\mathbf{P}$ act on bases e^a as follows;

$$\mathbf{E}^{ab}\,e^c = \delta^{bc}\,e^a,$$

$$\mathbf{P}\,e^a \otimes e^b = (-1)^{p(a)\,p(b)}\,e^b \otimes e^a.$$

We note that the permutation operator is also represented as [32]

$$\mathbf{P}_{ij} = \sum_{a,b=1}^{M+N} c^{\dagger}_{i,a}\,c^{\dagger}_{j,b}\,c_{i,b}\,c_{j,a}, \tag{3.3}$$

where $c_{i,a}$ is a bosonic (resp. fermionic) annihilation operator for $\epsilon_a \in \mathbf{B}_+$ (resp. $\epsilon_a \in \mathbf{B}_-$), and we have a constraint for each i;

$$\sum_{a=1}^{M+N} c^{\dagger}_{i,\alpha}\,c_{i,\alpha} = 1.$$

The representation (3.3) makes it clear a relationship between the Hamiltonians (3.1) and the supersymmetric t-J model with long-range interactions [33].

We comment on a definition of the super Yangian symmetry. It is defined from the fundamental commutation relation [34],

$$R^{12}(\xi_1 - \xi_2)\,\overset{1}{\mathbf{T}}(\xi_1)\,\overset{2}{\mathbf{T}}(\xi_2) = \overset{2}{\mathbf{T}}(\xi_2)\,\overset{1}{\mathbf{T}}(\xi_1)\,R^{12}(\xi_1 - \xi_2), \tag{3.4}$$

where $R(\xi)$ is the rational solution of the graded Yang–Baxter equation [35],

$$R(\xi) = \xi + \hbar\,\mathbf{P}. \tag{3.5}$$

Here $\hbar$ is a parameter. The super Yangian algebra is then constructed from a set of generators T_k^{ab}, which are defined by

$$\mathbf{T}(\xi) = \sum_{a,b=1}^{M+N} (-1)^{p(b)}\, T^{ab}(\xi) \otimes \mathbf{E}^{ba}, \tag{3.6}$$

$$T^{ab}(\xi) = \delta^{ab}\,(-1)^{p(b)} - \hbar \sum_{k=0}^{\infty} T_k^{ab}\, \xi^{-k-1}. \tag{3.7}$$

3.2 Path Configuration Sum

We shall show that the SRS polynomial, which plays a role of the restricted partition function (3.2), can be constructed from the path configuration sum. This correspondence was pointed out for the $s\ell_M$ case [36] based on a result of Ref. 29 where a representation of "motif" was given recursively. We define the energy function $H : \mathbf{B} \times \mathbf{B} \to \{0, 1\}$ as [37]

$$H(\epsilon_a, \epsilon_b) = \begin{cases} 0, & \text{if } \epsilon_a \prec \epsilon_b, \text{ or } \epsilon_a = \epsilon_b \in \mathbf{B}_-, \\ 1, & \text{if } \epsilon_a \succ \epsilon_b, \text{ or } \epsilon_a = \epsilon_b \in \mathbf{B}_+. \end{cases} \tag{3.8}$$

This is related to the order of the spectral parameter of the supersymmetric solution of the YBE [35]. It is noted that the energy function H has a symmetry,

$$H^{s\ell_{M,N}}(\epsilon_a, \epsilon_b) = 1 - H^{s\ell_{N,M}}(\epsilon_{M+N-b}, \epsilon_{M+N-a}).$$

We define a path configuration with finite length;

$$\underline{s} = (s_1, s_2, \ldots, s_{L+1}), \tag{3.9}$$

where $s_i \in \mathbf{B}$, and we denote the length of $\underline{s}$ as $\ell(\underline{s}) = L + 1$. The energy and the weight of the path configuration $\underline{s}$ are defined respectively as

$$E(\underline{s}) = \sum_{j=1}^{\ell(\underline{s})-1} j \cdot H(s_j, s_{j+1}), \tag{3.10}$$

$$\mathrm{wt}(\underline{s}) = \sum_{j=1}^{\ell(\underline{s})} s_j. \tag{3.11}$$

With these notations, we define the character of the path configuration by

$$\mathrm{ch}\, \mathcal{S}_L = \sum_{\underline{s}:\ell(\underline{s})=L+1} q^{E(\underline{s})}\, \mathrm{e}^{\mathrm{wt}(\underline{s})}. \tag{3.12}$$

This path configuration sum first appeared in calculation of the one-point function of the integrable lattice model by the Baxter corner transfer matrix method (see *e.g.* Ref. 38), and its importance was recognized by the correspondence with the character of the affine Lie algebra [10].

We now give the explicit form of these characters. The character $\mathrm{ch}\,\mathcal{S}_L$ can be naturally decomposed by a boundary condition as

$$\mathrm{ch}\,\mathcal{S}_L = \bigsqcup_{a=1}^{M+N} \mathrm{ch}\,\mathcal{S}_L^{(a)},$$

where

$$\mathrm{ch}\,\mathcal{S}_L^{(a)} = \sum_{\substack{\underline{s}:\ell(\underline{s})=L+1 \\ s_{L+1}=\epsilon_a}} q^{E(\underline{s})}\,\mathrm{e}^{\mathrm{wt}(\underline{s})}. \tag{3.13}$$

When we introduce a set of the q-polynomials $\underline{\chi}_L = (f_L^{(1)},\ldots,f_L^{(M)},b_L^{(1)},\ldots,b_L^{(N)})$ by

$$\mathrm{ch}\,\mathcal{S}_L^{(a)} = \begin{cases} q^{\frac{1}{2}L(L+1)}\cdot b_L^{(a)}(x,y;q^{-1}), & \text{for } \epsilon_a\in\mathbf{B}_+, \\ q^{\frac{1}{2}L(L+1)}\cdot f_L^{(a-M)}(x,y;q^{-1}), & \text{for } \epsilon_a\in\mathbf{B}_-, \end{cases}$$

we see by definition that $\underline{\chi}_L$ satisfies the recursion relation,

$$\underline{\chi}_{L+1} = \underline{\chi}_L\cdot\mathbf{M}\cdot\mathrm{diag}(x_1,\ldots,x_M,y_1,\ldots,y_N),$$

where each element of $(M+N)\times(M+N)$ matrix $\mathbf{M}$ is given by

$$M_{ab} = q^{1-H(\epsilon_a,\epsilon_b)}.$$

This recursion relation is solved, and we have a generating function for $\underline{\chi}_L$ as follows;

$$\sum_{L=0}^{\infty}\frac{b_L^{(1)}(x,y;q)}{(q;q)_L}\cdot t^L = G^{(M,N)}(t),$$

$$\sum_{L=0}^{\infty}\frac{b_L^{(a+1)}(x,y;q)}{(q;q)_L}\cdot t^L = \frac{x_{a+1}}{x_1}\cdot\left(\prod_{i=1}^{a}(1-t\,x_i)\right)\cdot G^{(M,N)}(t),$$

$$\sum_{L=0}^{\infty}\frac{f_L^{(N)}(x,y;q)}{(q;q)_L}\cdot t^L = \frac{y_N}{x_1}\cdot G^{(M,N)}(q\,t),$$

$$\sum_{L=0}^{\infty}\frac{f_L^{(a)}(x,y;q)}{(q;q)_L}\cdot t^L = \frac{y_a}{x_1}\cdot\left(\prod_{i=a+1}^{N}(1+t\,y_i)\right)\cdot G^{(M,N)}(q\,t).$$

Here the function $G^{(M,N)}(t)$ is defined in Eq. (A.14). This set of identities can be derived by identifying the recursion relation of the character $b_L^{(1)}(x,y;q)$ with Eq. (A.17). From these equalities, we can conclude that the character is written in terms of the super Rogers–Szegö (SRS) polynomial (A.15), and especially we have

$$\mathrm{ch}\,\mathcal{S}_L^{(1)} = q^{\frac{1}{2}L(L+1)} \cdot H_L^{(M,N)}(x,y;q^{-1}). \tag{3.14}$$

The motif, eigenstates of the super Yangian invariant spin system, can be constructed from this identification of the character with the SRS polynomial which is known to be connected with the partition function of the PF spin chain. We define the motif $\underline{m} = (m_1, \ldots, m_L)$ from the path $\underline{s}$ with $\ell(\underline{s}) = L+1$ and $s_{L+1} = \epsilon_1$ by

$$m_i = H(s_i, s_{i+1}) \in \{0, 1\}. \tag{3.15}$$

The energy (3.10) of the path is rewritten in terms of the motif $\underline{m}$ as

$$E(\underline{m}) = \sum_{j=1}^{L} j \cdot m_j. \tag{3.16}$$

As the character for this decomposed path is identified with the SRS polynomial (3.14) and the SRS polynomial is represented by Eq. (A.19), we find that the degeneracy of the motif $\underline{m}$ is given by the skew super Schur function for the border strip $\langle k_1+1, k_2+1, \ldots, k_r+1\rangle$ with $\sum_{i=1}^{r}(k_i+1) = L$ boxes where $r-1$ denotes a number of 0's in $\underline{m}$ and

$$\underline{m} = (\underbrace{1\ldots1}_{k_1}0\underbrace{1\ldots1}_{k_2}0\cdots0\underbrace{1\ldots1}_{k_r}).$$

We give examples in the following tables. Compare them with the explicit expression of the SRS polynomial $H_L(x,y;q)$ in Appendix.

$\underline{m}$	$E(\underline{m})$	border strip
0	0	$\langle 1,1\rangle$
1	1	$\langle 2\rangle$

$\underline{m}$	$E(\underline{m})$	border strip
00	0	$\langle 1,1,1\rangle$
01	2	$\langle 1,2\rangle$
10	1	$\langle 2,1\rangle$
11	3	$\langle 3\rangle$

$\underline{m}$	$E(\underline{m})$	border strip
000	0	$\langle 1,1,1,1\rangle$
100	1	$\langle 2,1,1\rangle$
010	2	$\langle 1,2,1\rangle$
001	3	$\langle 1,1,2\rangle$
110	3	$\langle 3,1\rangle$
101	4	$\langle 2,2\rangle$
011	5	$\langle 1,3\rangle$
111	6	$\langle 4\rangle$

3.3 Thermodynamics

We study the thermodynamics of quasi-particles for the super Yangian invariant system. From the fact that the Yangian symmetry preserves for spin chains with arbitrary L spins, the partition function $Z_L^{(M,N)}(q)$ of the PF spin chain with finite L spins can be regarded as the restricted partition function φ_L. In this interpretation 1's in each motif $\underline{m}$ indicates the presence of the quasi-particle (see Eq. (3.16)), and we can obtain the thermodynamic properties of the quasi-particles from the recursion relation of the SRS polynomial following a strategy in Section 2. By assuming the asymptotic form $\varphi_L \sim w^{-L}$, we obtain from the recursion relation (A.17) the algebraic equation for w [30];

$$V\,(2-z)^N = z^M, \qquad z = 1-(1-V)\,w, \tag{3.17}$$

where $V = \exp\bigl(-\beta\,(\varepsilon-\mu)\bigr)$ and ε is the energy. As was shown in the case of the ideal g-on gas, we can compute the central charge as

$$\begin{aligned} c &= -\frac{6}{\pi^2}\int_0^1 \mathrm{d}V\,\frac{1}{V}\log w(V) \\ &= M-1+\frac{N}{2}. \end{aligned} \tag{3.18}$$

The average occupation number of the quasi-particle is also given in terms of the solution of Eq. (3.17) by

$$\begin{aligned} \langle n_{\text{av}}\rangle &= -\frac{\partial}{\partial\,(\beta\,\mu)}\log w \\ &= -\frac{V}{1-V}+\frac{z}{1-z}\cdot\frac{1}{M+\frac{z}{2-z}N}. \end{aligned} \tag{3.19}$$

See that we have the particle-hole duality between $s\ell_{M,N}$ and $s\ell_{N,M}$ cases;

$$\langle n_{\text{av}}^{s\ell_{M,N}}(V)\rangle + \langle n_{\text{av}}^{s\ell_{N,M}}(V^{-1})\rangle = 1. \tag{3.20}$$

We list some interesting properties below.

1. $s\ell_{1,1}$;

   We have $c = 1/2$, and from Eq. (3.19) we get

$$\langle n_{\text{av}}\rangle = \frac{\partial}{\partial\,(\beta\,\mu)}\log(1+V). \tag{3.21}$$

We thus see that the theory coincides with the free (chiral) fermion theory. This can also be supported from the character;

$$\mathrm{ch}^{(1,1)}(q) = \lim_{L\to\infty} H_L^{(1,1)}(x,y;q) = \prod_{n=0}^{\infty} (x + q^n\, y)\,. \tag{3.22}$$

2. $s\ell_M$;

   We have $c = M - 1$, which exactly coincides with the central charge of the level-1 $\widehat{s\ell_M}$ WZW theory. In fact we recover the character of the level-1 WZW model in the large L limit as [29]

$$\mathrm{ch}_{\Lambda_\ell}(q) = \lim_{\substack{L\to\infty \\ L\equiv\ell \mod M}} q^{\frac{M-1}{2M}L^2}\, H_L^{(M,0)}(x;q^{-1}). \tag{3.23}$$

   The average occupation number (3.19) of the quasi-particle is given by

$$\langle n_{\mathrm{av}}\rangle = \frac{\partial}{\partial(\beta\,\mu)} \log\left(1 + V^{\frac{1}{M}} + \cdots + V^{\frac{M-1}{M}}\right). \tag{3.24}$$

   This distribution function denotes that of the Gentile statistics [39], in which at most $M-1$ quasi-particles can occupy the same energy levels though the energy is scaled by $1/M$. We have $\langle n_{\mathrm{av}}\rangle|_{\varepsilon\to-\infty} = (M-1)/M$.

3. $s\ell_{1,N}$;

   The central charge is computed as $c = N/2$, which indicates that the theory can be described by the N sets of the chiral fermions. Indeed it can be checked from the character,

$$\mathrm{ch}^{(1,N)}(q) = \lim_{L\to\infty} H_L^{(1,N)}(x,y;q) = \prod_{i=1}^{N}\Big(\prod_{n=0}^{\infty}(x + y_i\, q^n)\Big). \tag{3.25}$$

   Although, the distribution function of quasi-particles differs from that of the fermion case, and we have, *e.g.*, in $N = 2$ case,

$$\langle n_{\mathrm{av}}^{s\ell_{1,2}}\rangle = \frac{\partial}{\partial(\beta\,\mu)} \log\left(\frac{1 + 2\,V + \sqrt{1 + 8\,V}}{2}\right).$$

4. $s\ell_{M,M}$;

   We have the central charge $c = \frac{3}{2}M - 1$, and the distribution function of the quasi-particle is given by

$$\langle n_{\mathrm{av}}\rangle = \frac{\partial}{\partial(\beta\mu)} \log\left(\left(1 + V^{\frac{1}{M}}\right)\left(1 + V^{\frac{1}{M}} + \cdots + V^{\frac{M-1}{M}}\right)\right), \tag{3.26}$$

which is a new statistics as a supersymmetric extension of the Gentile statistics (3.24). This result also supports a validity of our results as a supersymmetric algebra. We have $\langle n_{\rm av}\rangle|_{\varepsilon\to-\infty} = 1$. In $M = 1$ case we recover the free fermion theory (3.21), and in $M = 2$ case we realize 2 sets of fermions, $\langle n_{\rm av}^{s\ell_{2.2}}\rangle = 2\,\langle n_{\rm av}^{s\ell_2}\rangle$.

We plot an average occupation of states at a temperature $T > 0$ for $s\ell_{M,N}$ cases $(M, N \leq 2)$ and the free fermion case in Fig. 3. See Ref. 40 for application of these symmetry to the edge states of the quantum Hall states.

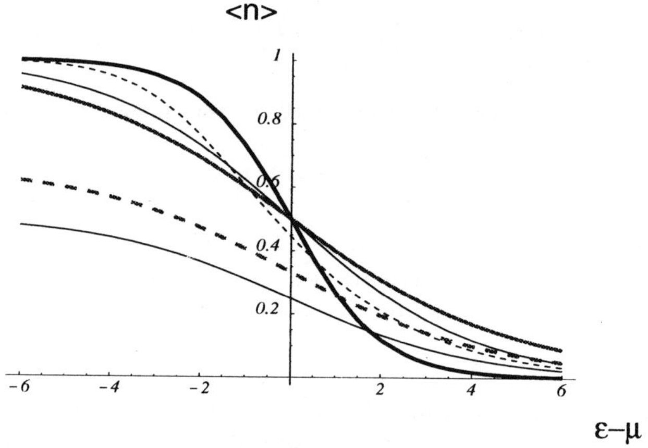

Figure 3. Average occupation number of quasi-particles is plotted for some of the super $s\ell_{M,N}$ cases. A bold line is the distribution function for the free fermion, $\langle n_{\rm av}\rangle = V/(1+V)$, which corresponds to the $s\ell_{1,1}$ case. Two solid lines denote those for $s\ell_2$ and $s\ell_{2,2}$. A dotted line shows for $s\ell_{2,1}$ case. A gray solid and dotted line are respectively for the $s\ell_{3,3}$ and $s\ell_3$ cases. The energy is given in units β^{-1}.

4 Concluding Remarks

We have studied the chiral partition function and the thermodynamics of quasi-particle which obeys the exclusion statistics. In a one-component case, we have computed explicitly the chiral partition function, and established the relationship between the statistical interaction and the chiral partition function. Based on this simple example, we have reviewed how to study the thermodynamic properties only from the recursion relation of the restricted partition function without an explicit form of the partition function. As an application to the multi-component case, we have adopted the super Yangian invariant system; for example we have a spin model such as the supersymmetric extension of the Haldane–Shastry model and the Heisenberg model with infinite spins. The restricted partition function of this system is given in terms of the super Rogers–Szegö polynomial, and we have reconstructed the SRS polynomial in terms of the path configuration sum. This representation of the SRS polynomial is shown to give the degeneracy of the motif, which is the eigenstates of the Yangian invariant system. As we have the correspondence between the $s\ell_M$ Rogers–Szegö polynomial and the character of the $\widehat{s\ell_M}$ affine Lie algebra, we can expect that our formula is useful in studying the character of the affine super Lie algebra (see Ref. 41 for recent work). Using the recursion relation of the SRS polynomial, we have revealed the thermodynamic properties of quasi-particles in the super Yangian invariant system. We have computed the central charge and the distribution function explicitly. These properties can be applicable for future studies of the physical problems such as the edge states of the fractional quantum Hall states and the disordered system [40, 42–44].

Acknowledgement

The author would like to thank B. Basu-Mallick for collaboration in Ref. 30. He also thanks A. N. Kirillov and F. Hirashita for kind hospitality during Workshop.

Appendix

A Super Polynomial

A.1 Super Schur Polynomial

We consider the super Schur polynomial associated with the supersymmetric $s\ell_{M,N}$ Lie algebra. We suppose $M \geq 1$. We set $\mathbf{B} = \mathbf{B}_+ \sqcup \mathbf{B}_-$, where

$$\mathbf{B}_+ = \{\epsilon_1, \ldots, \epsilon_M\}, \qquad \mathbf{B}_- = \{\epsilon_{M+1}, \epsilon_{M+2}, \ldots, \epsilon_{M+N}\}.$$

Two sets $\mathbf{B}_+$ and $\mathbf{B}_-$ respectively denote the even and odd set. To get the $s\ell_M$ case, we should treat $\mathbf{B}_-$ as an empty set; $\mathbf{B}_- = \phi$. We fix an ordering in $\mathbf{B}$ as

$$\epsilon_1 \prec \epsilon_2 \prec \cdots \prec \epsilon_{M+N}. \tag{A.1}$$

The super Schur function appears in the character for the $s\ell_{M,N}$ super Lie algebra [31], and it is defined combinatorially as follows [45]. For a given Young diagram λ, the Young tableaux T is given by filling a number $1, 2, \ldots, M+N$ in λ by the following rules;

- The entries in each row are increasing, allowing the repetition of elements in $\{i|\epsilon_i \in \mathbf{B}_+\}$, but not permitting the repetition of elements in $\{i|\epsilon_i \in \mathbf{B}_-\}$,
- The entries in each column are increasing, allowing the repetition of elements in $\{i|\epsilon_i \in \mathbf{B}_-\}$, but not permitting the repetition of elements in $\{i|\epsilon_i \in \mathbf{B}_+\}$.

To define the super Schur polynomial, we set the weight for the Young tableaux T as

$$\mathrm{wt}(T) = \sum_{a=1}^{M+N} m_a\, \epsilon_a, \tag{A.2}$$

where m_a is the number of a's in T. In terms of the weight of T, the super Schur polynomial $S_\lambda(x,y)$ is defined as

$$S_\lambda(x,y) = \sum_{\text{tableaux } T \text{ of shape } \lambda} \mathrm{e}^{\mathrm{wt}(T)}, \tag{A.3}$$

where we use

$$\begin{cases} x_i = \mathrm{e}^{\epsilon_i}, & \text{for } \epsilon_i \in \mathbf{B}_+, \\ y_i = \mathrm{e}^{\epsilon_{M+i}}, & \text{for } \epsilon_{M+i} \in \mathbf{B}_-. \end{cases}$$

Due to the rule of filling a number, the Young diagram λ must be contained by the (M,N) hook (see Fig. 4). By this reason the super Schur function $S_\lambda(x,y)$ is often called as the *hook* Schur function. We note that the super Schur polynomial can be rewritten as

$$S_\lambda(x,y) = \det\bigl(c_{\lambda_i+j-i}\bigr)_{1\leq i,j\leq \mathrm{length}(\lambda)}, \tag{A.4}$$

where $c_n = c_n(x,y)$ is defined by

$$\frac{\prod_{j=1}^{N}(1+t\,y_j)}{\prod_{i=1}^{M}(1-t\,x_i)} = \sum_{n=0}^{\infty} c_n\, t^n.$$

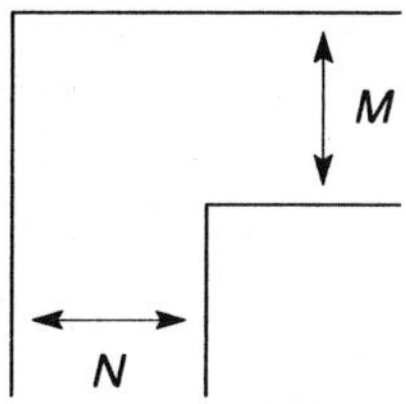

Figure 4. (M, N) hook Young diagram.

Hereafter we denote $s_\lambda(x)$ as the usual Schur function. For our convention we use the elementary polynomial $e_L(x)$, complete symmetric polynomial $h_L(x)$, and elementary super polynomial $E_L(x)$ defined respectively by

$$e_L(x) = s_{[1^L]}(x), \qquad h_L(x) = s_{[L]}(x), \qquad E_L(x,y) = S_{[1^L]}(x,y). \qquad \text{(A.5)}$$

Note that

$$\prod_{i=1}^{M}(1 + t\, x_i) = \sum_{k=0}^{M} e_k(x)\, t^k, \qquad \text{(A.6)}$$

$$\prod_{i=1}^{M} \frac{1}{1 - t\, x_i} = \sum_{k=0}^{\infty} h_k(x)\, t^k, \qquad \text{(A.7)}$$

$$\frac{\prod_{i=1}^{M}(1 - t\, x_i)}{\prod_{j=1}^{N}(1 + t\, y_j)} = \sum_{k=0}^{\infty}(-1)^k E_k(x,y)\, t^k. \qquad \text{(A.8)}$$

The skew super Schur function $S_{\lambda/\mu}(x,y)$ can be also defined in a same way by filling numbers in the skew Young diagram λ/μ following the rules before. It is easy to see that we have [46]

$$S_{\lambda/\mu}(x,y) = \sum_{\mu\subset\nu\subset\lambda} s_{\lambda/\nu}(x)\, s_{\nu'/\mu'}(y), \qquad \text{(A.9)}$$

where μ' and ν' denote partitions conjugate to μ and ν respectively. For the super skew Schur function, we have a Jacobi–Trudi type formula,

$$S_{\lambda/\mu}(x,y) = \det\big(E_{\lambda'_i-\mu'_j-i+j}(x,y)\big)_{1\le i,j\le r}. \qquad \text{(A.10)}$$

Throughout this paper, we denote $\langle m_1, m_2, \ldots, m_r\rangle$ as the border strip (or, rim hook, or ribbon), which is the connected skew Young diagram with no

2×2 square;

$\langle m_1, m_2, \ldots, m_r \rangle =$ 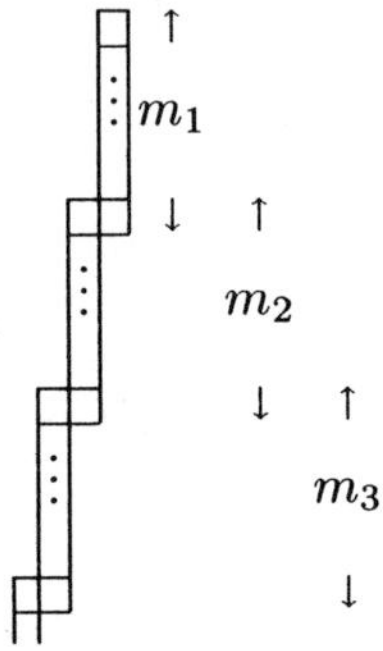 (A.11)

For instance, $\langle 2,1,4,1,1 \rangle =$ 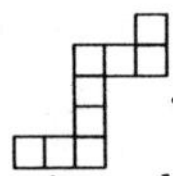. Due to Eq. (A.10), the skew super Schur function for $\langle m_1, \ldots, m_r \rangle$ is given by

$$S_{\langle m_1,\ldots,m_r \rangle}(x,y) = \begin{vmatrix} E_{m_r} & E_{m_r+m_{r-1}} & \cdots & \cdots & E_{m_r+\cdots+m_1} \\ 1 & E_{m_{r-1}} & E_{m_{r-1}+m_{r-2}} & \cdots & E_{m_{r-1}+\cdots+m_1} \\ 0 & 1 & E_{m_{r-2}} & \cdots & \vdots \\ \vdots & \ddots & \ddots & \ddots & \vdots \\ 0 & \cdots & 0 & 1 & E_{m_1} \end{vmatrix}. \tag{A.12}$$

A.2 Super Rogers–Szegö Polynomial

We define the super Rogers–Szegö polynomial $H_L^{(M,N)}(x,y;q)$ as [30]

$$\begin{aligned} H_L^{(M,N)}(x,y;q) &= \sum_{\substack{\sum_{i=1}^M a_i + \sum_{j=1}^N b_j = L \\ a_i \geq 0, b_j \geq 0}} \frac{(q;q)_L}{\prod_{i=1}^M (q;q)_{a_i} \cdot \prod_{j=1}^N (q^{-1};q^{-1})_{b_j}} \cdot \prod_{i=1}^M {x_i}^{a_i} \cdot \prod_{j=1}^N \left(-\frac{y_j}{q}\right)^{b_j} \\ &= \sum_{\sum_{i=1}^{M+N} a_i = L} \frac{(q;q)_L}{\prod_{i=1}^{M+N} (q;q)_{a_i}} \cdot q^{\frac{1}{2}\sum_{j=M+1}^{M+N} a_j\,(a_j-1)} \cdot e^{\sum_{i=1}^{M+N} a_i\,\epsilon_i}. \end{aligned} \tag{A.13}$$

The SRS polynomial has a generating function; when we set $G^{(M,N)}(t) \equiv G^{(M,N)}(t;x,y;q)$ by

$$G^{(M,N)}(t) = \frac{1}{\prod_{i=1}^M (t\,x_i;q)_\infty \cdot \prod_{j=1}^N (-t\,y_j\,q^{-1};q^{-1})_\infty}, \tag{A.14}$$

we have

$$G^{(M,N)}(t) = \sum_{L=0}^{\infty} \frac{H_L^{(M,N)}(x,y;q)}{(q;q)_L} \cdot t^L. \tag{A.15}$$

It is straightforward to see that the function $G^{(M,N)}(t)$ satisfies the q-difference equation,

$$\left(\prod_{j=1}^{N}(1+t\,y_j)\right) \cdot G^{(M,N)}(q\,t) = \left(\prod_{i=1}^{M}(1-t\,x_i)\right) \cdot G^{(M,N)}(t), \tag{A.16}$$

which gives the recursion relation of the SRS polynomial as

$$\begin{aligned} &H_L^{(M,N)}(x,y;q) \\ &= \sum_{k=1}^{\max(M,N)} (-1)^{k-1} \frac{(q;q)_{L-1}}{(q;q)_{L-k}} \cdot \left(e_k(x) - (-1)^k q^{L-k} e_k(y)\right) \cdot H_{L-k}^{(M,N)}(x,y;q). \end{aligned} \tag{A.17}$$

Here we have used the generating function of the elementary polynomials (A.6). Recalling Eq. (A.8), we can rewrite the q-difference equation (A.16) as

$$G^{(M,N)}(q\,t) = \left(\sum_{L=0}^{\infty}(-1)^L E_L(x,y)\,t^L\right) \cdot G^{(M,N)}(t),$$

and we obtain another type of recursion relation of the SRS polynomial;

$$H_L^{(M,N)}(x,y;q) = \sum_{k=1}^{L}(-1)^{k+1}\frac{(q;q)_{L-1}}{(q;q)_{L-k}} \cdot E_k(x,y) \cdot H_{L-k}^{(M,N)}(x,y;q). \tag{A.18}$$

As was proved in Ref. 30 the SRS polynomial can be rewritten as

$$\begin{aligned} &H_L^{(M,N)}(x,y;q) \\ &\quad = \sum_{r=1}^{L} \sum_{\substack{\sum_{i=1}^{r} m_i = L \\ m_i \geq 1}} q^{\frac{1}{2}L\,(L+1) - \sum_{i=1}^{r}(m_1+\cdots m_i)} \cdot S_{\langle m_1,\ldots,m_r\rangle}(x,y), \end{aligned} \tag{A.19}$$

where $S_{\langle m_1,\ldots,m_r\rangle}(x,y)$ is the skew super Schur polynomial (A.12). The essential point of the proof of above equality is that the skew super Schur polynomial (A.12) fulfills

$$S_{\langle m_1,\ldots,m_r\rangle}(x,y) = \sum_{i=1}^{r}(-1)^{i+1}\, E_{m_r+\cdots+m_{r-i+1}}(x,y) \cdot S_{\langle m_1,\ldots,m_{r-i}\rangle}(x,y).$$

Then we see that the expression (A.19) also satisfies the recursion relation (A.17). We give some explicit form of the SRS polynomials below;

$$H_0(x,y;q) = 1,$$

$$H_1(x,y;q) = S_{(1)}(x,y),$$

$$H_2(x,y;q) = q\cdot S_{(1,1)}(x,y) + S_{(2)}(x,y),$$

$$H_3(x,y;q) = q^3\cdot S_{(1,1,1)}(x,y) + q^2\cdot S_{(2,1)}(x,y) + q\cdot S_{(2,2)/(1)}(x,y) + S_{(3)}(x,y),$$

$$\begin{aligned} H_4(x,y;q) = {} & q^6\cdot S_{(1,1,1,1)}(x,y) + q^5\cdot S_{(2,1,1)}(x,y) + q^4\cdot S_{(2,2,1)/(1)}(x,y) + q^3\cdot S_{(2,2,2)/(1,1)}(x,y) \\ & + q^3\cdot S_{(3,1)}(x,y) + q^2\cdot S_{(3,2)/(1)}(x,y) + q\cdot S_{(3,3)/(2)}(x,y) \\ & + S_{(4)}(x,y). \end{aligned}$$

A.3 (q, t)-Deformation of the Super Schur Polynomial

In the $s\ell_M$ case, the Rogers–Szegö polynomial $H_L^{(M)}(x;q)$ coincides with a limit $t \to 0$ of the Macdonald polynomial with one-row Young diagram $P_{[L]}(x;q,t)$. It is therefore natural to define the super Macdonald polynomial for the one-row Young diagram by a t-deformation of the SRS polynomial $H_L^{(M,N)}(x,y;q)$;

$$\prod_{i=1}^{M}\frac{(t\,z\,x_i;q)_\infty}{(z\,x_i;q)_\infty}\cdot\prod_{j=1}^{N}\frac{(-t\,z\,y_j\,q^{-1};q^{-1})_\infty}{(-z\,y_j\,q^{-1};q^{-1})_\infty} = \sum_{L=0}^{\infty}\frac{(t;q)_L}{(q;q)_L}\cdot P_{[L]}^{(M,N)}(x,y;q,t)\cdot z^L. \tag{A.20}$$

This polynomial has a following property;

$$P_{[L]}^{(M,N)}(x,y;q,t) = \sum_\lambda u_{[L],\lambda}\cdot S_\lambda(x,y) \longleftrightarrow P_{[L]}(x;q,t) = \sum_\lambda u_{[L],\lambda}\cdot s_\lambda(x). \tag{A.21}$$

Further studies should be required on these polynomials.

References

1. F. D. M. Haldane: Phys. Rev. Lett. **67**, 937 (1991).
2. A. Shapere and F. Wilczek, eds.: *Geometric Phases in Physics, Advanced Series in Mathematical Physics* **5** (World Scientific, Singapore, 1989).

3. Y.-S. Wu: Phys. Rev. Lett. **73**, 922 (1994).
4. S. B. Isakov: Phys. Rev. Lett. **73**, 2150 (1994).
5. C. Nayak and F. Wilczek: Phys. Rev. Lett. **73**, 2740 (1994).
6. D. Bernard and Y.-S. Wu: In *New Developments of Integrable Systems and Long-Ranged Interaction Models* (Edited by M.-L. Ge and Y.-S. Wu) (World Scientific, Singapore, 1995) pp. 10–20.
7. T. Fukui and N. Kawakami: Phys. Rev. B **51**, 5239 (1995).
8. K. Hikami: Phys. Lett. A **205**, 364 (1995).
9. A. Berkovich and B. M. McCoy: In *Statistical Physics on the Eve of the 21st Century* (Edited by M. T. Batchelor and L. T. Wille) (World Scientific, Singapore, 1999) pp. 240–256.
10. E. Date, M. Jimbo, A. Kuniba, T. Miwa, and M. Okado: Lett. Math. Phys. **17**, 69 (1987).
11. O. J. Heilmann and E. H. Lieb: Phys. Rev. Lett. **24**, 1412 (1970).
12. F. Calogero: J. Math. Phys. **12**, 419 (1971).
13. G. E. Andrews: *q-Series: Their Development and Application in Analysis, Number Theory, Combinatorics, Physics, and Computer Algebra* (AMS, Providence, 1986).
14. N. Kawakami: J. Phys. Soc. Jpn **62**, 4163 (1993).
15. W. Nahm, A. Recknagel, and M. Terhoeven: Mod. Phys. Lett. A **8**, 1835 (1993).
16. B. Richmond and G. Szekeres: J. Austral. Math. Soc. (Sereis A) **31**, 362 (1981).
17. A. N. Kirillov: Prog. Theor. Phys. Suppl. **118**, 61 (1995).
18. A. K. Rajagopal: Phys. Rev. Lett. **74**, 1048 (1995).
19. K. Hikami: Phys. Rev. Lett. **80**, 4374 (1998).
20. P. Bouwknegt and K. Schoutens: Nucl. Phys. B **547**, 501 (1999).
21. V. G. Drinfeld: Sov. Math. Dokl. **36**, 212 (1988).
22. F. D. M. Haldane, Z. N. C. Ha, J. C. Talstra, D. Bernard, and V. Pasquier: Phys. Rev. Lett. **69**, 2021 (1992).
23. D. Bernard, V. Pasquier, and D. Serban: Nucl. Phys. B **428**, 612 (1994).
24. K. Schoutens: Phys. Lett. B **331**, 335 (1994).
25. F. D. M. Haldane: Phys. Rev. Lett. **60**, 635 (1988).
26. B. S. Shastry: Phys. Rev. Lett. **60**, 639 (1988).
27. A. P. Polychronakos: Phys. Rev. Lett. **70**, 2329 (1993).
28. H. Frahm: J. Phys. A: Math. Gen. **26**, L473 (1993).
29. K. Hikami: Nucl. Phys. B **441**, 530 (1995); J. Phys. Soc. Jpn. **64**, 1047 (1995).
30. K. Hikami and B. Basu-Mallick: Nucl. Phys. B **566**, 511 (2000).
31. V. G. Kac: Commun. Math. Phys. **53**, 31 (1977).

32. F. D. M. Haldane: In *Correlation Effects in Low Dimensional Electron Systems* (Edited by A. Okiji and N. Kawakami) (Springer, Berlin, 1994) pp. 3–20.
33. Y. Kuramoto and H. Yokoyama: Phys. Rev. Lett. **67**, 1338 (1991).
34. M. L. Nazarov: Lett. Math. Phys. **21**, 123 (1991).
35. P. P. Kulish and E. K. Sklyanin: J. Sov. Math. **19**, 1596 (1982).
36. A. N. Kirillov, A. Kuniba, and T. Nakanishi: Commun. Math. Phys. **185**, 441 (1997).
37. K. Hikami and R. Inoue: J. Phys. A: Math. Gen. **33**, 4081 (2000).
38. R. J. Baxter: *Exactly Solved Models in Statistical Mechanics* (Academic Press, London, 1982).
39. G. Gentile: Nuovo Cim. **17**, 493 (1940).
40. K. Hikami: Phys. Rev. B **61**, 4473 (1999).
41. V. G. Kac and M. Wakimoto: math-ph/0006007 .
42. R. A. J. van Elburg and K. Schoutens: Phys. Rev. B **58**, 15704 (1998).
43. S. B. Isakov, T. Martin, and S. Ouvry: Phys. Rev. Lett. **80**, 580 (1999).
44. L. G. C. Rego and G. Kirczenow: Phys. Rev. B **59**, 13080 (1999).
45. A. Berele and A. Regev: Adv. Math. **64**, 118 (1987).
46. A. Molev: In *Kirillov's Seminar on Representation Theory* (Edited by G. I. Olshanski) (Amer. Math. Soc., Providence, 1998), *AMS Translations Series 2* **181**, pp. 109–137.

FORRESTER'S CONSTANT TERM CONJECTURE AND ITS q–ANALOGUE

JYOICHI KANEKO

Graduate School of Mathematics, Kyushu University
Ropponmatsu Fukuoka 810-8560, Japan
E-mail: kaneko@math.kyushu-u.ac.jp

We describe our approach to the Forrester's conjectured constant term identity which is equivalent to a new kind of generalization of the Selberg integral. We also deal with the q–analogue of the conjecture in a parallel way and give a simple proof in the case $N_1 = 2$

1 Introduction

Suppose $a, b, k \in \mathbb{Z}_{\geq 0}$ and let

$$H(x_1, \dots, x_N; a, b, k) := \prod_{l=1}^{N}(1 - x_l)^a(1 - \frac{1}{x_l})^b \prod_{i \neq j}^{N}(1 - \frac{x_i}{x_j})^k. \tag{1.1}$$

The Morris constant term identity [13] states that

$$\mathrm{CT}_{\{x\}} H(x_1, \dots, x_N; a, b, k) = M_N(a, b, k) \tag{1.2}$$

where $\mathrm{CT}_{\{x\}}$ denotes the constant term in the Laurent polynomial and

$$M_N(a, b, k) := \prod_{l=0}^{N-1} \frac{\Gamma(a + b + 1 + kl)\Gamma(1 + k(l + 1))}{\Gamma(a + 1 + kl)\Gamma(b + 1 + kl)\Gamma(1 + k)}. \tag{1.3}$$

In [4], after some numerical computations, Forrester suggested the following remarkable generalization of this identity and proved the special cases $a = b = 0$ (general k, N_0, N_1) and $k = 1$ (general a, b, N_0, N_1).

Conjecture

$$\mathrm{CT}_{\{x\}} \prod_{i,j=N_0+1, i \neq j}^{N} (1 - \frac{x_i}{x_j}) H(x_1, \dots, x_N; a, b, k) = M_{N_0}(a, b, k)$$
$$\times \prod_{j=0}^{N_1-1} \frac{(j + 1)\Gamma((k + 1)j + a + b + kN_0 + 1)\Gamma((k + 1)(j + 1) + kN_0)}{\Gamma(k + 1)\Gamma((k + 1)j + a + kN_0 + 1)\Gamma((k + 1)j + b + kN_0 + 1)}, \tag{1.4}$$

where $N = N_0 + N_1$.

Subsequently a q–analogue of the conjecture was formulated by Baker and Forrester [1] where several cases including the $N_1 = 2$ case were proved. All these proofs are strongly of combinatorial nature. It should be noted here that Forrester formulated a more general constant term conjecture associated to a partition $N = N_0 + N_1 + \cdots + N_m$, on which not much has been known.

In this paper we shall briefly describe our approach to the conjecture invoking the integration formula of Jack polynomials in [9,10]. We also deal with the q–analogue in a parallel way and give a simple proof in the case $N_1 = 2$.

2 Forrester's conjecture

It is known [4] that the conjecture (1.4) is equivalent to the following integral evaluation.

$$\int_{[0,1]^N} \prod_{i=1}^{N} t_i^{x-1}(1-t_i)^{y-1} \prod_{j=N_0+1}^{N} t_j^{-N_1+1} \times \prod_{N_0+1\le i<j\le N} (t_i - t_j)^2 \prod_{1\le i<j\le N} (t_i - t_j)^{2k} dt_1 \cdots dt_N$$

$$= \prod_{l=0}^{N_0-1} \frac{\Gamma(y+kl)\Gamma(x+k(N-1-l))\Gamma(1+k(l+1))}{\Gamma(x+y+k(N-1+l))\Gamma(1+k)}$$

$$\times \prod_{j=0}^{N_1-1} \frac{(j+1)\Gamma(y+(k+1)j+kN_0))\Gamma(x+k(N-1-N_0-j)-j)}{\Gamma(x+y+k(N-1+N_0+j)+j)}$$

$$\times \prod_{j=0}^{N_1-1} \frac{\Gamma((k+1)(j+1)+kN_0)}{\Gamma(1+k)}. \tag{2.1}$$

This is clearly equivalent to the symmetrized form:

$$\int_{[0,1]^N} \prod_{i=1}^{N} t_i^{x-N_1}(1-t_i)^{y-1} f_{N,N_0}(t_1,\ldots,t_N) \prod_{1\le i<j\le N} (t_i - t_j)^{2k} dt_1 \cdots dt_N$$

$$= \frac{N!}{N_0!N_1!}(\text{RHS of } (2.1)), \tag{2.2}$$

where

$$f_{N,N_0}(t_1,\ldots,t_N) := \sum_{1\le i_1<\cdots<i_{N_0}\le N} (t_{i_1}\cdots t_{i_{N_0}})^{N_1-1} \times \prod_{j_1<j_2,\{j_1,j_2\}\subset\{1,\ldots,N\}\setminus\{i_1,\ldots,i_{N_0}\}} (t_{j_1}-t_{j_2})^2. \tag{2.3}$$

We rewrite (2.3) by invoking on the integration formula of Jack polynomials.

Let $\lambda = (\lambda_1,\ldots,\lambda_N)$ be a partition of length $\le N$ and denote by $|\lambda|$ its weight: $|\lambda| = \lambda_1+\cdots+\lambda_N$. For a parameter α, define

$$(a)_\lambda^{(\alpha)} = \prod_{i=1}^{N}(a-\frac{1}{\alpha}(i-1))_{\lambda_i} \tag{2.4}$$

where $(a)_n = a(a+1)\cdots(a+n-1)$ and $(a)_0 = 1$. Let $J_\lambda^{(\alpha)}(t)$ be the Jack polynomial corresponding to the partition λ with the parameter α [12,14]. They form a basis o f symmetric polynomials in $t_1,\ldots,t_N$ with coefficients in $\mathbb{Q}(\alpha)$, the rational function field in α. Moreover we have [6,7]:

$$\int_{[0,1]^N} J_\lambda^{(\frac{1}{k})}(t)\prod_{i=1}^{N} t_i^{x-1}(1-t_i)^{y-1}\prod_{1\le i<j\le N}(t_i-t_j)^{2k}dt_1\cdots dt_N = k^{-|\lambda|}S_{N,0}(x,y,k)\frac{(kN)_{\lambda a}^{(\frac{1}{k})}(k(N-1)+x)_\lambda^{(\frac{1}{k})}}{(2k(N-1)+x+y)_\lambda^{(\frac{1}{k})}}, \tag{2.5}$$

where

$$S_{N,0}(x,y,k) = \prod_{i=1}^{N}\frac{\Gamma(1+ki)\Gamma(x+k(i-1))\Gamma(y+k(i-1))}{\Gamma(1+k)\Gamma(x+y+k(N+i-2))}. \tag{2.6}$$

Suppose the symmetric polynomial $f_{N,N_0}(t)$ is expressed in terms of Jack polynomials:

$$f_{N,N_0}(t) = \sum_\lambda C_\lambda J_\lambda^{(\frac{1}{k})}(t). \tag{2.7}$$

Here we notice that the same kind of expansion, namely, the square of the difference product in terms of Schur polynomials, was extensively investigated in [3]. Substituting (2.7) into (2.2), we see that the integral formula (2.2) is equivalent to

$$\sum_\lambda C_\lambda \frac{(kN)_\lambda^{(\frac{1}{k})}(k(N-1)+x-N_1+1)_\lambda^{(\frac{1}{k})}}{(2k(N-1)+x+y-N_1+1)_\lambda^{(\frac{1}{k})}}$$

$$= k^{N(N_1-1)}\frac{N!}{N_0!}\prod_{l=0}^{N_0-1}\frac{(x-N_1+1+k(N-1-l))_{N_1-1}}{(x+y-N_1+1+k(N-1+l))_{N_1-1}} \tag{2.8}$$

$$\times \prod_{j=0}^{N_1-1}\frac{(x-N_1+1+k(N_1-1-j))_{N_1-1-j}(y+k(N_0+j))_j}{(x+y-N_1+1+k(N+N_0+j-1))_{N_1+j-1}}$$

$$\times \prod_{j=0}^{N_1-1}(1+k(N_0+j+1))_j.$$

Now define partitions κ, μ, ν by

$$\begin{aligned}\kappa &= (2N_1-2, 2N_1-3, \ldots, N_1, \overbrace{N_1-1, \ldots, N_1-1}^{N_0+1}),\\ \mu &= (\overbrace{N_1-1, \ldots, N_1-1}^{N_0+1}, N_1-2, \ldots, 1),\\ \nu &= (N_1-1, N_1-2, \ldots, 1).\end{aligned} \tag{2.9}$$

and write

$\widetilde{C_\lambda} = C_\lambda (kN)_\lambda^{(\frac{1}{k})}, X = x-N_1+1+k(N-1), Y = x+y-N_1+1+2k(N-1).$

Then it is straightforward to rewrite (2.8) to get

$$\sum_\lambda \widetilde{C_\lambda}\frac{(X)_\lambda^{(\frac{1}{k})}(Y)_\kappa^{(\frac{1}{k})}}{(X)_\mu^{(\frac{1}{k})}(Y)_\lambda^{(\frac{1}{k})}} = C_{N,N_0}(Y-X)_\nu^{(\frac{1}{k})}, \tag{2.10}$$

where $C_{N,N_0} = k^{N(N_1-1)}\frac{N!}{N_0!}\prod_{j=0}^{N_1-1}(1+k(N_0+j+1))_j$.

We verified (2.10) in the cases $N_1 = 2, 3$ by calculating C_λ's [9], and the case $N_1 = N-1$ by using the Chu–Vandermonde formula for the generalized binomial coefficients [10], but it still seems difficult to give a full proof and we content ourselves with showing a partial factorization associated to the first column of ν (Theorem 2.1 below).

One needs to prove the LHS of (2.10) is actually a polynomial and is divisible by $(Y-X)_{\nu}^{(\frac{1}{k})}$, i.e., the LHS vanishes when one substitutes $X = Y-(i-1)k+j-1$ for all (i,j) with $1 \le i \le N_1-1, 1 \le j \le \nu_i = N_1-i$:

$$\sum_{\lambda} \widetilde{C_\lambda} \frac{(Y-(i-1)k+j-1)_{\lambda}^{(\frac{1}{k})}}{(Y)_{\lambda}^{(\frac{1}{k})}} = 0. \tag{2.11}$$

The case $i=j=1$ of (2.11) is clear because substituting $t_1 = \cdots = t_N = 1$ in (2.7) yields $\sum_\lambda \widetilde{C_\lambda} = 0$ in view of (2.13) below. We can rewrite the equality (2.11) by using the Chu–Vandermonde formula for the generalized binomial coefficients which we now recall.

The *generalized binomial coefficient* $\binom{\lambda}{\mu}$ is defined by

$$J_{\lambda}^{(\alpha)}(1+t_1,\ldots,1+t_n) = J_{\lambda}^{(\alpha)}(1^n) \sum_{\mu} \binom{\lambda}{\mu} \frac{J_{\mu}^{(\alpha)}(t_1,\ldots,t_n)}{J_{\mu}^{(\alpha)}(1^n)} \tag{2.12}$$

where $J_{\lambda}^{(\alpha)}(1^n) := J_{\lambda}^{(\alpha)}(\overbrace{1,ts,1}^{n})$ has the following explicit formula [14, Theorem 5.4]:

$$J_{\lambda}^{(\alpha)}(1^n) = \alpha^{|\lambda|} \left(\frac{n}{\alpha}\right)_{\lambda}^{(\alpha)}. \tag{2.13}$$

It holds that $\binom{\lambda}{\mu}$ is independent of the dimension n in the definition (2.12) and $\binom{\lambda}{\mu} = 0$ unless $\mu \subset \lambda$ [7, Theorem 3], where $\mu \subset \lambda$ means $\mu_i \le \lambda_i$ for every i.

The *Chu–Vandermonde formula* is ([14, Theorem 12.1]):

$$\sum_{\mu \subset \lambda} (-1)^{|\mu|} \frac{(a)_{\mu}^{(\alpha)}}{(b)_{\mu}^{(\alpha)}} \binom{\lambda}{\mu} = \frac{(b-a)_{\lambda}^{(\alpha)}}{(b)_{\lambda}^{(\alpha)}} \tag{2.14}$$

This follows from change of variables t_i with $1-t_i, 1 \le i \le N$ in (2.5) and the definition of the generalized binomial coefficients. We apply this formula to the left–hand side of (2.11):

$$\frac{(Y-(i-1)k+j-1)_{\lambda}^{(\frac{1}{k})}}{(Y)_{\lambda}^{(\frac{1}{k})}} = \sum_{\tau \subset \lambda} (-1)^{|\tau|} \frac{((i-1)k-j+1)_{\tau}^{(\frac{1}{k})}}{(Y)_{\tau}^{(\frac{1}{k})}} \binom{\lambda}{\tau},$$

so that (2.11) is equivalent to

$$\sum_{\tau} (-1)^{|\tau|} \frac{((i-1)k-j+1)_{\tau}^{(\frac{1}{k})}}{(Y)_{\tau}^{(\frac{1}{k})}} \left(\sum_{\lambda \supset \tau} \widetilde{C_\lambda} \binom{\lambda}{\tau} \right) = 0, \tag{2.15}$$

where, since $((i-1)k-j+1)_\tau^{(\frac{1}{k})} = 0$ for a partition τ of $\tau_i \geq N_1 - i$, the summation is taken over τ with $\tau_i < N_1 - i$. We observe that the sum $\sum_{\lambda \supset \tau} \widetilde{C_\lambda}\binom{\lambda}{\tau}$ (τ being fixed) is the coefficient of $k^{-N(N_1-1)} \frac{J_\tau^{(\frac{1}{k})}(t_1,\ldots,t_N)}{J_\tau^{(\frac{1}{k})}(1^N)}$ in the expansion of $f_{N,N_0}(1+t_1,\ldots,1+t_N)$ in terms of Jack polynomials. By this rewriting, it is now easy to show

Theorem 2.1 *The equality (2.11) holds in the cases* $1 \leq i \leq N_1 - 1, j = 1$.
Proof. Let $f_m(t_1,\ldots,t_N)$ be the homogeneous part of degree m of the inhomogeneous polynomial $f_{N,N_0}(1+t_1,\ldots,1+t_N)$, so that $f_m(t_1,\ldots,t_N)$ is identically zero if $m < N_1(N_1-1)$. Let

$$\begin{aligned} f_m(t_1,\ldots,t_N) &= \textstyle\sum_\tau d_\tau J_\tau^{(\frac{1}{k})}(t_1,\dot{s},t_N) \\ &= \textstyle\sum_\tau \widetilde{d_\tau} k^{-m} \frac{J_\tau^{(\frac{1}{k})}(t_1,\ldots,t_N)}{J_\tau^{(\frac{1}{k})}(1^N)}. \end{aligned} \tag{2.16}$$

For the proof of (2.15) with $j=1$ it suffices to show that

$$\sum_{|\tau|=m} \frac{((i-1)k)_\tau^{(\frac{1}{k})}}{(Y)_\tau^{(\frac{1}{k})}} \widetilde{d_\tau} = 0. \tag{2.17}$$

Since $((i-1)k)_\tau^{(\frac{1}{k})} = 0$ when $\tau_i > 0$, the summation is over the partitions with $\tau_i = 0$. Substituting $t_i = \cdots = t_N = 0$ in (2.16) gives

$$f_m(t_1,\ldots,t_{i-1},0,\ldots,0) = \sum_\tau d_\tau J_\tau^{(\frac{1}{k})}(t_1,\ldots,t_{i-1}).$$

Since $\prod_{N_0+1 \leq i < j \leq N}(t_i - t_j)^2$ is translation invariant, we see

$$f_m(t_1,\ldots,t_{i-1},0,\ldots,0) = 0,$$

provided $i \leq N_1 - 1$, and this implies that $d_\tau = 0$ if $\tau_i = 0$. We have thus proved the theorem. □

We note that our proof of the conjecture in the case $N_1 = N - 1$ [10] was carried out by comparison of expansion of $f_m(t_1,\ldots,t_N)$ in terms of monomial symmetric polynomials and in terms of Jack polynomials.

3 q–analogue

Fix q with $0 < q < 1$ and set $(x)_\infty = (x;q)_\infty = \prod_{i=0}^{\infty}(1 - xq^i)$, $(x)_a = (x;q)_a = (x)_\infty/(xq^a)_\infty$. Define

$$\Gamma_q(x) := \frac{(q;q)_\infty}{(q^x;q)_\infty}(1-q)^{1-x},$$
$$[x]_q := \frac{1-q^x}{1-q} = \frac{\Gamma_q(x+1)}{\Gamma_q(x)}.$$

We shall denote by ${}_Q\Delta_N(t)$ the product $\prod_{1\le i<j\le N}(t_i - Qt_j)$ and by $\Delta_N(t)$ the product $\prod_{1\le i<j\le N}(t_i - t_j)$.

Suppose $a, b, k \in \mathbb{Z}_{\ge 0}$ and let

$$H_q(x_1,\ldots,x_n;a,b,k) := \prod_{l=1}^{N}(x_l;q)_a\left(\frac{q}{x_l};q\right)_b \prod_{1\le i<j\le N}\left(q\frac{x_j}{x_i};q\right)_k\left(\frac{x_i}{x_j};q\right)_k. \quad (3.1)$$

The q–Morris constant term identity [13,6,5] states that

$$\mathrm{CT}_{\{x\}}H_q(x_1,\ldots,x_N;a,b,k) = M_N(a,b,k;q) \quad (3.2)$$

where

$$M_N(a,b,k;q) := \prod_{l=0}^{N-1}\frac{\Gamma_q(a+b+1+kl)\Gamma_q(1+k(l+1))}{\Gamma_q(a+1+kl)\Gamma_q(b+1+kl)\Gamma_q(1+k)}. \quad (3.3)$$

In [1] Baker and Forrester have given a conjecture extending this identity to read:

Conjecture 3.1 *For $a, b, k \in \mathbb{Z}_{\ge 0}$ we have*

$$\mathrm{CT}_{\{x\}}\prod_{N_0+1\le i<j\le N}\left(1-q^k\frac{x_i}{x_j}\right)\left(1-q^{k+1}\frac{x_j}{x_i}\right)H_q(x_1,\ldots,x_N;a,b,k)$$

$$= M_{N_0}(a,b,k;q)\prod_{j=0}^{N_1-1}\frac{\Gamma_q((k+1)j+a+b+kN_0+1)\Gamma((k+1)(j+1)+kN_0)}{\Gamma_q((k+1)j+a+kN_0+1)\Gamma_q((k+1)j+b+kN_0+1)}$$

$$\times\prod_{j=0}^{N_1-1}\frac{[(k+1)(j+1)]_q}{\Gamma_q(k+2)}, \quad (3.4)$$

where $N = N_0 + N_1$.

They proved this conjecture in the special cases $a = \lambda$ and $a = b = 0$ by using a theorem of Bressoud and Goulden [2]. Also the special case $N_1 = 2$ was proved by adapting the method of Zeilberger [15]. In this section we formulate the conjecture in an analogous way as in the case of Forrester's conjecture by employing the q–integration formula of Macdonald polynomial $P_\lambda(t_1, \dots, t_N; q, t)$.

Define the q–integral of a function $f(t_1, \dots, t_N)$ over $[0,1]^N$ by

$$\int_{[0,1]^N} f(t_1, \dots, t_N) d_q t_1 \cdots d_q t_N = (1-q)^N \sum_{s_i \in \mathbb{Z}_{\geq 0}} f(q^{s_1}, \dots, q^{s_N}) q^{s_1} \cdots q^{s_N}.$$

It is known [1] that the conjectured constant term identity (3.4) is equivalent to the following q–Selberg type integral evaluation

$$\begin{aligned} \left(\frac{\Gamma_q(x+y)}{\Gamma_q(x)\Gamma_q(y)}\right)^N \int_{[0,1]^N} \prod_{j=1}^N t_j^{x-1} \frac{(qt_j)_\infty}{(q^y t_j)_\infty} f_{N_0,N}(t) d_q t_1 \cdots d_q t_N \\ = \left(\frac{(q)_a (q)_b}{(q)_{a+b}}\right)^N (\text{RHS of } (3.4)) \end{aligned} \tag{3.5}$$

where $b = -x, a = x + y - 1$ and

$$\begin{aligned} f_{N_0,N}(t) &:= \prod_{N_0+1 \leq i < j \leq N} \left(1 - q^k \frac{t_i}{t_j}\right)\left(1 - q^{k+1}\frac{t_j}{t_i}\right) f_N \\ f_N &:= \prod_{1 \leq i < j \leq N} (q\frac{t_j}{t_i}; q)_k (\frac{t_i}{t_j}; q)_k . \end{aligned} \tag{3.6}$$

Let us first recall the q–integration formula of Macdonald polynomials [8, Proposition 5.2]. Let $\lambda = (\lambda_1, \dots, \lambda_N)$ be a partition of length $\leq N$ and λ' be the conjugate partition. For each square $s = (i,j)$ in the diagram of λ, let

$$\begin{aligned} a(s) &= \lambda_i - j, \quad a'(s) = j - 1 \\ l(s) &= \lambda'_j - i, \quad l'(s) = i - 1 \end{aligned}$$

and put

$$\begin{aligned} h_\lambda(q,t) &= \prod_{s \in \lambda} (1 - q^{a(s)} t^{l(s)+1}), \\ (a)_\lambda^{(q,t)} &= \prod_{s \in \lambda} (t^{l'(s)} - q^{a'(s)} a) = t^{n(\lambda)} \prod_{i \geq 1} (at^{1-i})_{\lambda_i}, \end{aligned}$$

where

$$n(\lambda) := \sum_{i\geq 1}(i-1)\lambda_i.$$

For a Macdonald polynomial $P_\lambda(t_1,\ldots,t_N;q,t)$ we have

$$\int_{[0,1]^N} P_\lambda(t;q,q^k)\prod_{j=1}^N t_j^{x-1}\frac{(qt_j)_\infty}{(q^y t_j)_\infty}\prod_{1\leq i<j\leq N} t_i^{2k}(q^{1-k}t_j/t_i)_{2k}\,d_qt_1\cdots d_qt_N$$

$$= q^{kx\binom{N}{2}+2k^2\binom{N}{3}}\frac{(q^{Nk})_\lambda^{(q,q^k)}}{h_\lambda(q,q^k)}$$

$$\times\prod_{i=1}^N\frac{\Gamma_q(ik+1)\Gamma_q(x+(N-i)k+\lambda_i)\Gamma_q(y+(N-i)k)}{\Gamma_q(k+1)\Gamma_q(x+y+(2N-i-1)k+\lambda_i)}$$

$$= {}_qS_N(x,y;k)\frac{(q^{Nk})_\lambda^{(q,q^k)}}{h_\lambda(q,q^k)}\prod_{i=1}^N\frac{(q^{x+(N-i)k})_{\lambda_i}}{(q^{x+y+(N-i)k})_{\lambda_i}}, \tag{3.7}$$

where

$${}_qS_N(x,y;k) := \int_{[0,1]^N}\prod_{j=1}^N t_j^{x-1}\frac{(qt_j)_\infty}{(q^y t_j)_\infty}\prod_{1\leq i<j\leq N} t_i^{2k}(q^{1-k}t_j/t_i)_{2k}\,d_qt_1\cdots d_qt_N$$

$$= q^{kx\binom{N}{2}+2k^2\binom{N}{3}}\prod_{i=1}^N\frac{\Gamma_q(ik+1)\Gamma_q(x+(N-i)k)\Gamma_q(y+(N-i)k)}{\Gamma_q(k+1)\Gamma_q(x+y+(2N-i-1)k)}. \tag{3.8}$$

Note that $\prod_{1\leq i<j\leq N} t_i^{2k-1}(q^{1-k}t_j/t_i)_{2k-1}$ is antisymmetric and

$$\prod_{1\leq i<j\leq N} t_i^{2k}(q^{1-k}t_j/t_i)_{2k} = {}_{q^k}\Delta_N(t)\prod_{1\leq i<j\leq N} t_i^{2k-1}(q^{1-k}t_j/t_i)_{2k-1}.$$

Let A denotes the antisymmetrizer:

$$Af(t_1,\ldots,t_N) = \sum_{\sigma\in S_N}\varepsilon(\sigma)f(t_{\sigma(1)},\ldots.t_{\sigma(N)}),$$

where S_N is the symmetric group of degree N and $\varepsilon(\sigma)$ denotes the sign of the permutation σ. Note that (cf. [12, p. 207])

$$A({}_{q^k}\Delta_N(t)) = \frac{(q^k;q^k)_N}{(1-q^k)^N}\Delta_N(t).$$

Hence symmetrization of the integrand of LHS of (3.7) gives

$$\int_{[0,1]^N} P_\lambda(t;q,q^k)\prod_{j=1}^N t_j^{x-1}\frac{(qt_j)_\infty}{(q^y t_j)_\infty}\Delta_N(t)t_i^{2k-1}(q^{1-k}t_j/t_i)_{2k-1}\,d_qt_1\cdots d_qt_N$$

$$= N!\frac{(1-q^k)^N}{(q^k;q^k)_N}\,{}_qS_N(x,y;k)\frac{(q^{Nk})_\lambda^{(q,q^k)}}{h_\lambda(q,q^k)}\prod_{i=1}^N\frac{(q^{x+(N-i)k})_{\lambda_i}}{(q^{x+y+(N-i)k})_{\lambda_i}}.$$

Since

$$f_N=(-1)^{kN(N-1)/2}q^{N(N-1)k(k-1)/4}\Big(\prod_{j=1}^N t_j^{-k(N-1)}\Big)\prod_{1\le i<j\le N} t_i^{2k}\big(q^{1-k}t_j/t_i;q\big)_{2k},$$

the integral evaluation (3.5) is equivalent to

$$\int_{[0,1]^N}\prod_{j=1}^N t_j^{x-k(N-1)-N_1}\frac{(qt_j)_\infty}{(q^yt_j)_\infty}\prod_{1\le i<j\le N} t_i^{2k}(q^{1-k}t_j/t_i)_{2k}$$

$$\times\prod_{i=1}^{N_0}t_i^{N_1-1}\prod_{N_0+1\le i<j\le N}(t_i-q^{-k}t_j)(t_i-q^{k+1}t_j)d_qt_1\cdots d_qt_N$$

$$= A_N\Big(\frac{\Gamma_q(x)\Gamma_q(y)}{\Gamma_q(x+y)}\Big)^N\Big(\frac{(q)_a(q)_b}{(q)_{a+b}}\Big)^N(\text{RHS of (3.4)}),$$

where

$$A_N:=(-1)^{kN(N-1)/2+N_1(N_1-1)/2}q^{-k(k-1)N(N-1)/4-kN_1(N_1-1)/2}.$$

Therefore if we write

$$F_{N,N_0}(t_1,\ldots,t_N)=\Delta_N(t)^{-1}$$

$$\times A\Big({}_{q^k}\Delta_N(t)\prod_{i=1}^{N_0}t_i^{N_1-1}\prod_{N_0+1\le i<j\le N}(t_i-q^{-k}t_j)(t_i-q^{k+1}t_j)\Big),\qquad(3.9)$$

the symmetrized form of this evaluation is

$$\int_{[0,1]^N} \prod_{j=1}^{N} t_j^{x-k(N-1)-N_1} \frac{(qt_j)_\infty}{(q^y t_j)_\infty} \Delta_N(t)$$

$$\times \prod_{1\le i<j\le N} t_i^{2k-1} (q^{1-k} t_j/t_i)_{2k-1} F_{N,N_0}(t_1,\dots,t_N) d_q t_1 \cdots d_q t_N \tag{3.10}$$

$$= N! A_N \Big(\frac{\Gamma_q(x)\Gamma_q(y)}{\Gamma_q(x+y)}\Big)^N \Big(\frac{(q)_a (q)_b}{(q)_{a+b}}\Big)^N \text{(RHS of (3.4))}.$$

Let us expand the symmetric homogeneous polynomial $F_{N,N_0}(t_1,\dots,t_N)$ in terms of Macdonald polynomials $P_\lambda(t_1,\dots,t_N;q,q^k)$

$$F_{N,N_0}(t_1,\dots,t_N) = \sum_\lambda C_\lambda P_\lambda(t;q,q^k), \tag{3.11}$$

and substitute this into (3.10). Then by straightforward (but tedious) computation we can rewrite (3.10) as

$$\sum_\lambda C_\lambda \frac{(q^{Nk})_\lambda^{(q,q^k)}}{h_\lambda(q,q^k)} \prod_{i=1}^{N} \frac{(q^{x+k(i-1)-N_1+1})_{\lambda_i}}{(q^{x+y+k(i-1)-N_1+1})_{\lambda_i}} = q^{\frac{N_1(N_1-1)x}{2}+B_N} (1-q)^{\frac{N_1(N_1-1)}{2}}$$

$$\times \frac{(q^k;q^k)_N}{(1-q^k)^N} \prod_{j=0}^{N_1-1} \frac{[(k+1)(j+1)]_q}{[k+1]_q} \prod_{j=0}^{N_1-1} \frac{(q^{k(j+N_0+1)+1})_j}{(1-q)^j}$$

$$\times \prod_{j=0}^{N_1-1} (q^{y+k(N_0+j)})_j \prod_{l=0}^{N_0-1} (q^{x-kl-N_1+1})_{N_1-1} \prod_{j=0}^{N_1-1} (q^{x-k(N_0+j)-N_1+1})_{N_1-j-1}$$

$$\times \prod_{l=0}^{N_0-1} (q^{x+y+kl-N_1+1})_{N_1-1}^{-1} \prod_{j=0}^{N_1-1} (q^{x+y+k(N_0+j)-N_1+1})_{N_1-j+1}^{-1}, \tag{3.12}$$

where

$$B_N = k\Big(\frac{1}{2}N(N_1-1)(N-N_1-1) + \frac{1}{6}N_1(N_1-1)(N_1-2)\Big) - \frac{1}{6}N_1(N_1-1)(N_1+1). \tag{3.13}$$

Put

$$\widetilde{C_\lambda} = C_\lambda \frac{(q^{Nk})_\lambda^{(q,q^k)}}{h_\lambda(q,q^k)}, \quad X = q^{x-N_1+1}, \quad Y = q^{x+y-N_1+1+k(N-1)}, \tag{3.14}$$

so that (3.12) becomes

$$\sum_\lambda \widetilde{C_\lambda} \frac{(X)_\lambda^{(\frac{1}{k})}(Y)_\kappa^{(\frac{1}{k})}}{(X)_\mu^{(\frac{1}{k})}(Y)_\lambda^{(\frac{1}{k})}} = C_{N,N_0}(q) X^{N_1(N_1-1)/2}(YX^{-1})_\nu^{(\frac{1}{k})} \tag{3.15}$$

where

$$C_{N,N_0}(q) = q^{C_N} \frac{(q^k;q^k)_N}{(1-q^k)^N} \prod_{j=0}^{N_1-1} \frac{[(k+1)(j+1)]_q}{[k+1]_q} \prod_{j=0}^{N_1-1} (q^{k(j+N_0+1)+1})_j, \tag{3.16}$$

and

$$C_N = \frac{1}{2}k(N_1-1)((N(N-1)-N_1) + \frac{1}{3}N_1(N_1-1)(N_1-2) \tag{3.17}$$

Once the conjecture is formulated in this way, it is rather easy to prove the case $N_1 = 2$. In fact, one can readily show that the partitions appearing in the left–hand side of (3.15) are

$$\lambda^{(1)} = (2, 1^{N-2}), \lambda^{(2)} = (1^N),$$

so that the left–hand side of (3.15) is a polynomial in X and Y of degree at most one in each variable. Since

$$X^{N_1(N_1-1)/2}(YX^{-1})_\nu^{(\frac{1}{k})} = X - Y,$$

it suffices to show that the left–hand side of (3.15) is divisible by $X - Y$ and the coefficient of X is $C_{N,N_0}(q) = C_{N,N-2}(q)$. The divisibility is equivalent to

$$\sum_\lambda \widetilde{C_\lambda} = 0. \tag{3.18}$$

This is implied by

Proposition 3.2 *We have*

$$F_{N,N_0}(1, q^k, \ldots, q^{k(N-1)}) = 0. \tag{3.19}$$

Proof. The term corresponding to the unit element of the symmetric group in the alternating sum is zero because of the factor $t_{N-1} - q^{-k}t_N$. The vanishing of the remaining terms is a consequence of the factor ${}_{q^k}\Delta_N(t_{\sigma(1)}, \ldots, t_{\sigma(N)})$ in view of the following lemma.

Lemma 3.3 *For any element $\sigma \neq$ the unit in the symmetric group S_N, there exists a pair (i,j) such that $1 \le i < j \le N$ and $\sigma(i) - \sigma(j) = 1$*

Proof. This is easily verified by induction on N.

To obtain the formula of $C_{\lambda(1)}$, recall that the Macdonald polynomial $P_\lambda(x_1,\ldots,x_N;q,t)$ is an eigenfunction of the operator $D_1 = D_1(q,t)$ with the eigenvalue $e_\lambda = e_\lambda(q,t) = \sum_{i=1}^N t^{N-i}q^{\lambda_i}$ [12, Chapter 6, (4.15)]:

$$D_1 P_\lambda(x_1,\ldots,x_N;q,t) = e_\lambda P_\lambda(x_1,\ldots,x_N;q,t),$$

where

$$D_1 = \sum_{i=1}^N A_i(x;t)T_{q,x_i}, \quad A_i(x;t) = \prod_{j\neq i}\frac{tx_i - x_j}{x_i - x_j}$$

and T_{q,x_i} is the shift operator:

$$(T_{q,x_i}f)(x_1,\ldots,x_N) = f(x_1,\ldots,qx_i,\ldots,x_N).$$

Apply D_1 with $t = q^k$ to both sides of (3.11) ($N_0 = N-2$) and then substitute $t_i = q^{k(i-1)}, i = 1.\ldots,N$. By the lemma above we have

$$\begin{aligned}
&(D_1F_{N,N-2})(1,q^k,\ldots,q^{k(N-1)})\\
&= A_N(1,q^k,\ldots,q^{k(N-1)};q^k)(T_{q,t_N}F_{N,N-2})(1,q^k,\ldots,q^{k(N-1)})\\
&= q^{\frac{k(N+1)(N-2)}{2}}\frac{(1-q)(1-q^{2k+2})(1-q^{kN+1})(q^k;q^k)_N}{(1-q^{k+1})(1-q^k)^N}
\end{aligned}$$

The right-hand side is

$$e_{\lambda(1)}\widetilde{C_{\lambda(1)}} + e_{\lambda(2)}\widetilde{C_{\lambda(2)}} = (e_{\lambda(1)} - e_{\lambda(2)})\widetilde{C_{\lambda(1)}} = (1-q)(1-q^{k(N-1)+1})\widetilde{C_{\lambda(1)}},$$

so that

$$\widetilde{C_{\lambda(1)}} = q^{\frac{k(N+1)(N-2)}{2}}\frac{(1-q^{kN+1})(1+q^{k+1})(q^k;q^k)_N}{(1-q^{k(N-1)+1})(1-q^k)^N}.$$

Now the coefficient of X in the left–hand side of (3.15) is

$$\begin{aligned}
-q^{k(N-1)+1}\widetilde{C_{\lambda(1)}} - \widetilde{C_{\lambda(2)}} &= (1-q^{k(N-1)+1})\widetilde{C_{\lambda(1)}}\\
&= q^{\frac{k(N+1)(N-2)}{2}}\frac{(1-q^{kN+1})(1+q^{k+1})(q^k;q^k)_N}{(1-q^k)^N}.
\end{aligned}$$

This coincides with $C_{N,N-2}(q)$ in view of (3.16) and (3.17), thereby completing the proof of the conjecture in the case $N_1 = 2$.

References

1. Baker T. H. and Forrester P. J., *Generalizations of the q–Morris constant term identity*, J. Comb. Theory Ser. A. **81** (1998), 69-87.
2. Bressoud D. M. and Goulden I. P., *Constant term identities extending the q–Dyson theorem*, Trans. Amer. Math. Soc. **291** (1985), 203-228.
3. Di Francesco P., Gaudin M., Itzykson C. and Lesage F., *Laughlin's wave functions, Coulomb gases and expansions of the discriminant*, Internat. J. Modern Phys. **A9** (1994), 4257-4351.
4. Forrester P. J., *Normalization of the wavefunction for the Calogero–Sutherland model with internal degree of freedom*, Internat. J. Modern Phys. **B9** (1995), 1243-1261.
5. Habsieger L., *Une q–intégrale de Selberg et Askey,* SIAM J. Math. Anal. **19** (1988), 969-986.
6. Kadell K., *The Selberg–Jack symmetric functions*, Adv. Math. **130** (1997), 33-102.
7. Kaneko J., *Selberg integrals and hypergeometric functions associated with Jack polynomials*, SIAM J. Math. Anal. **24** (1993), 1086-1110.
8. Kaneko J., *q–Selberg integrals and Macdonald polynomials*, Ann. Sci. École. Norm. Sup., 4^{e}série, **29** (1996), 583-637.
9. Kaneko J., *On Forrester's Generalization of Morris constant term conjecture*, Contemp. Math. **254** (2000), 271-282.
10. Kaneko J., *Forrester's conjectured constant term identity* II, preprint, 2000.
11. Lassalle M., *Some combinatorial conjectures for shifted Jack polynomials*, Ann. Combin. **2** (1998), 145-163.
12. Macdonald I. G., *Symmetric Functions and Hall Polynomials*, Second Edition, Clarendon Press, Oxford, 1995.
13. Morris W. G., *Constant term identities for finite and affine root system*, Ph. D. thesis, University of Wisconsin, Madison, 1982.
14. Stanley R. P., *Some combinatorial properties of Jack symmetric functions*, Adv. Math. **77** (1989), 76-115.
15. Zeilberger D., *A Stembridge–Stanton style elementary proof of the Habsieger–Kadell q–Morris identity,* Discrete Math. **79** (1989), 313-322.

ON THE SPECTRUM OF DEHN TWISTS IN QUANTUM TEICHMÜLLER THEORY

R. M. KASHAEV

Steklov Mathematical Institute at St. Petersburg
27 Fontanka, St. Petersburg 191011, Russia
and
Helsinki Institute of Physics, P.O. Box 9, (Siltavuorenpenger 20C)
FIN-00014 University of Helsinki, Finland
E-mail: kashaev@pdmi.ras.ru

The operator realizing a Dehn twist in quantum Teichmüller theory is diagonalized and continuous spectrum is obtained. This result is in agreement with the expected spectrum of conformal weights in quantum Liouville theory at $c > 1$. The completeness condition of the eigenvectors includes the integration measure which appeared in the representation theoretic approach to quantum Liouville theory by Ponsot and Teschner. The underlying quantum group structure is also revealed.

1 Introduction

The quantization problem of Teichmüller spaces of punctured surfaces is connected to quantum Chern–Simons theory with non-compact gauge groups (and thereby to 2+1-dimensional quantum gravity [1,22]) and non-rational quantum conformal field theory in two dimensions. In particular, quantum Teichmüller theory is expected to describe the space of conformal blocks in quantum Liouville theory [21]. From the mathematical point of view this can be useful in three-dimensional topology, in particular for construction of new link and three-manifold invariants.

The Teichmüller spaces of punctured surfaces have been quantized in two different but essentially equivalent ways: as (degenerate) Poisson manifolds [8,3], and as simplectic manifolds with Weil–Petersson simplectic structure [13]. The both approaches employ the real analytic coordinates closely related to the Penner parameterization of the decorated Teichmüller spaces [17]. The main outcome of the theory is the construction of a particular projective (infinite dimensional) representation of the mapping class groups of punctured surfaces. The projective factor has been shown to be related to the Virasoro central charge in quantum Liouville theory [14].

In this paper we describe the spectrum of Dehn twists in the quantum Teichmüller theory, derive an explicit formula for the associated with braidings R-matrix (square of which is a Dehn twist), and uncover the underlying infinite dimensional representations of the quantum group $\mathcal{U}_q(\mathfrak{sl}(2))$. Those representations were considered in papers [20,18,19].

The paper is organized as follows. In section 2 we briefly remind the main properties of the non-compact quantum dilogarithm (closely related to the double sine function, see for example [9]). This is the main technical object entering practically all the constructions of the paper. In section 3 we describe the algebraic system underlying the quantum Teichmüller theory and its realization in terms of the quantum dilogarithm. Section 4 is a survey of the part of papers [13,14] which describes the mapping class group representation in terms of the basic algebraic system of section 3. Solution of the Dehn twist spectral problem is described in section 5 (proof of Theorem 1 is to appear in a separate publication). Section 6 contains derivation of the R-matrix and revealing the associated quantum group structure.

2 Quantum dilogarithm

Let complex b have a nonzero real part $\Re b \neq 0$. The *non-compact quantum dilogarithm*, $\mathrm{e}_b(z)$, is defined by the integral formula [5]

$$\mathrm{e}_b(z) \equiv \exp\left(\frac{1}{4}\int_{\mathrm{i}0-\infty}^{\mathrm{i}0+\infty} \frac{e^{-\mathrm{i}2zw}\,dw}{\sinh(wb)\sinh(w/b)w}\right) \tag{1}$$

in the strip $|\Im z| < |\Im c_b|$, where

$$c_b \equiv \mathrm{i}(b+b^{-1})/2.$$

In particular, when $\Im b^2 > 0$, we have the following product formula:

$$\mathrm{e}_b(z) = (e^{2\pi(z+c_b)b}; q^2)_\infty / (e^{2\pi(z-c_b)b^{-1}}; \bar{q}^2)_\infty, \tag{2}$$

where

$$q \equiv e^{\mathrm{i}\pi b^2}, \quad \bar{q} \equiv e^{-\mathrm{i}\pi b^{-2}}.$$

Using the symmetry properties

$$\mathrm{e}_b(z) = \mathrm{e}_{-b}(z) = \mathrm{e}_{1/b}(z),$$

we assume in what follows

$$\Re b > 0, \quad \Im b \geq 0.$$

Function $\mathrm{e}_b(z)$ can be analytically continued in variable z to the entire complex plane as a meromorphic function with essential singularity at infinity, and with the following characteristic properties:

poles and zeros

$$(\mathrm{e}_b(z))^{\pm 1} = 0 \;\Leftrightarrow\; z = \mp(c_b + m\mathrm{i}b + n\mathrm{i}b^{-1}),\; m,n \in \mathbb{Z}_{\geq 0}; \tag{3}$$

behavior at infinity depending on the direction along which the limit is taken,

$$\mathrm{e}_b(z)\Big|_{|z|\to\infty} \approx \begin{cases} 1 & |\arg z| > \frac{\pi}{2} + \arg b; \\ e^{\mathrm{i}\pi z^2 - \mathrm{i}\pi(1+2c_b^2)/6} & |\arg z| < \frac{\pi}{2} - \arg b; \\ \frac{(\bar{q}^2;\bar{q}^2)_\infty}{\Theta(\mathrm{i}b^{-1}z;-b^{-2})} & |\arg z - \frac{\pi}{2}| < \arg b; \\ \frac{\Theta(\mathrm{i}bz;b^2)}{(q^2;q^2)_\infty} & |\arg z + \frac{\pi}{2}| < \arg b, \end{cases} \tag{4}$$

where

$$\Theta(z;\tau) \equiv \sum_{n\in\mathbb{Z}} e^{\mathrm{i}\pi\tau n^2 + \mathrm{i}2\pi z n}, \quad \Im\tau > 0;$$

inversion relation

$$\mathrm{e}_b(z)\,\mathrm{e}_b(-z) = e^{\mathrm{i}\pi z^2 - \mathrm{i}\pi(1+2c_b^2)/6}; \tag{5}$$

functional equations

$$\mathrm{e}_b(z - \mathrm{i}b^{\pm 1}/2) = (1 + e^{2\pi b^{\pm 1} z})\,\mathrm{e}_b(z + \mathrm{i}b^{\pm 1}/2); \tag{6}$$

unitarity when b is either real or on unit circle,

$$(1-|b|)\Im b = 0 \;\Rightarrow\; \overline{\mathrm{e}_b(z)} = 1/\,\mathrm{e}_b(\bar{z}); \tag{7}$$

pentagon equation

$$\mathrm{e}_b(\mathsf{p})\,\mathrm{e}_b(\mathsf{q}) = \mathrm{e}_b(\mathsf{q})\,\mathrm{e}_b(\mathsf{p}+\mathsf{q})\,\mathrm{e}_b(\mathsf{p}), \tag{8}$$

if self-adjoint operators p and q in $L^2(\mathbb{R})$ satisfy the Heisenberg commutation relation $[\mathsf{p},\mathsf{q}] = (2\pi\mathrm{i})^{-1}$;

integral analogue of Ramanujan's summation formula

$$\begin{aligned}\int_{\mathbb{R}} &\frac{\mathrm{e}_b(x+u)}{\mathrm{e}_b(x+v)} e^{2\pi\mathrm{i}wx}\,dx \\ &= \frac{\mathrm{e}_b(u-v-c_b)\,\mathrm{e}_b(w+c_b)}{\mathrm{e}_b(u-v+w-c_b)} e^{-2\pi\mathrm{i}w(v+c_b)+\mathrm{i}\pi(1-4c_b^2)/12} \\ &= \frac{\mathrm{e}_b(v-u-w+c_b)}{\mathrm{e}_b(v-u+c_b)\,\mathrm{e}_b(-w-c_b)} e^{-2\pi\mathrm{i}w(u-c_b)-\mathrm{i}\pi(1-4c_b^2)/12},\end{aligned} \tag{9}$$

where

$$\Im(v+c_b) > 0, \quad \Im(-u+c_b) > 0, \quad \Im(v-u) < \Im w < 0. \tag{10}$$

Restrictions (10) can actually be relaxed by deforming the integration path in the complex x plane, keeping the asymptotic directions of the two ends within the sectors $\pm(|\arg x| - \pi/2) > \arg b$. The enlarged in this way domain for the variables u, v, w in eqn (9) has the form:

$$|\arg(\mathrm{i}z)| < \pi - \arg b, \quad z \in \{w, v-u-w, u-v-2c_b\}. \tag{11}$$

As the matter of fact the pentagon equation (8) is equivalent to the integral Ramanujan formula (9), see [7] for the proof.

3 The basic algebraic system

Introduce an important notation. Let V be a vector space. For any natural $1 \le i \le m$ we define embeddings

$$\iota_i\colon \operatorname{End} V \ni \mathsf{a} \mapsto \mathsf{a}_i = \underbrace{id \otimes \cdots \otimes id}_{i-1 \text{ times}} \otimes \mathsf{a} \otimes id \otimes \cdots \otimes id \in \operatorname{End} V^{\otimes m},$$

ie a stands in the i-th position. If $\mathsf{b} \in \operatorname{End} V^{\otimes k}$ for some $1 \le k \le m$ and $\{i_1, i_2, \dots, i_k\} \subset \{1, 2, \dots, m\}$, we write

$$\mathsf{b}_{i_1 i_2 \dots i_2} \equiv \iota_{i_1} \otimes \iota_{i_2} \otimes \cdots \otimes \iota_{i_k}(\mathsf{b}).$$

Note also that the permutation group $\mathbb{S}_m$ is naturally represented in $V^{\otimes m}$:

$$\mathsf{P}_\sigma(x_1 \otimes \cdots \otimes x_i \otimes \cdots) = x_{\sigma^{-1}(1)} \otimes \cdots \otimes x_{\sigma^{-1}(i)} \otimes \dots, \quad \sigma \in \mathbb{S}_m. \tag{12}$$

The projective representation of the mapping class groups of punctured surfaces, associated with the quantum Teichmüller theory, is based on the concrete realization of the following algebraic system of equations on two invertible elements $\mathsf{A} \in \operatorname{End} V$ and $\mathsf{T} \in \operatorname{End} V^{\otimes 2}$,

$$\mathsf{A}^3 = id, \tag{13}$$
$$\mathsf{T}_{12}\mathsf{T}_{13}\mathsf{T}_{23} = \mathsf{T}_{23}\mathsf{T}_{12}, \tag{14}$$
$$\mathsf{A}_1\mathsf{T}_{12}\mathsf{A}_2 = \mathsf{A}_2\mathsf{T}_{21}\mathsf{A}_1, \tag{15}$$
$$\mathsf{T}_{12}\mathsf{A}_1\mathsf{T}_{21} = \zeta\mathsf{A}_1\mathsf{A}_2\mathsf{P}_{(12)}, \tag{16}$$

where ζ is a (nonzero) complex number, $\mathsf{P}_{(12)}$ being defined in eqn (12) with

$$\mathbb{S}_2 \ni \sigma = (12)\colon 1 \mapsto 2 \mapsto 1.$$

3.1 Useful notation

In practical calculations it is very convenient to use the following notation which permits to avoid writing explicitly operator A. For any $\mathsf{a} \in \operatorname{End} V$ we denote

$$\mathsf{a}_{\hat{k}} \equiv \mathsf{A}_k \mathsf{a}_k \mathsf{A}_k^{-1}, \quad \mathsf{a}_{\check{k}} \equiv \mathsf{A}_k^{-1} \mathsf{a}_k \mathsf{A}_k. \tag{17}$$

Obviously

$$\mathsf{a}_{\check{\hat{k}}} = \mathsf{a}_{\hat{\check{k}}} = \mathsf{a}_k, \quad \mathsf{a}_{\hat{\hat{k}}} = \mathsf{a}_{\check{k}}, \quad \mathsf{a}_{\check{\check{k}}} = \mathsf{a}_{\hat{k}},$$

where the last two equations follow from eqn (13). Besides, it is useful to denote

$$\mathsf{P}_{(kl\ldots m\hat{k})} \equiv \mathsf{A}_k \mathsf{P}_{(kl\ldots m)}, \quad \mathsf{P}_{(kl\ldots m\check{k})} \equiv \mathsf{A}_k^{-1} \mathsf{P}_{(kl\ldots m)}, \tag{18}$$

where $(kl\ldots m)$ is cyclic permutation,

$$(kl\ldots m)\colon k \mapsto l \mapsto \ldots \mapsto m \mapsto k.$$

Eqns (15), (16) in this notation take very compact form

$$\mathsf{T}_{12} = \mathsf{T}_{\hat{2}\check{1}}, \tag{19}$$

$$\mathsf{T}_{12}\mathsf{T}_{2\hat{1}} = \zeta \mathsf{P}_{(12\hat{1})}. \tag{20}$$

3.2 Realization through the quantum dilogarithm

Take

$$V = L^2(\mathbb{R}).$$

Let self-adjoint operators p, q satisfy the Heisenberg commutation relation

$$[\mathsf{p}, \mathsf{q}] = (2\pi \mathsf{i})^{-1}.$$

Then operators

$$\mathsf{A} \equiv e^{-\mathsf{i}\pi/3} e^{\mathsf{i}3\pi \mathsf{q}^2} e^{\mathsf{i}\pi(\mathsf{p}+\mathsf{q})^2} \in L^2(\mathbb{R}), \tag{21}$$

$$\mathsf{T} \equiv e^{\mathsf{i}2\pi \mathsf{p}_1 \mathsf{q}_2} (\mathrm{e}_b(\mathsf{q}_1 + \mathsf{p}_2 - \mathsf{q}_2))^{-1} \in L^2(\mathbb{R}^2), \tag{22}$$

do satisfy equations (13) – (16) with $\zeta = e^{\mathsf{i}\pi c_b^2/3}$. Note that operator A here is unitary and is characterized by the equations (up to a normalization factor)

$$\mathsf{A}\mathsf{q}\mathsf{A}^{-1} = \mathsf{p} - \mathsf{q}, \quad \mathsf{A}\mathsf{p}\mathsf{A}^{-1} = -\mathsf{q},$$

while operator T is unitary if $(1 - |b|)\Im b = 0$.

4 The mapping class group representation

Here we recall how the mapping class groups of punctured surfaces are represented in terms of system (13) – (16).

4.1 Decorated ideal triangulations

We call a two-cell in a cell complex (or CW-complex, see [4]) *triangle* if exactly three boundary points of the corresponding two-disk are mapped to the zero-skeleton. We shall also call zero-cells and one-cells *vertices* and *edges*, respectively.

Let Σ be a compact oriented (possibly with boundary) surface with finite non-empty set V_Σ of marked points called also *punctures.*
Definition 1 *A cell complex decomposition of* Σ *is called* ideal triangulation *if*

- *the zero-skeleton coincides with* V_Σ*;*
- *all two-cells are triangles.*

Two ideal triangulations are considered equivalent *if there exists an isotopy of* Σ *fixed on* $\partial\Sigma \cup V_\Sigma$ *deforming one into another.*
We suppose that Σ admits ideal triangulations (for example, each boundary component must contain at least one marked point). Any ideal triangulation of Σ has one and the same number of triangles n_Σ. Denote $T(\tau)$ the set of triangles in ideal triangulation τ.
Definition 2 *Ideal triangulation* τ *is called* decorated *if*

- *each triangle is provided by a marked corner;*
- *all triangles are numbered, ie a bijective* numbering *mapping*

  $$\bar{\tau}\colon \{1, \dots, n_\Sigma\} \to T(\tau)$$

  is fixed.

Graphically (see below) in a triangle we put the corresponding integer inside of it, and asterisk at the marked corner. The set of all decorated ideal triangulations of Σ will be denoted Δ_Σ. For terminological convenience and if no confusion is possible, in what follows we sometimes will use the term triangulation as a substitute for decorated ideal triangulation.

Action of the permutation group. The permutation group $\mathbb{S}_{n_\Sigma}$ naturally acts in Δ_Σ from the right,

$$\Delta_\Sigma \times \mathbb{S}_{n_\Sigma} \ni (\tau, \sigma) \mapsto \tau\sigma \in \Delta_\Sigma,$$

by changing the numbering mapping:

$$\overline{\tau\sigma} = \bar{\tau} \circ \sigma.$$

Changing of a marked corner. If $\tau \in \Delta_\Sigma$ and $1 \le i \le n_\Sigma$, then triangulation $\rho_i(\tau)$ is obtained from τ by changing the marked corner of triangle $\bar{\tau}(i)$ as is shown in figure 1.

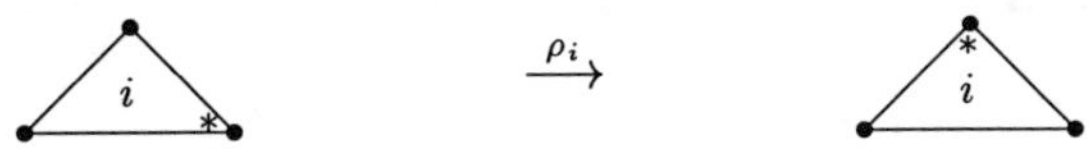

Figure 1. Transformation ρ_i changes the marked corner of triangle $\bar{\tau}(i)$.

The flip transformation. Let two distinct triangles $\bar{\tau}(i)$, $\bar{\tau}(j)$ have a common edge and their marked corners be as in the lhs of figure 2, so the common edge is a diagonal of a quadrilateral combined of the two triangles. Then tri-

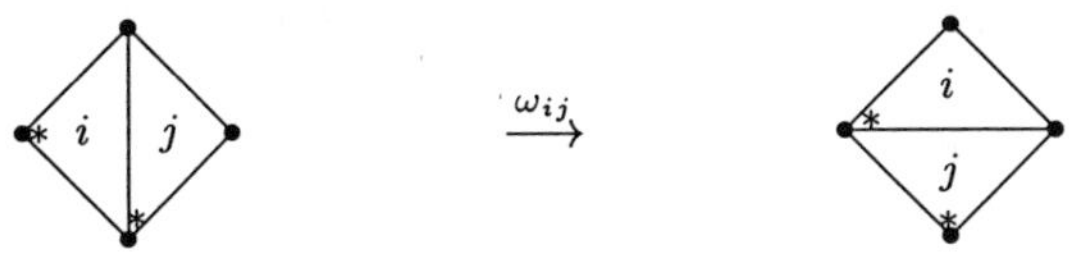

Figure 2. The flip transformation ω_{ij} clockwise "rotates" one diagonal of the quadrilateral until it matches another diagonal.

angulation $\omega_{ij}(\tau)$ is obtained from τ by replacing the common edge by the opposite diagonal of the quadrilateral, assigning the numbers and marked corners to new triangles as is shown in the rhs of figure 2. Note that this *flip* transformation ω_{ij} implicitly depends on the triangulation it transforms as at fixed i and j it is not defined on all triangulations.

The properties. We have the following properties:

$$\rho_i \circ \rho_i \circ \rho_i = id, \tag{23}$$

$$\omega_{jk} \circ \omega_{ik} \circ \omega_{ij} = \omega_{ij} \circ \omega_{jk}, \tag{24}$$

$$(\rho_i^{-1} \times \rho_j) \circ \omega_{ij} = \omega_{ji} \circ (\rho_i^{-1} \times \rho_j), \tag{25}$$

$$\omega_{ji} \circ \rho_i \circ \omega_{ij} = (ij) \circ (\rho_i \times \rho_j). \tag{26}$$

The first equation is evident since there are only three possibilities to mark a corner in a triangle. The other three equations are proved pictorially in figures 3 – 5.

Any two (decorated ideal) triangulations can be transformed to each other by a (finite) composition of elementary transformations, ie ρ_i, ω_{ij} and permutations. This follows from the known fact that any two ideal triangulations

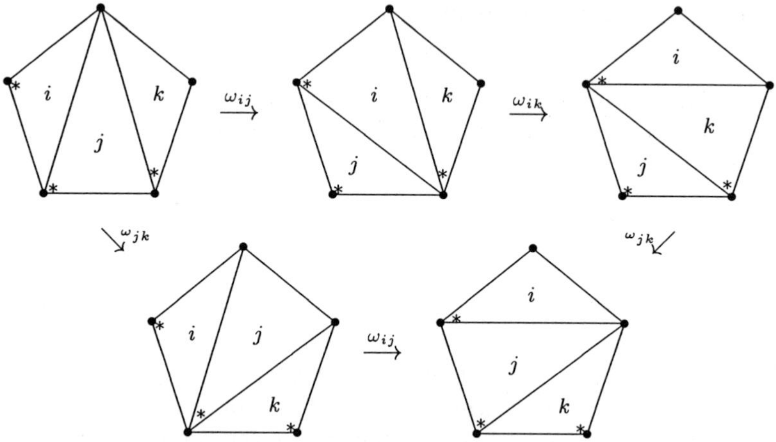

Figure 3. Proof of the pentagon equation (24).

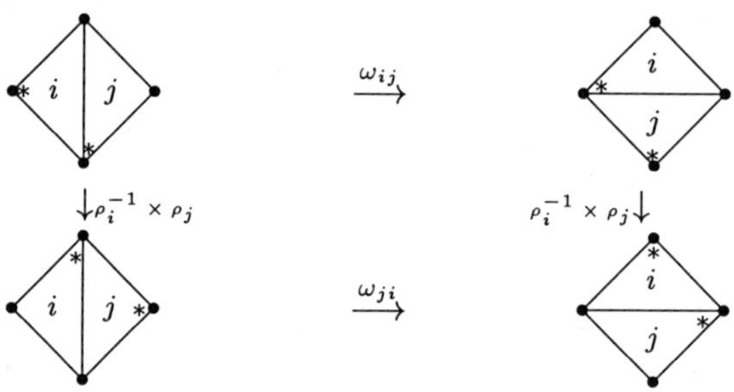

Figure 4. Proof of the symmetry relation (25).

(without decoration) can be transformed one into another by a composition of flips [15], and that all possible decorations of a fixed ideal triangulation are transitively acted upon by compositions of permutations and ρ_i transformations.

4.2 Quantum theory

Suppose we are given a solution to eqns (13) – (16). For each $\tau \in \Delta_\Sigma$ and $1 \leq i \leq n_\Sigma$ assign

$$\mathsf{F}(\tau, \rho_i(\tau)) \equiv \mathsf{A}_i \in \operatorname{End} V^{\otimes n_\Sigma}. \tag{27}$$

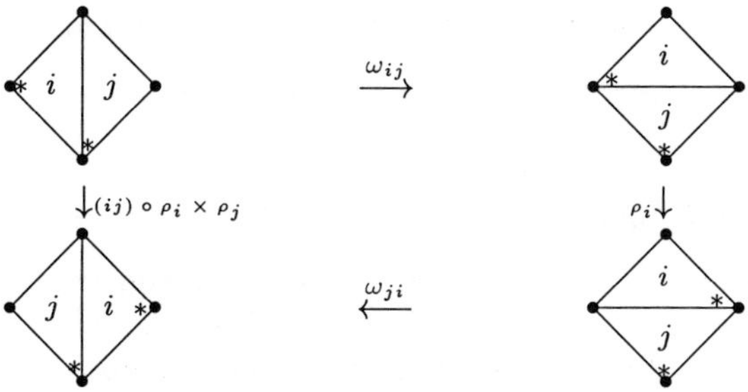

Figure 5. Proof of the inversion relation (26).

Let $i \neq j$ be such that triangles $\bar{\tau}(i)$ and $\bar{\tau}(j)$ are as in the lhs of figure 2. Then we put

$$\mathsf{F}(\tau, \omega_{ij}(\tau)) \equiv \mathsf{T}_{ij} \in \operatorname{End} V^{\otimes n_\Sigma}. \tag{28}$$

Finally, for any permutation $\sigma \in \mathbb{S}_{n_\Sigma}$ set

$$\mathsf{F}(\tau, \tau\sigma) \equiv \mathsf{P}_\sigma \in \operatorname{End} V^{\otimes n_\Sigma},$$

where operator P_σ is defined by eqn (12). Ensured by the consistency of eqns (13) – (16) with eqns (23) – (26), mapping F can be extended to an operator valued function $\mathsf{F}(\tau, \tau')$ on $\Delta_\Sigma \times \Delta_\Sigma$ such that for any $\tau, \tau', \tau'' \in \Delta_\Sigma$

$$\mathsf{F}(\tau, \tau) = 1, \quad \mathsf{F}(\tau, \tau')\mathsf{F}(\tau', \tau'')\mathsf{F}(\tau'', \tau) \in \mathbb{C} - \{0\}. \tag{29}$$

In particular (when $\tau'' = \tau$),

$$\mathsf{F}(\tau, \tau')\mathsf{F}(\tau', \tau) \in \mathbb{C} - \{0\}. \tag{30}$$

As an example, deduce operator $\mathsf{F}(\tau, \omega_{ij}^{-1}(\tau))$. Denoting $\tau' \equiv \omega_{ij}^{-1}(\tau)$ and employing eqn (30) as well as definition (28), we have

$$\mathsf{F}(\tau, \omega_{ij}^{-1}(\tau)) = \mathsf{F}(\omega_{ij}(\tau'), \tau') \simeq (\mathsf{F}(\tau', \omega_{ij}(\tau')))^{-1} = \mathsf{T}_{ij}^{-1}, \tag{31}$$

where we denote by $\simeq$ an equality up to a numerical factor.

The mapping class or modular group[a] $\mathcal{M}_\Sigma$ of Σ naturally acts in Δ_Σ. By construction we have the following invariance property of function F:

$$\mathsf{F}(f(\tau), f(\tau')) = \mathsf{F}(\tau, \tau'), \qquad \forall f \in \mathcal{M}_\Sigma.$$

[a]The group of homeomorphisms permuting the marked points, factored wrt the connected component of the identical homeomorphism.

This enables us to construct a projective representation of $\mathcal{M}_\Sigma$:

$$\mathcal{M}_\Sigma \ni f \mapsto \mathsf{F}(\tau, f(\tau)) \in \mathrm{End}\, V^{\otimes n_\Sigma}.$$

Indeed,

$$\mathsf{F}(\tau, f(\tau))\mathsf{F}(\tau, h(\tau)) = \mathsf{F}(\tau, f(\tau))\mathsf{F}(f(\tau), f(h(\tau))) \simeq \mathsf{F}(\tau, fh(\tau)).$$

Any Dehn twist, at least if it is along a non-separating contour, is equivalent to the Dehn twist of an annulus (with two marked points on the two boundary components) along its only non-contractible loop denoted α in figure 6. As an operator it is given in terms of operator T. Indeed, from figure 6 it follows that $\omega_{12} \circ D_\alpha(\tau) = \tau$. So, using eqn (31), we obtain

$$\mathsf{F}(\tau, D_\alpha(\tau)) = \mathsf{F}(\tau, \omega_{12}^{-1}(\tau)) \simeq \mathsf{T}_{12}^{-1}, \tag{32}$$

where the normalization is to be fixed.

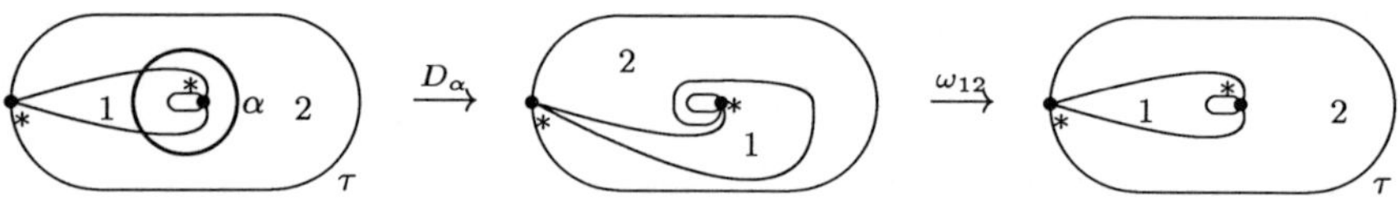

Figure 6. The Dehn twist along contour α followed by the flip transformation ω_{12} does not change the initial triangulation.

5 Diagonalizing the Dehn twist

In this section we work within the quantum Teichmüller theory, ie we consider solution (21), (22) of system (13) – (16). We assume that $(1 - |b|)\Im b = 0$, so the range of mapping F in this case is given by unitary operators.

Properly normalized quantum Dehn twist of the annulus has the form:

$$\mathsf{D}_\alpha = \zeta^{-6}\mathsf{T}_{12}^{-1} e^{\mathrm{i}2\pi \mathsf{z}_\alpha^2}, \quad \mathsf{z}_\alpha = (\mathsf{p}_1 + \mathsf{q}_2)/2, \quad \zeta = e^{\mathrm{i}\pi c_b^2/3},$$

where the exponential factor removes the spurious degree of freedom, see [13,14]. Using formula (22) we rewrite it equivalently

$$\mathsf{D}_\alpha = e^{\mathrm{i}2\pi(\mathsf{q}_\alpha^2 - c_b^2)}\, e_b(\mathsf{p}_\alpha + \mathsf{q}_\alpha),$$

where the self-adjoint operators

$$\mathsf{p}_\alpha \equiv \mathsf{q}_1 + \mathsf{p}_2 - (\mathsf{q}_2 + \mathsf{p}_1)/2, \quad \mathsf{q}_\alpha \equiv (\mathsf{q}_2 - \mathsf{p}_1)/2$$

satisfy the Heisenberg commutation relation

$$[\mathsf{p}_\alpha, \mathsf{q}_\alpha] = (2\pi\mathrm{i})^{-1}.$$

To diagonalize D_α, we use the fact that mutually commuting (and conjugate to each other in the case when $|b|=1$) unbounded operators

$$\mathsf{L}_\alpha^\pm \equiv 2\cosh(2\pi b^{\pm 1}\mathsf{q}_\alpha) + e^{2\pi b^{\pm 1}\mathsf{p}_\alpha}$$

also commute with D_α:

$$[\mathsf{L}_\alpha^-, \mathsf{L}_\alpha^+] = [\mathsf{D}_\alpha, \mathsf{L}_\alpha^\pm] = 0.$$

Geometrically operator L_α^+ is nothing else but the quantized geodesic length (hyperbolic cosine thereof) of contour α, see [3,8].

Let us consider the "coordinate" basis $\langle x|$, $x \in \mathbb{R}$, where q_α is diagonal and p_α is a differentiation operator,

$$\langle x|\mathsf{q}_\alpha = x\langle x|, \quad \langle x|\mathsf{p}_\alpha = \frac{1}{2\pi \mathrm{i}}\frac{\partial}{\partial x}\langle x|,$$

and define one parameter family of vectors:

$$\langle x|\alpha_s\rangle = \frac{\mathrm{e}_b(s+x+c_b-\mathrm{i}0)}{\mathrm{e}_b(s-x-c_b+\mathrm{i}0)} e^{-\mathrm{i}2\pi(x+c_b)s} = \langle x|\alpha_{-s}\rangle, \quad s\in\mathbb{R}. \tag{33}$$

Theorem 1 *Vectors $|\alpha_s\rangle$ are eigenvectors of operators $\mathsf{L}_\alpha^\pm, \mathsf{D}_\alpha$:*

$$\mathsf{L}_\alpha^\pm|\alpha_s\rangle = |\alpha_s\rangle 2\cosh(2\pi b^{\pm 1}s), \quad \mathsf{D}_\alpha|\alpha_s\rangle = |\alpha_s\rangle e^{\mathrm{i}2\pi(s^2-c_b^2)}. \tag{34}$$

They are orthogonal to each other:

$$\langle\alpha_r|\alpha_s\rangle = \nu(s)^{-1}\delta(|r|-|s|), \quad \nu(s) \equiv 4\sinh(2\pi bs)\sinh(2\pi b^{-1}s), \tag{35}$$

and complete in $L^2(\mathbb{R})$:

$$\int_0^\infty |\alpha_s\rangle\nu(s)\,ds\langle\alpha_s| = 1. \tag{36}$$

The continuous spectrum of Dehn twists in quantum Teichmüller theory has also been observed by V. Fock[b]. It is worth noting that measure $\nu(s)ds$ in eqn (36) also appears in the representation theory of the non-compact quantum group $\mathcal{U}_q(\mathfrak{sl}(2,\mathbb{R}))$ [18,19]. In the next section we reveal that this is not accidental.

---

[b]Private communication.

6 Braiding and R-matrix

Braiding of a disk with two vertices is naturally associated with the contour (isotopy class thereof) surrounding the vertices. Moreover, square of the braiding is Dehn twist along the associated contour.

Figure 7 shows how the braiding of triangulation τ of a disk with two interior and two boundary marked points can be removed by a sequence of elementary transformations:

$$\tau = (13)(24) \circ (\rho_3 \times \rho_1^{-1}) \circ \omega_{14} \circ (\omega_{13} \times \omega_{24}) \circ (\rho_1 \times \omega_{23}) \circ \rho_3^{-1} \circ B_\alpha^{-1}(\tau).$$

Using the construction of subsection 4.2, the corresponding quantum braiding

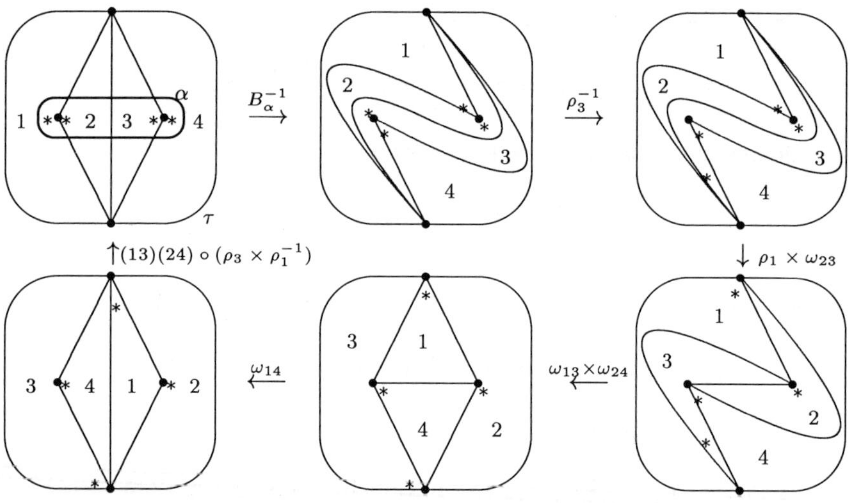

Figure 7. Braiding along contour α followed by the indicated sequence of transformations brings back to the initial triangulation τ.

operator has the form (up to a normalization factor)

$$\mathsf{B}_\alpha \equiv \mathsf{F}(\tau, B_\alpha(\tau)) \simeq \mathsf{P}_{(13)(24)}\mathsf{R}_{1234},$$

with

$$\mathsf{R} = \mathsf{R}_{1234} \equiv \mathsf{A}_1^{-1}\mathsf{A}_3\mathsf{T}_{41}\mathsf{T}_{31}\mathsf{T}_{42}\mathsf{T}_{32}\mathsf{A}_1\mathsf{A}_3^{-1} = \mathsf{T}_{1\check{4}}\mathsf{T}_{13}\mathsf{T}_{42}\mathsf{T}_{\hat{3}2}, \qquad (37)$$

where in the second equation we have used notation (17) and eqn (19). As a consequence of relations (13) – (16) operator $\mathsf{R} \in \operatorname{End} V^4$ solves the following

Yang–Baxter equation

$$\mathsf{R}_{1234}\mathsf{R}_{1256}\mathsf{R}_{3456} = \mathsf{R}_{3456}\mathsf{R}_{1256}\mathsf{R}_{1234}, \tag{38}$$

and in the case corresponding to realization through the quantum dilogarithm it is in fact (as will be shown below) $\mathcal{U}_q(\mathfrak{sl}(2))$ (more precisely, the modular double thereof [6]) universal R-matrix evaluated at a reducible infinite dimensional representation acting in $L^2(\mathbb{R}^2)$. Note that eqn (37) is a particular case of the realization of universal R-matrices in the Drinfeld doubles of Hopf algebras in terms of solutions of the pentagon equation associated with the Heisenberg doubles [12].

6.1 Relation to Hopf algebras

To identify the generators of the quantum group, we first rewrite the R-matrix in the form

$$\mathsf{R} = \mathrm{Ad}(\mathsf{T}_{1\hat{2}}^{-1}\mathsf{T}_{\check{4}3})\mathsf{T}_{\hat{2}\check{4}} = \mathrm{Ad}(\mathsf{A}_2\mathsf{T}_{12}^{-1}\mathsf{A}_4^{-1}\mathsf{T}_{43})\mathsf{T}_{24}, \tag{39}$$

where

$$\mathrm{Ad}(\mathsf{a})\mathsf{x} \equiv \mathsf{a}\mathsf{x}\mathsf{a}^{-1},$$

and we have used the pentagon equation (14). Let now operator T be written as a sum

$$\mathsf{T} = \sum_a \mathsf{e}_a \otimes \mathsf{e}^a,$$

where two operator sets $\{\mathsf{e}_a\}$ and $\{\mathsf{e}^a\}$ are realizations of mutually dual linear bases of two (mutually dual) Hopf sub-algebras A_T and A_T^* in the associated to T Heisenberg double $H(A_\mathsf{T})$. Their co-products are given by the formulae

$$\Delta(\mathsf{e}_a) = \mathrm{Ad}(\mathsf{T}^{-1})(1 \otimes \mathsf{e}_a), \quad \Delta(\mathsf{e}^a) = \mathrm{Ad}(\mathsf{T})(\mathsf{e}^a \otimes 1).$$

From eqn (39) we conclude that the R-matrix is also decomposed into a sum

$$\mathsf{R} = \sum_a \mathsf{E}_a \otimes \mathsf{E}^a$$

with

$$\mathsf{E}_a = \mathrm{Ad}(\mathsf{A}_2\mathsf{T}^{-1})(1 \otimes \mathsf{e}_a) = \mathrm{Ad}(\mathsf{A}_2)\Delta(\mathsf{e}_a), \tag{40}$$

$$\mathsf{E}^a = \mathrm{Ad}(\mathsf{A}_2^{-1}\mathsf{T}_{21})(1 \otimes \mathsf{e}^a) = \mathrm{Ad}(\mathsf{A}_2^{-1})\Delta'(\mathsf{e}^a), \tag{41}$$

where Δ' is the opposite co-product. These are also realizations of the Hopf algebras A_T and A_T^*, but this time within the Drinfeld double $D(A_\mathsf{T})$.

6.2 Quantum Teichmüller theory and $\mathcal{U}_q(\mathfrak{sl}(2))$

Here we will see what quantum group and in which representation does correspond to our solution (21), (22).

From eqns (22), (2) we see that all elements $\{\mathsf{e}_a\}$ can be thought to be generated by operators p and $e^{2\pi b^{\pm1}\mathsf{q}}$ with the co-product

$$\Delta(\mathsf{p}) = \mathsf{p}_1 + \mathsf{p}_2, \quad \Delta(e^{2\pi b^{\pm1}\mathsf{q}}) = e^{2\pi b^{\pm1}(\mathsf{q}_1+\mathsf{p}_2)} + e^{2\pi b^{\pm1}\mathsf{q}_2}.$$

Similarly dual elements $\{\mathsf{e}^a\}$ are generated by q and $e^{2\pi b^{\pm1}(\mathsf{p}-\mathsf{q})}$ with the co-product

$$\Delta(\mathsf{q}) = \mathsf{q}_1 + \mathsf{q}_2, \quad \Delta(e^{2\pi b^{\pm1}(\mathsf{p}-\mathsf{q})}) = e^{2\pi b^{\pm1}(\mathsf{p}_1-\mathsf{q}_1-\mathsf{q}_2)} + e^{2\pi b^{\pm1}(\mathsf{p}_2-\mathsf{q}_2)}.$$

From eqn (40) we deduce that operators

$$\mathsf{g}_{12} \equiv \mathsf{p}_1 - \mathsf{q}_2, \quad \mathsf{f}_{12}^{\pm} \equiv e^{2\pi b^{\pm1}(\mathsf{q}_1-\mathsf{q}_2)} + e^{2\pi b^{\pm1}(\mathsf{p}_2-\mathsf{q}_2)} \tag{42}$$

generate the set $\{\mathsf{E}_a\}$, while according to eqn (41) the generators of the dual set $\{\mathsf{E}^a\}$ are the operators

$$\mathsf{q}_1 - \mathsf{p}_2 = -\mathsf{g}_{21}, \quad e^{2\pi b^{\pm1}(\mathsf{q}_2-\mathsf{q}_1)} + e^{2\pi b^{\pm1}(\mathsf{p}_1-\mathsf{q}_1)} = \mathsf{f}_{21}^{\pm}. \tag{43}$$

In what follows we restrict our attention to the sub-algebra corresponding to the positive exponent of parameter b.

Proposition 1 *Operators $\mathsf{f}_{mn} \equiv \mathsf{f}_{mn}^{+}$, g_{mn} ($mn = 12$ or 21) with the relations*

$$[\mathsf{g}_{12}, \mathsf{g}_{21}] = 0, \quad [\mathsf{g}_{mn}, \mathsf{f}_{mn}] = -\mathrm{i}b\mathsf{f}_{mn}, \quad [\mathsf{g}_{nm}, \mathsf{f}_{mn}] = \mathrm{i}b\mathsf{f}_{mn},$$
$$[\mathsf{f}_{12}, \mathsf{f}_{21}] = (q - q^{-1})(e^{2\pi b\mathsf{g}_{12}} - e^{2\pi b\mathsf{g}_{21}}), \quad q \equiv e^{\mathrm{i}\pi b^2};$$

and the twisted co-product

$$\Delta_\varphi = \mathrm{Ad}(e^{\mathrm{i}\varphi(\mathsf{g}_{21}\otimes\mathsf{g}_{12}-\mathsf{g}_{12}\otimes\mathsf{g}_{21})}) \circ \Delta, \quad \varphi \in \mathbb{R},$$

$$\Delta(\mathsf{g}_{mn}) = \mathsf{g}_{mn} \otimes 1 + 1 \otimes \mathsf{g}_{mn},$$
$$\Delta(\mathsf{f}_{12}) = \mathsf{f}_{12} \otimes e^{2\pi b\mathsf{g}_{12}} + 1 \otimes \mathsf{f}_{12}, \quad \Delta(\mathsf{f}_{21}) = e^{2\pi b\mathsf{g}_{21}} \otimes \mathsf{f}_{21} + \mathsf{f}_{21} \otimes 1.$$

generate a Hopf algebra $\mathcal{G}_\varphi$.

Algebra $\mathcal{G}_{\pi/2}$ is closely related with $\mathcal{U}_q(\mathfrak{sl}(2))$.

Definition 3 [c] *Quantum group $\mathcal{U}_q(\mathfrak{sl}(2))$ is a Hopf algebra with*

- *generators E, F, K, K^{-1};*

[c] This definition is that of [19] without concretizing the star-structure, since the latter depends on a solution to the equation $(1 - |b|)\Im b = 0$.

- *relations*

$$KE = qEK, \quad KF = q^{-1}FK, \quad [E,F] = -(K^2 - K^{-2})/(q - q^{-1});$$

- *co-product*

$$\Delta(K) = K \otimes K, \quad \Delta(X) = X \otimes K + K^{-1} \otimes X, \quad X = E \text{ or } F.$$

Proposition 2 *There exists a faithful Hopf algebra homomorphism*

$$\eta\colon \mathcal{U}_q(\mathfrak{sl}(2)) \hookrightarrow \mathcal{G}_{\pi/2}$$

such that

$$K \mapsto e^{\pi b(\mathsf{g}_{12} - \mathsf{g}_{21})/2},$$

$$E \mapsto e^{-\pi b(c_b + \mathsf{g}_{21})} \frac{\mathsf{f}_{21}}{q - q^{-1}}, \quad F \mapsto \frac{\mathsf{f}_{12}}{q - q^{-1}} e^{\pi b(c_b - \mathsf{g}_{12})}.$$

The generators of algebra $\mathcal{G}_\varphi$ depend on only three combinations of the Heisenberg operators $\mathsf{p}_i, \mathsf{q}_i$ $(i = 1, 2)$

$$\mathsf{z}_\beta \equiv (\mathsf{p}_1 - \mathsf{q}_1 + \mathsf{p}_2 - \mathsf{q}_2)/2,$$

$$\mathsf{p}_\beta \equiv (\mathsf{p}_1 + \mathsf{q}_1 - \mathsf{p}_2 - \mathsf{q}_2)/2, \quad \mathsf{q}_\beta \equiv (-\mathsf{p}_1 + \mathsf{q}_1 + \mathsf{p}_2 - \mathsf{q}_2)/2$$

which satisfy the commutation relations

$$[\mathsf{z}_\beta, \mathsf{p}_\beta] = [\mathsf{z}_\beta, \mathsf{q}_\beta] = 0, \quad [\mathsf{p}_\beta, \mathsf{q}_\beta] = (2\pi \mathsf{i})^{-1}.$$

Subscript β here refers to the associated to our operators contour around the interior vertex of a triangulated disk (with one interior vertex) in figure 8. In particular, in quantum Teichmüller theory operator z_β describes the spurious degree of freedom associated with contour β. Substituting these definitions

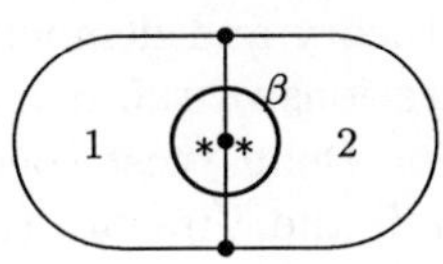

Figure 8. Triangulated disk with one interior vertex.

into eqns (42), (43) we obtain

$$\mathsf{g}_{12} = \mathsf{z}_\beta + \mathsf{p}_\beta, \quad \mathsf{g}_{21} = \mathsf{z}_\beta - \mathsf{p}_\beta,$$

$$\mathsf{f}_{12} = 2e^{2\pi b \mathsf{q}_\beta} e^{\pi b(\mathsf{z}_\beta + \mathsf{p}_\beta - c_b)} \sinh(\pi b(\mathsf{z}_\beta - \mathsf{p}_\beta + c_b)),$$

$$\mathsf{f}_{21} = -2e^{\pi b(\mathsf{z}_\beta - \mathsf{p}_\beta + c_b)} e^{-2\pi b \mathsf{q}_\beta} \sinh(\pi b(\mathsf{z}_\beta + \mathsf{p}_\beta + c_b)),$$

while R-matrix (37) takes the form

$$\mathsf{R} = e^{\mathrm{i}2\pi(\mathsf{p}_{\beta_1}+\mathsf{z}_{\beta_1})(\mathsf{p}_{\beta_2}-\mathsf{z}_{\beta_2})}\,\mathrm{Ad}\left(\frac{\mathrm{e}_b(\mathsf{p}_{\beta_1}-\mathsf{z}_{\beta_1})}{\mathrm{e}_b(\mathsf{p}_{\beta_2}+\mathsf{z}_{\beta_2})}\right)(\mathrm{e}_b(\mathsf{q}_{\beta_1}+\mathsf{z}_{\beta_1}-\mathsf{q}_{\beta_2}-\mathsf{p}_{\beta_2}))^{-1}. \tag{44}$$

On the eigenvectors $\langle\xi, x|$ of operators z_β and q_β,

$$\langle\xi, x|\mathsf{z}_\beta = \xi\langle\xi, x|, \quad \langle\xi, x|\mathsf{q}_\beta = x\langle\xi, x|, \quad \langle\xi, x|\mathsf{p}_\beta = \frac{1}{2\pi\mathrm{i}}\frac{\partial}{\partial x}\langle\xi, x|,$$

algebra $\mathcal{G}_\varphi$ is irreducibly represented for each fixed $\xi \in \mathbb{R}$, and these are exactly the representations from Schmüdgen's classification list [20] considered by Ponsot and Teschner in [18,19].

Theorem 2 *If $\Im b = 0$, the following equation holds*

$$\langle\xi, x|\eta(a) = \pi_{\mathrm{i}(\xi-c_b)}(a)\langle\xi, x|, \quad \forall a \in \mathcal{U}_q(\mathfrak{sl}(2)),$$

where representations π_α, $\alpha \in \mathrm{i}(\mathbb{R} - c_b)$, are defined in eqn (12) of [19].

Now it becomes clear the relevance of the result of Ponsot and Teschner on the decomposition of the tensor products of the representations under consideration to the spectral problem of Dehn twists in quantum Teichmüller theory considered in section 5: the decomposition of tensor products of representations into irreducibles is essentially equivalent to the spectral problem of the R-matrix, while square of the latter is nothing else but a Dehn twist. Elaboration of this connection is to be published elsewhere.

7 Conclusion

In this paper we have described the spectrum of Dehn twists (Theorem 1) in the quantum Teichmüller theory and demonstrated appearance of certain infinite dimensional representations of the quantum group $\mathcal{U}_q(\mathfrak{sl}(2))$ (Theorem 2) studied in [20,18,19]. The Dehn twist spectrum coincides with that of the operator $e^{\mathrm{i}2\pi L_0}$, where L_0 is the Virasoro (infinitesimal light-cone translation) operator in quantum Liouville theory. The quantum group structure indicates a relationship between the two approaches to quantum Liouville theory: one through quantization of the Teichmüller spaces and another through representation theory of a non-compact quantum group. The result of [2] on the representation theory of the quantum Lorentz group is also likely to be relevant here.

Quantum Teichmüller theory should be connected to quantum Liouville theory on a space-time lattice [7] where the non-compact quantum dilogarithm plays the central role too. The first step in this direction could be the following observation. For any integers j and t with odd sum, operators $\chi_{j,t}$ defined through the relations

$$\chi_{j,t} = \mathsf{D}^{-t} e^{-2\pi b(\mathsf{p}+\mathsf{q})} \mathsf{D}^{t}, \quad j+t = 1 \pmod 2,$$

satisfy the quantum lattice Liouville equation

$$\chi_{j,t+1}\chi_{j,t-1} = (1 + q\chi_{j+1,t})(1 + q^3\chi_{j-1,t}), \quad q \equiv e^{\mathrm{i}\pi b^2},$$

with periodicity property $\chi_{j,t} = \chi_{j+2,t}$. Here pair of self-adjoint operators p and q satisfy the Heisenberg commutation relation $[\mathsf{p}, \mathsf{q}] = (\mathrm{i}2\pi)^{-1}$ and $\mathsf{D} = e^{\mathrm{i}2\pi \mathsf{q}^2} \mathrm{e}_b(\mathsf{p}+\mathsf{q})$ is the quantum Dehn twist operator (up to normalization) considered in section 5. Clearly, the quantum Dehn twist in this case is the "light-cone" traslation operator

$$\mathsf{D}^{-1}\chi_{j,t}\mathsf{D} = \chi_{j-1,t+1}.$$

This again is in agreement with formula $\mathsf{D} \simeq e^{\mathrm{i}2\pi L_0}$. It is thus possible that such simple quantum evolution model describes the zero modes (responsible for the spectrum of conformal weights) in quantum lattice Liouville model of arbitrary (even) length.

We have derived the R-matrix (associated with braidings in the mapping class groups) in terms of the non-compact quantum dilogarithm, formula (44), which first has been suggested by Faddeev in [6] as the universal $\mathcal{U}_q(\mathfrak{sl}(2))$ R-matrix for the corresponding modular double. Note that more general formula (37) directly follows from the canonical embedding of the Drinfeld doubles of Hopf algebras into tensor product of two Heisenberg doubles [12]. Thus the explanation of section 6 can be considered as a geometrical view on the pure algebraic construction of [12]. It is also worth mentioning that this R-matrix is the direct non-compact analogue of the finite dimensional R-matrix introduced in [10] which leads to a specialization of the colored Jones link invariants (polynomials) [16] with the particularly remarkable asymptotic behavior [11]. In this light it would be interesting to study the corresponding non-compact analogue of the Jones invariants.

Acknowledgments

I wish to thank L. D. Faddeev for valuable discussions and encouragement. I am grateful to the organizers of the International Workshop on Physics and Combinatorics, Nagoya, August 21-26, 2000 for giving me the opportunity to report about these results. This work is supported by RFBR grant 99-01-00101, INTAS grant 99-01705, and Finnish Academy.

References

1. A. Achúcarro, P. K. Townsend: Chern–Simons actions for three dimensional anti-De Sitter supergravity theories. *Phys. Lett.* B **180** (1986), 89–92.
2. E. Buffenoir, Ph. Roche: Tensor products of principal unitary representations of quantum Lorentz group and Askey–Wilson polynomials. Preprint math.QA/9910147.
3. L. Chekhov, V. V. Fock: Quantum Teichmüller space. *Theor. Math. Phys.* **120** (1999) 1245–1259, math.QA/9908165.
4. B. A. Dubrovin, A. T. Fomenko, S. P. Novikov: *Modern Geometry — Methods and Applications. Part III. Introduction to Homology Theory.* Springer–Verlag 1990.
5. L. D. Faddeev: Discrete Heisenberg–Weyl group and modular group. *Lett. Math. Phys.* **34** (1995), 249–254, hep-th/9504111.
6. ———: Modular double of quantum group. Preprint math.QA/9912078.
7. L. D. Faddeev, R. M. Kashaev, A. Yu. Volkov: Strongly coupled quantum discrete Liouville theory. I: Algebraic approach and duality. Preprint hep-th/0006156.
8. V. V. Fock: Dual Teichmüller spaces. Preprint dg-ga/9702018.
9. M. Jimbo, T. Miwa: Quantum KZ equation with $|q| = 1$ and correlation functions of the XXZ model in the gapless regime. *J. Phys. A: Math. Gen.* **29** (1996), 2923–2958.
10. R. M. Kashaev: A link invariant from quantum dilogarithm. *Mod. Phys. Lett.* A, Vol. 10, No. 19 (1995), 1409-1418, q-alg/9504020.
11. ———: The hyperbolic volume of knots from the quantum dilogarithm. *Lett. Math. Phys.* **39** (1997), 269–275, q-alg/9601025.
12. ———: The Heisenberg double and the Pentagon relation. St. Petersburg Math. J. Vol. 8, No. 4 (1997), 585–592, q-alg/9504020.
13. ———: Quantization of Teichmüller spaces and the quantum dilogarithm. *Lett. Math. Phys.* **43** (1998), 105–115, q-alg/9705021.
14. ———: Liouville central charge in quantum Teichmüller theory. *Proc. of the Steklov Inst. of Math.* Vol. 226 (1999), 63–71, hep-th/9811203.
15. L. Mosher: Mapping class groups are automatic. *Annals of Math.*, Vol. **142** (1995), 303–384.
16. H. Murakami, J. Murakami: The colored Jones polynomials and the simplicial volume of a knot. Preprint math.GT/9905075.
17. R. C. Penner: The decorated Teichmüller space of punctured surfaces. *Commun. Math. Phys.* **113** (1987), 299–339.
18. P. Ponsot, J. Teschner: Liouville bootstrap via harmonic analysis on a

noncompact quantum group. Preprints DIAS-STR-99-14, ESI-791, LPM-99/46, hep-th/9911110.

19. ———: Clebsch–Gordan and Racah–Wigner coefficients for a continuous series of representations of $\mathcal{U}_q(\mathfrak{sl}(2,\mathbb{R}))$. Preprint DIAS-STP-00-15, LPM-00/21, math.QA/0007097.
20. K. Schmüdgen: Operator representations of $\mathcal{U}_q(\mathfrak{sl}(2,\mathbb{R}))$. *Lett. Math. Phys.* **37** (1996), 211–222.
21. H. Verlinde: Conformal field theory, two-dimensional quantum gravity and quantization of Teichmüller space. *Nuclear Phys.* B **337** (1990), 652–680.
22. E. Witten: 2 + 1-dimensional gravity as an exactly soluble system. *Nuclear Phys.* B **311** (1988/89), 46–78.

INTRODUCTION TO TROPICAL COMBINATORICS

ANATOL N. KIRILLOV
Graduate School of Mathematics, Nagoya University
Chikusa-ku, Nagoya 486-8602, Japan
and
Steklov Mathematical Institute at St.Petersburg
Fontanka 27, St.Petersburg 191011, Russia
E-mail: kirillov@math.nagoya-u.ac.jp

We construct birational representations of the symmetric, affine symmetric and extended affine symmetric groups on the affine spaces $\mathbb{A}^{n^2}$ and $\mathbb{A}^{n(n+1)/2}$ which are *tropical* versions of the action of these groups on the set of transportation matrices and Lascoux–Schützenberger's action of the symmetric group on the set of semistandard Young tableaux. We show that *tropical* version of the Robinson–Schensted–Knuth correspondence (*RSK* for short) is a birational automorphism of the affine space $\mathbb{A}^{n^2}$ which intertwines two birational actions of the symmetric group on the space $\mathbb{A}^{n^2}$. We show that *tropical* version of *RSK* has many properties which are similar to those of the classical *RSK*. We study *tropical* version of statistics cocharge $\bar{c}_n$ and proof, in particular, that it is invariant with respect to the birational action of the symmetric group on the space $\mathbb{A}^{n(n+1)/2}$ constructed in this paper. We prove that *tropical* version of Schützenberger's involution satisfies the discrete Hirota–Miwa type equations (the Plücker relations for the 9-th variation of Schur functions). This result allows to find *explicit* formula for the tropical version of Schützenberger's involution as a generating function for certain weighted non–crossing lattice paths.

CONTENTS

1 Introduction

This notes presents Introduction to the subject which we call by *Tropical Combinatorics.* In a few words, by Tropical Combinatorics we mean a branch of Mathematics dealing with the properties and applications of *tropical image* (tropification) of the Piece–wise Linear Combinatorics. By *tropical image* of a piece–wise linear function we mean a rational function constructed from the former by certain rules, see Section 4. It happens that *tropification* of certain piece–wise linear functions which have appeared in Combinatorics and Representation Theory [4,5,6,7,14,15,21,24,25,31,32], have many striking and unexpected properties, and closely connected to some famous discrete integrable equations of Statistical Physics [17,22,35,40]. For example, the tropical images of Schützenberger's involution q_{n-1} and Robinson–Schensted–Knuth's correspondence RSK are closely connected to the 9-th variation of Schur functions introduced by I.G. Macdonald, see e.g. [34], Chapter I, Section 3, Exercise 21, and discrete Hirota–Miwa's type equations. Such a connection not only allows to obtain explicit piece–wise linear formulae for the latter, but also throw new light on their combinatorial, algebraic and algebra–geometric properties. For example, the Robinson–Schensted–Knuth correspondence is a remarkable birational automorphism of the n^2–dimensional affine space which intertwines two birational representations of the symmetric group S_n on the affine space $\mathbb{A}^{n^2}$.

The Piece–wise Linear Combinatorics takes its origin in a series of papers by I.M. Gelfand and A.V. Zelevinsky [14,15] about *good bases* in irreducible finite dimensional representations of the general linear algebra $\mathfrak{gl}(n)$, and has been further developed by A.D. Berenstein and A.V. Zelevinsky [6,7]. A.D. Berenstein and A.N. Kirillov [4,23,24], H. Knight and A.V. Zelevinsky [25], M. Kashiwara [20], M. Kashiwara and T. Nakashima [21], G. Lusztig [31,32,33] and many others. One of the main subject of the Piece–wise Linear Combinatorics is to study relationships between different parameterizations of *bases* in an irreducible finite dimensional representation of a Lie algebra. Recall that even if the canonical basis of an irreducible finite dimensional representation of a Lie algebra $\mathfrak{g}$ is defined uniquely, its parameterizations *depend* on a choice of reduced decomposition of the longest element of the Weyl group $W(\mathfrak{g})$. It is one of the most challenging open problem of Piece–wise Linear Combinatorics to find for any semisimple finite dimensional Lie algebra $\mathfrak{g}$ an explicit formula for the transition map that relates parameterizations of canonical basis corresponding two reduced decompositions of the longest element of the Weyl group $W(\mathfrak{g})$. Essential progress to the solution of this problem for the Lie algebra of type A was made in [3].

As far as I am aware, the tropification map, i.e. a map which transfers the maximum $\max(a, b)$ to the sum $a+b$ and the sum $a+b$ to the product $a \cdot b$, was invented and widely used since the early sixties by the "French School" led by Marcel–Paul Schützenberger, in connection with the theory of formal languages, finite automata, operations research and theoretical computer science. We refer the reader to [1,9,18] for further details about these subjects.

According to J.-E. Pin [37], the name "tropical semiring" for a commonly known $(\max, +)$–semiring was invented by Dominique Perrin in honor of the pioneering works in this field by brazilian mathematician Imre Simon. The $(\max, +)$–semirings were widely used for algebraic description of Discrete Event Dynamical Systems, see e.g. [1]. Recently, the tropical semirings were used for investigation of Soliton Cellular Automata, see e.g. [17,23,39,40].

The name "Tropical Combinatorics" was invented by the author in honor of Marcel–Paul Schützenberger (1920–1996), one of the most creative and influential combinatorialists of 20th century.

In this notes we will consider only the case of the Lie algebra of type A. There exist several bases in irreducible finite dimensional representation of

$\mathfrak{gl}(n)$. For example, Schur–Weyl–Young basis, Gelfand–Tsetlin bases, Canonical basis, Crystal bases, PBW–bases, String basis, Verma basis and many others. The main purpose of this notes is to study combinatorial relationships between the Schur–Weyl–Young basis and the Gelfand–Tsetlin one.

There exists a simple and well–known bijection between the set of (integral) Gelfand–Tsetlin patterns (GT–patterns for short) $K^\lambda \cap \mathbb{Z}^{n(n+1)/2}$ and that of semistandard Young tableaux $STY(\lambda, \leq n)$, see Section 2.3. This bijection allows to transfer certain combinatorial structures given on the set of semistandard Young tableaux to those given on the set of integral GT–patterns. It turns out that in many cases combinatorial structures on the set of integral GT–patterns which came from the set of semistandard Young tableaux, have a natural extension to the whole Gelfand–Tsetlin polytope $K^\lambda(\beta)$ and even to the whole space of triangles X_n. This is a basic idea of the present paper. Among such combinatorial structures on the set of semistandard Young tableaux, we are going to study in the present paper, are the following:

- exponents of Young tableau;
- Bender–Knuth transformations;
- Lascoux–Schützenberger's action of the symmetric group on the set of Young tableaux $STY(\lambda, \leq n)$;
- Schützenberger transformation;
- Lascoux–Schẗzenberger's statistics cocharge and charge.

Let us say a few words about the content of our paper. In Section 2 we recall some basic notation, definitions and results which we are going to use throughout the paper, and tropical images of which we are going to study in Section 4. This material is standard except, probably, Section 2.12 and Theorem 2.28. In Section 2.12 we define the action of the symmetric group on the space of matrices which is compatible with the Robinson–Schensted–Knuth correspondence, its inverse and the upper action of the symmetric group on the "space of plane partitions" introduced by A. Lascoux and M.-P. Schützenberger [30].

In order to illustrate the power of *tropical* approach, let us give some historical remarks. In the paper by the author and A.D. Berenstein [24] a certain piece–wise linear action of the affine symmetric group $W(A_{n-1}^{(1)})$ on the space of triangles X_n was introduced and studied. This action had originated from the familiar action of the symmetric group on the set of Young tableaux introduced by A. Lascoux and M.-P. Schützenberger [30]. Recall [24] that the generators s_j, $1 \leq j \leq n-1$, of the symmetric group S_n are defined as a product of Bender–Knuth's involutions:

$$s_j = (t_1t_2\cdots t_{n-1})^j t_1 (t_1t_2\cdots t_{n-1})^{-j}. \qquad (1.1)$$

It is easy to check that $s_j^2 = 1$. One can show, see [24], that the Coxeter relation

$$(s_js_{j+1})^3 = 1, \quad 1 \le j \le n-2,$$

is a corollary of the following relation $(t_1t_2)^6 = 1$ between Bender–Knuth's involutions t_1 and t_2. The main problem is to show that s_i and s_j commute if $|i-j| \ge 2$. In [24] we had proved more precise statement, namely, that the action of involution s_j changes only the entries of the j-th row of triangle x, and depends only on the entries of the $(j-1)$-th, j-th and $(j+1)$-th rows of the latter. The fact that s_i and s_j commute if $|i-j| \ge 2$ follows from the above statement immediately. The proof of this more precise statement given in [24] is based on yet another product formula for involution s_j:

$$s_j = R_{j-1,j}R_{j-2,j}\cdots R_{1,j}\theta_j R_{1,j}\cdots R_{j-2,j}R_{j-1,j}, \quad 1 \le j \le n-1. \qquad (1.2)$$

Precise definitions of piece–wise linear involutions R_{jk} and (linear) transformation θ_j may be found in [24], p.92, or in Section 2, Exercises 6 and 7. Essential point is that the action of involutions R_{ij} and that of transformation θ_j change only the j-th row of a triangle x, and depends only on the entries of the $(j-1)$-th, j-th and $(j+1)$-th rows of the latter. This observation and formula (1.2) imply the more precise statement about involution s_j mentioned above, and, therefore, imply that involutions s_i and s_j commute if $|i-j| \ge 2$.

Both formulae (1.1) and (1.2) for the involutions s_j were used by the author to obtain an explicit (if complicated) expression for the piece–wise linear action of s_j on the space of triangles. In the max–min form the latter expression appeared to be messy and untractable even for $j=3$, see e.g. [24], (3.14). As a big surprise for the author was an observation that after *tropification* the messy max–min expression for the action of s_3 on the space of triangles X_4 has drastically simplified. We give a simple explicit formula for the action of *tropical* version of the involution s_j in Section 4.3, together with a sketch of the proof that transformations s_j, $1 \le j \le n-1$, defined by (4.11), *indeed* satisfy the defining relations of the symmetric group S_n.

A similar situation had happened with the author's attempts to find explicit formulae for the action of the symmetric group S_n on the space of matrices M_n. Our construction of the symmetric group action on the space of matrices had appeared first as the transform by means of the Robinson–Schensted–Knuth correspondence and its inverse of the upper action of the symmetric group on the set of plane partitions:

$$\sigma_j = (RSK)^{-1} \circ s_j^{up} \circ RSK.$$

Namely, let $d \in M_{n\times n}(\mathbb{Z} \geq 0)$ be a matrix with non-negative integer entries, and (P, Q) be a pair of semistandard Young tableaux which corresponds to d according to the Robinson–Schensted–Knuth correspondence

$$d \xrightarrow{RSK} (P, Q),$$

then

$$\sigma_j(d) = (RSK)^{-1}(P, s_j(Q)).$$

Note that *explicit* piece–wise linear formulae for the transformations RSK, RSK^{-1} and s_j had been known for some time [4,24]. To compute the action of σ_j explicitly, it remained only to plug one explicit formula to another. The formula for σ_j obtained in this way was messy and untractable. However after *tropification* of the corresponding formulae for RSK, $(RSK)^{-1}$ and s_j some unexpected cancelations had happened, and the final formulae for σ_j has drastically simplified, see Section 2.12. In my opinion, the main advantage of the tropical approach to the piece–wise linear combinatorics is a possibility to use the "advanced determinantal calculus" including applications of the non–crossing lattice paths theory, see e.g. [16,27], and Plücker's relations for the 9-th variation of Schur functions introduced by I.G. Macdonald, see e.g. [34].

In Section 2.11 we recall [24] the definition of the affine symmetric group action on the set of Gelfand–Tsetlin patterns and that on the space of triangles. We also construct on the space of triangles an action of the extended affine symmetric group. Having in mind many recent applications of piece–wise linear and birational representations of extended affine Weyl groups to the theory of discrete dynamical systems [36], Cellular Automata and Ultradiscrete systems, see e.g. [17,40] and the literature quoted therein, birational and piece–wise linear representations of the (extended) affine symmetric group constructed in the present paper are hopefully of some use.

In Section 3 we review several piece–wise linear formulae:

• for the action of involutions s_k on the space of triangles [24].

Recall that involutions s_k being restricted to the set of Gelfand–Tsetlin patterns and after that transformed to the set of Young tableaux, correspond to the symmetric group action on the former set, introduced by A. Lascoux and M.-P. Schützenberger [29];

• for the Schützenberger involution [4,25].

• for the Robinson–Schensted–Knuth correspondence [4].

In the same paper [4] the piece–wise linear formulae for the inverse Robinson–Schensted–Knuth correspondence and for the Hillman–Grassl bijection which is a generalization of the RSK, have been obtained and proved.

We give *tropical* version for the inverse RSK in Section 4. The piece–wise linear formula for the latter can be obtained by means of the ultradiscretizaton map; *tropical* versions of the Hillman–Grassl bijection and its inverse will be considered in a separate publication.

The main body of our paper is Section 4 which is devoted to the description and study of the *tropical* version for

- Schützenberger involution;
- Lascoux–Schützenberger's action of the symmetric group on the set of Young tableaux;
- Robinson–Schensted–Knuth correspondence and its inverse;
- action of the symmetric group on the space of matrices;
- cocharge and charge.

The main purpose of Section 4 is to show that after passing to *tropical version*, the items mentioned above still have "nice combinatorial" properties which are similar to those they had before. For example, see Theorems 4.27 and 4.29, let

$$\mathbb{A}^{n^2} \to \mathbb{A}^{n^2},$$

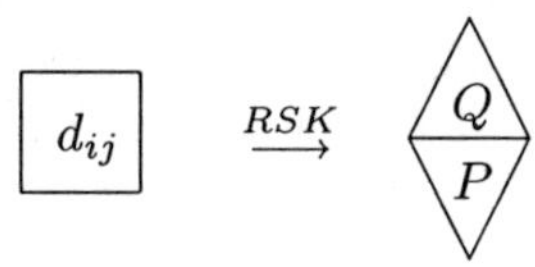

be the tropical version of the Robinson–Schensted–Knuth correspondence (RSK), then

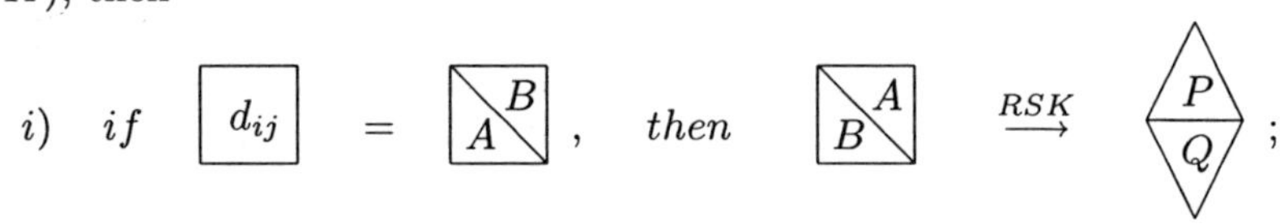

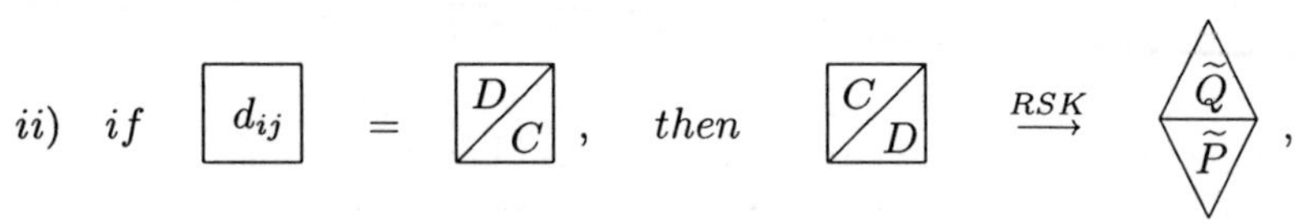

where $\widetilde{P} = \mathbb{S}P$ and $\widetilde{Q} = \mathbb{S}Q$, and $\mathbb{S} = q_{n-1} : \mathbb{A}^{n(n+1)/2} \to \mathbb{A}^{n(n+1)/2}$ denotes the tropical image of Schützenberger's involution;

iii) $RSK(h_j^{(0)}(\mathbf{d})) = Q_{jj}(y(Q))$, if $1 \le j \le n-1$,

where $h_j^{(0)}(\mathbf{d})$ and $Q_{jj}(y)$ stands for the tropical versions of the row exponents $h_j^{(0)}(d)$ of a matrix $d \in M_{n\times n}(\mathbb{R})$, see Section 2.10, and the exponents $\kappa_j(x)$ of a triangle $x \in X_n$, see Section 2.4, correspondingly;

iv) $s_j^{up}(RSK(\mathbf{d})) = RSK(\sigma_j(\mathbf{d}))$ if $1 \le j \le n-1$,

see Theorem 4.27 for explanation of notation we have used.

Another example which shows fruitfulness of the tropical approach to Piece–wise Linear Combinatorics is familiar statistics *charge* and *cocharge*, introduced by A. Lascoux and M.-P. Schützenberger [29]. In [24] we had suggested conjectural piece–wise linear formula for the statistics cocharge $\overline{c}_n$ on the space of triangle X_n. In particular, we made a conjecture that statistics cocharge $\overline{c}_n$ is invariant with respect to the action of the affine symmetric group $W(A_{n-1}^{(1)})$ on the space of triangles X_n. Here we prove this conjecture. Let us briefly explain the main steps of the proof. First of all we compute the tropical image $\overline{c}_n(y)$, called by tropical cocharge, of the formula for statistics cocharge, suggested in [24]. The next step is to study the ratio $\widetilde{c}_n(y) = \overline{c}_n(y)/\overline{c}_{n-1}(y)$. It is easy to see that $\widetilde{c}_1(y) = \widetilde{c}_2(y) = 1$, and $\overline{c}_n(y) = \prod_{j=1}^{n} \widetilde{c}_j(y)$. The main tool of our proof is two *inductive* formulae for the Laurent polynomial $\widetilde{c}_n(y)$, namely, (4.26) and (4.27). We state here only their corollary, namely, the following inductive formula:

$$\overline{c}_{n+1}(y) = \frac{\overline{c}_n(y)\overline{c}_n(s_n(y))}{\overline{c}_{n-1}(y)} \left\{ \frac{y_{n+1,n+1}Q_{n,n}(y)\beta_n(y)}{\beta_{n-1}(y)\beta_{n+1}(y)} \prod_{k=1}^{n-1} (s_n(y))_{k,n} \right\}. \quad (1.3)$$

It is not difficult to see from (4.26) and (4.27) that if $n \geq 2$ be an integer, then Laurent's polynomial $\widetilde{c}_{n+1}(y)\widetilde{c}_n(y)$ is a s_n–invariant, and furthermore, the function $\widetilde{c}_{n+1}(y)$ is a s_j–invariant for all j, $1 \leq j \leq n-1$. These two statements imply that statistics *tropical* cocharge $\overline{c}_n(y)$ is a S_n–invariant. Because it is clear from the very definition of statistics tropical cocharge $\overline{c}_n(y)$ that the latter is invariant with respect to the action of affine shift operator T_ϵ, see Section 4.2, we conclude that Laurent's polynomial $\overline{c}(y)$ is a $W(A_{n-1}^{(1)})$–invariant.

We also introduce the tropical version of statistics *charge* $c_n(y)$. It follows from [8,28] that the ultradiscretization of the statistics charge $c_n(y)$ being restricted to the set dominant weight semistandard Young tableaux $STY_+(\leq n)$, coincides with statistics *charge* $c_n(\bullet)$ on the set $STY_+(\leq n)$, introduced by A. Lascoux and M.-P. Schützenberger. Combining this result with the one about relationships between (tropical) statistics $c_n(y)$ and $\overline{c}_n(y)$, one can deduce that the ultradiscritization of statistics cocharge $\overline{c}_n(y)$ being restricted to the set of semistandard Young tableaux, coincides with statistics *cocharge* $\overline{c}_n(\bullet)$, introduced by A. Lascoux and M.-P. Schützenberger. The main tool to obtain inductive formulae for statistics tropical cocharge $\overline{c}_{n+1}(y)$ is to study the tropical versions for the exponents $\kappa_j(x)$of a triangle $x \in X_n$, and that for the row exponents $h_j^{(0)}(d)$ of matrix $d \in M_{n\times n}(\mathbb{R})$.

In this notes we consider mainly the algebraic and combinatorial aspects of *Tropical Combinatorics*. Very interesting connections of the latter with

Crystal bases, Canoniacal basis, Toric varieties, Birational Geometry, Discrete and Ultradiscrete dynamical systems will be considered in subsequent publications.

2 Basic Definitions

In this section we collect some basic notation, definitions, and results dealing with combinatorics of Young tableaux and Gelfand–Tsetlin patterns. For more details and proofs see [11,14,24,34].

2.1 Young tableaux

Let λ be a partition and β be a composition, denote by $STY(\lambda,\beta)$ the set of semistandard Young tableaux of shape λ and weight (or content) β. We refer the reader to [11], p.1–4, for a condensed explanation of the above notions. We also will use notation $STY(\lambda,\leq n)=\coprod\limits_{|\beta|=n} STY(\lambda,\beta)$ to denote the set of all semistandard Young tableaux of shape λ filled by the numbers from 1 to n; notation $STY_{+}(\lambda,\leq n)=\coprod\limits_{\mu\vdash n} STY(\lambda,\mu)$ to denote the set of semistandard Young tableaux which have a dominant weight $\mu=(\mu_1\geq\mu_2\geq\cdots\geq\mu_n)$; notation $STY(\leq n)=\coprod\limits_{\lambda\vdash n} STY(\lambda,\leq n)$ to denote the set of all semistandard Young tableaux filled by the numbers from 1 to n, and that $STY_{+}(\leq n)$ to denote the set of dominant weight semistandard Young tableaux filled by the numbers from 1 to n.

2.2 Triangles and Gelfand–Tsetlin patterns

Let n be a positive integer. By definition, the space of triangles X_n consists of all sequences $x=(x^{(n)},x^{(n-1)},\ldots,x^{(1)})$, where $x^{(j)}=(x_{1j},\ldots,x_{jj})\in\mathbb{R}^j$ for $j=1,\ldots,n$. As a vector space, $X_n\simeq\mathbb{R}^{\frac{n(n+1)}{n}}$. In the sequel we will display a triangle $x=(x^{(n)},\ldots,x^{(1)})\in\mathbb{R}^{\frac{n(n+1)}{n}}$ either follow to a traditional notation for Gelfand–Tsetlin patterns:

$$x:=\begin{array}{ccccccc} x_{1n} & & x_{2n} & \cdots & x_{n-1,n} & & x_{nn}\\ & x_{1,n-1} & & \cdots & & x_{n-1,n-1} & \\ & & & \vdots & & & \\ & & & x_{12}\ x_{22} & & & \\ & & & x_{11} & & & \end{array},$$

or as a upper triangular matrix:

$$x := \begin{pmatrix} x_{11} & x_{12} & \cdots & x_{1,n-1} & x_{1n} \\ 0 & x_{22} & \cdots & x_{2,n-1} & x_{2n} \\ & & \vdots & & \\ 0 & 0 & \cdots & 0 & x_{nn} \end{pmatrix}$$

The vector $x^{(n)} \in \mathbb{R}^n$ is called by *highest weight* of a triangle $x \in X_n$ and denoted by $\lambda(x) := x^{(n)}$. The weight $\beta := \beta(x)$ of a triangle $x \in X_n$ is defined to be a vector $\beta = (\beta_1, \ldots, \beta_n) \in \mathbb{R}^n$ such that

$$\beta_j = |x^{(j)}| - |x^{(j-1)}|, \quad 1 \le j \le n,$$

where by definition $|x^{(j)}| = \sum_{i=1}^{j} x_{ij}$, and $|x^{(0)}| = 0$. For given vectors $\lambda \in \mathbb{R}^n$ and $\beta \in \mathbb{R}^n$ we introduce the following subspaces of X_n:

$$X^\lambda = \{x \in X_n | \lambda(x) = \lambda\}, \tag{2.1}$$

$$X^\lambda(\beta) = \{x \in X^\lambda | \beta(x) = \beta\}. \tag{2.2}$$

Now we are going to define the Gelfand–Tsetlin patterns, see [14,5,24]. By definition a triangle $x \subset X_n$ is called by *Gelfand–Tsetlin pattern* (GT–pattern for short) if the following inequalities are satisfied:

$$x_{ij} \ge x_{i+1,j+1}, \quad 1 \le i \le j \le n-1,$$
$$x_{ij} \ge x_{i,j-1}, \quad 1 \le i < j \le n,$$
$$x_{ij} \ge 0, \quad 1 \le i \le j \le n.$$

Follow [24] we denote by $K := K_n \subset X_n$ the set of all GT–patterns, and by K^λ and $K^\lambda(\beta)$ the subsets consisting of GT–patterns that belong to the subspaces (2.1) and (2.2) correspondingly. It is easy to see that K^λ and $K^\lambda(\beta)$ are convex compact polytopes in the space $\mathbb{R}_+^{n(n+1)/2}$. The latter polytope is commonly called by Gelfand–Tsetlin polytope. A GT–pattern $x = (x_{ij})$ is called *integral* if all its entries x_{ij}, $1 \le i \le j \le n$, are (non–negative) integer numbers. The following result goes back to the original paper by I.M. Gelfand and M.L. Tsetlin (Doklady Akad. Nauk SSSR (N.S.) **71** (1950), 825–828), and was rediscovered many times.

Proposition 2.1 *Let λ be a partition and β be a composition, then the number of integer points in the convex polytope K^λ (respectively in $K^\lambda(\beta)$) is equal to the dimension of irreducible highest weight λ representation V_λ of the Lie algebra $\mathfrak{gl}(n)$ (respectively, to the dimension of weight β subspace $V_\lambda(\beta) \subset V_\lambda$).*

2.3 Gelfand–Tsetlin patterns and Young tableaux

It is well–known, see e.g. [14], that the dimension of weight subspace $V_\lambda(\beta) \subset V_\lambda$ admits a pure combinatorial description as the number of semistandard Young tableaux of shape λ and weight (or content) β:

$$|K^\lambda \cap \mathbb{Z}^{n(n+1)/2}| = \dim V_\lambda = |STY(\lambda, \leq n)|, \tag{2.3}$$
$$|K^\lambda(\beta) \cap \mathbb{Z}^{n(n+1)/2}| = \dim V_\lambda(\beta) = |STY(\lambda, \beta)|. \tag{2.4}$$

One way to explain and prove the equalities (2.3), (2.4) is based on an observation that there exist several realizations of the irreducible highest weight λ representation V_λ of the Lie algebra $\mathfrak{gl}(n)$. One such a realization of the $\mathfrak{gl}(n)$–representation V_λ goes back to I. Schur and H. Weyl (see e.g., [12]). The representation space in this case is called by *Weyl module*, and its *basis* happens to be parameterized by the set of semistandard Young tableaux $STY(\lambda, \leq n)$. Another realization of the same representation V_λ is due to I.M. Gelfand and M.L. Tsetlin [*ibid*], and integral GT–patterns happen to be parameterize a *basis* in this representation (the so–called Gelfand–Tsetlin basis). Summarizing, integral Gelfand–Tsetlin's patterns $K^\lambda(\beta) \cap \mathbb{Z}^{n(n+1)/2}$ and semistandard Young tableaux $STY(\lambda, \beta)$ parameterize *bases* in the weight β subspace $V_\lambda(\beta)$ of the irreducible representation V_λ in the Gelfand–Tsetlin and Schur–Weyl realizations of the latter correspondingly. This is an explanation of the equalities (2.3) and (2.4) from the representation theory point of view. Equalities (2.4) have an interesting corollary: the number of integer points in the Gelfand–Tsetlin polytope $K^\lambda(\beta)$ is equal to the number of semistandard Young tableaux of shape λ and weight β. There exists a simple and well–known bijective proof of this corollary. Namely, let T be a semistandard Young tableaux of shape λ and weight $\beta = (\beta_1, \ldots, \beta_n)$. The tableau T defines (and is defined by) a sequence of partitions

$$0 = \mu^{(0)} \subset \mu^{(1)} \subset \cdots \subset \mu^{(n)} = \mu \tag{2.5}$$

such that each skew diagram $\theta^{(i)} = \mu^{(i)} \setminus \mu^{(i-1)}$ $(1 \leq i \leq n)$ is a horizontal strip, see e.g. [34], Chapter I, Section 1. Starting from the sequence of partitions (2.5), one can define a triangle $x := x(T) \in X_n$ by the following rule

$$x^{(i)}(T) = \text{shape}(\mu^{(i)}), \quad i = 1 \ldots, n.$$

It is easy to check that the resulting triangle $x(T)$ appears to be an integral GT–pattern and *vise versa*, i.e. if $x \in K^\lambda(\beta) \cap \mathbb{Z}^{n(n+1)/2}$ is a GT–pattern, then a sequence of partitions $\mu^{(i)} := x^{(i)}$, $1 \leq i \leq n$, defines a tableau T which is in fact a semistandard one. This simple bijection allows to transfer certain combinatorial structures given on the set of semistandard Young tableaux to those given on the set of integral Gelfand–Tsetlin patterns. Among such combinatorial structures on the set of semistandard Young tableaux we consider in subsequent Sections the following:

- exponents of a Young tableau;
- Bender–Knuth transformations;
- Lascoux–Schützenberger action of the symmetric group;
- Schützenberger transformation;
- Lascoux–Schützenberger's statistics *cocharge.*

It happens very often that the latter combinatorial structures have natural extensions to the whole Gelfand–Tsetlin polytope $K^{\lambda}(\beta)$ and even to the whole space of triangles X_n.

2.4 Exponents and Littlewood–Richardson numbers

Let $x \in X_n$ be a triangle. Follow [15] define *exponents* $\kappa_1, \ldots, \kappa_{n-1}$ of the triangle x by $\kappa_i = \max\{d_j^{(i)}(x) | 1 \le j \le i\}$, where

$$d_j^{(i)}(x) = \sum_{1 \le k < j} (x_{k,i+1} - 2x_{k,i} + x_{k,i-1}) + x_{j,i+1} - x_{j,i}. \tag{2.6}$$

In the case when a triangle x corresponds to a semistandard Young tableaux $x := x(T)$ under the bijection constructed in Section 2.3, the exponents $\kappa_i(x)$, $1 \le i \le n-1$, have a simple combinatorial interpretation in terms of the so-called *descent set* $DES(T)$ of tableau T. Namely, let T be a semistandard Young tableau of shape λ and weight β, $l(\beta) \le n$. For any integer m between 1 and $n-1$, denote by $d_m(T)$ the maximal number of quadruples of positive integers $((i_1, j_1), (i'_1, j'_1)), \ldots, ((i_k, j_k), (i'_k, j'_k))$ such that

i) all pairs (i_r, j_r) and (i'_r, j'_r), $1 \le r \le k$, are distinct;
ii) $i_r < i'_r$, $1 \le r \le k$;
iii) $T(i_r, j_r) = m$ and $T(i'_r, j'_r) = m + 1$.

Recall that $T(i, j)$ denotes the entry of the tableau T which is located in the box with coordinates (i, j).

Finally, let us introduce numbers $\kappa_i(T) = \beta_{i+1} - d_i(T)$, $1 \le i \le n-1$. The numbers $\kappa_1(T), \ldots, \kappa_{n-1}(T)$ are called by *exponents* of the tableau T. It is not difficult to see that $\kappa_i(x(T)) = \kappa_i(T)$ for any semistandard Young tableau T. We refer the reader to the paper by A.D. Berenstein and A.V. Zelevinsky [7], where interesting connections between *exponents* $\kappa_j(x)$ of GT–patterns, string bases and Lusztig's canonical basis are explained.

Let λ be a partition and μ be a composition. For any partition ν we denote by $STY(\lambda, \mu \parallel \nu)$ the following set

$$\{T \in STY(\lambda, \mu - \nu) | \kappa_i(x(T)) \le \nu_i - \nu_{i+1}\}.$$

Proposition 2.2 *Let λ, μ, ν be partitions such that $|\lambda| + |\mu| = |\nu|$. Then*

$$|STY(\lambda, \nu \parallel \mu)| = c^{\nu}_{\lambda\mu},$$

where the number $c^{\nu}_{\lambda\mu}$ denotes the Littlewood–Richardson number, i.e. the multiplicity Mult $[V_\nu : V_\lambda \otimes V_\mu]$.

Example 2.3 Let us take $\lambda = \mu = (321)$ and $\nu = (4332)$. Thus $\nu - \mu = (1122)$ and the set $STY(\lambda, \nu - \mu)$ consists of the following semistandard Young tableaux:

$$\begin{array}{lcccc} T: & \begin{matrix} 1\,2\,3 \\ 3\,4 \\ 4 \end{matrix} & \begin{matrix} 1\,2\,4 \\ 3\,3 \\ 4 \end{matrix} & \begin{matrix} 1\,3\,3 \\ 2\,4 \\ 4 \end{matrix} & \begin{matrix} 1\,3\,4 \\ 2\,4 \\ 3 \end{matrix} \\ \\ DES(T): & (233) & (23) & (133) & (123) \\ \\ \kappa(T): & (1,1,0) & (1,1,1) & (1,2,0) & (0,1,1) \end{array}$$

Below each tableau T we put, respectively, its descent set $DES(T)$ and a set of exponents $\kappa(T) = (\kappa_1(T), \ldots, \kappa_{n-1}(T))$. For example, let us compute $\kappa_2(x\begin{pmatrix} 1\,3\,3 \\ 2\,4 \\ 4 \end{pmatrix})$. It is easy to see that

$$T := \begin{matrix} 1\,3\,3 \\ 2\,4 \\ 4 \end{matrix} \quad \longrightarrow \quad x(T) := \begin{matrix} 3 & & 2 & & 1 & & 0 \\ & 3 & & 1 & & 0 & \\ & & 1 & & 1 & & \\ & & & 1 & & & \end{matrix},$$

and

$$\kappa_2(x\begin{pmatrix} 3 & & 2 & & 1 & & 0 \\ & 3 & & 1 & & 0 & \\ & & 1 & & 1 & & \\ & & & 1 & & & \end{pmatrix}) = \max(d_1^{(2)}(x), d_2^{(2)}(x))$$

$$= \max(x_{13} - x_{12}, x_{13} - 2x_{12} + x_{11} + x_{23} - x_{22}) = 2.$$

Therefore, the set

$$STY(\lambda, \nu \parallel \mu) = \{T \in STY(\lambda, \nu - \mu) | \kappa_i(T) \leq \mu_i - \mu_{i+1} = 1, \quad i = 1, 2, 3\}$$

consists of all the above semistandard tableaux, except that $\begin{matrix} 1\,3\,3 \\ 2\,4 \\ 4 \end{matrix}$. Hence

$c^{\nu}_{\lambda\mu} = 3$.

2.5 Bender–Knuth transformations

It is an easy consequence of the representation theory of the Lie algebra $\mathfrak{gl}(n)$ that the number of semistandard Young tableaux of a given shape λ and weight $\beta = (\beta_1, \dots, \beta_n)$ does not depend on the order of components of the weight β. In other words, if $\sigma \in S_n$ be a permutation, then

$$|STY(\lambda, \beta)| = |STY(\lambda, (\beta_{\sigma(1)}, \dots, \beta_{\sigma(n)}))|.$$

A pure combinatorial proof of this statement had been given for the first time by A. Bender and D. Knuth, [2]. The key step in their proof is a combinatorial construction of certain *transformations* on the set of semistandard Young tableaux. Follow [24] we will call these transformations by Bender–Knuth's involutions. In this Section we will recall the definition of Bender–Knuth's transformations, and, follow [24], we are going to rewrite the latter in terms of Gelfand–Tsetlin patterns. We recall also one of the basic results of the paper [24] which describes *relations* between Bender–Knuth's involutions.

Given a tableau $T \in STY(\lambda, \beta)$, consider the part T_j of tableau T filled by the letters j and $j+1$. It is clear that T_j is a disjoint union of the following fragments:

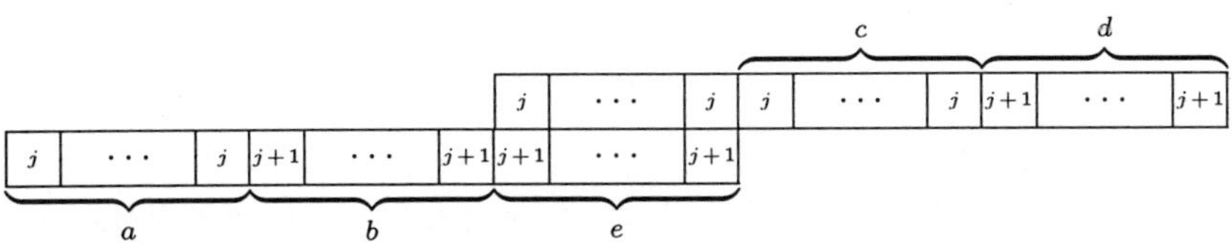

We replace every such fragment by the following one

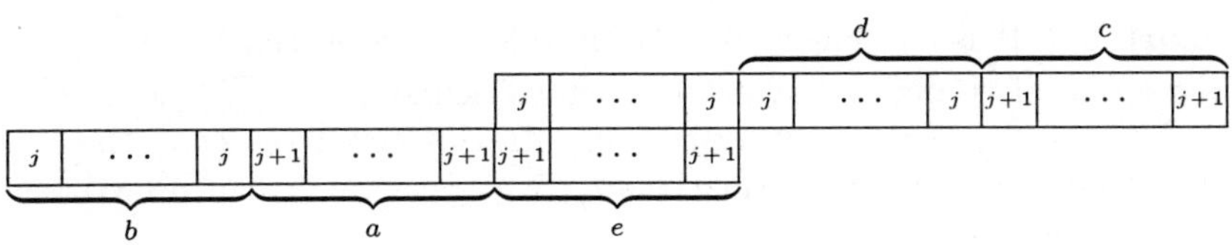

The part $T \setminus T_j$ of the tableau T remains unchanged. As a result, we obtain a new tableau $\widetilde{T} \in STY(\lambda, (j, j+1)\beta)$, where $(j, j+1)$ denotes the simple transposition that interchanges j and $j+1$, and fixes all other elements of $[1, n]$; for any sequence $\beta = (\beta_1, \dots, \beta_j, \beta_{j+1}, \dots, \beta_n)$, we put by definition

$$(j, j+1)\beta = (\beta_1, \dots, \beta_{j+1}, \beta_j, \dots, \beta_n).$$

Denote the new tableau $\widetilde{T}$ by $t_j(T)$. The transformation

$$t_j : STY(\lambda, \beta) \to STY(\lambda, (j, j+1)\beta)$$
$$T \to \widetilde{T}$$

is called by the Bender–Knuth transformation. It was introduced by A. Bender and D. Knuth [2].

Proposition 2.4 [24] *Let $T \in STY(\lambda, \beta)$ and $x = x(T)$ be the corresponding integral GT–pattern. Then the GT–pattern $\widetilde{x}$ corresponding to the tableau $\widetilde{T} = t_j(T)$ has the following coordinates*

$$\begin{aligned} \widetilde{x}_{ik} &= x_{ik}, \quad \text{if} \quad k \neq j; \\ \widetilde{x}_{ij} &= \min(x_{i,j+1}, x_{i-1,j-1}) + \max(x_{i,j-1}, x_{i+1,j+1}) - x_{ij}, \end{aligned} \tag{2.7}$$

where by definition we put $x_{0j} := +\infty$, $x_{j,j-1} = -\infty$ for all j, $1 \leq j \leq n-1$.

Definition 2.5 *The Bender–Knuth transformations t_i, $1 \leq i \leq n-1$, acting on the space of triangles X_n are defined by the formulae (2.7).*

Theorem 2.6 (A.D. Berenstein and A.N. Kirillov [24])

Transformations $t_i : X_n \to X_n$, $1 \leq i \leq n-1$, satisfy the following relations

1) $t_i^2 = 1, \quad t_i t_j = t_j t_i \;$ if $\; |i-j| \geq 2$;
2) $(t_1 t_2)^6 = 1$;
3) $(q_i t_1)^4 = 1, \;$ if $\; i \geq 3, \;$ where $\; q_i = t_1 \underbrace{t_2 t_1} \cdots \underbrace{t_i t_{i-1} \cdots t_2 t_1}$.

It is clear from the very definition that $q_i^2 = 1$, $1 \leq i \leq n-1$.

The transformation q_{n-1} of the space of triangles X_n is called by Schützenberger involution, and will denoted also by $\mathbb{S}_n$.

Remark 2.7 It is convenient to introduce a new basis $\{y_{ij} | 1 \leq i \leq j \leq n\}$ in the space of triangles X_n, namely, let us define $y_{ij} = \sum_{r=1}^{i} x_{rj}$, so that $x_{ij} = y_{ij} - y_{i-1,j}$, if $i \geq 2$, and $x_{1j} = y_{1j}$. In this new basis $\{y_{ij}\}$ the action of the Bender–Knuth transformations t_j, $1 \leq j \leq n-1$, can be written as follows: if $t_j(y) = \widetilde{y}$, then

- $\widetilde{y}_{ik} = y_{ik}$, if $\; k \neq j$;
- $\widetilde{y}_{ij} = \max(y_{i,j-1} + y_{i,j+1}, y_{i-1,j-1} + y_{i+1,j+1}) - y_{ij}$, if $2 \leq i < j \leq n-1$;
- $\widetilde{y}_{1j} = \max(y_{1,j-1} + y_{1,j+1}, y_{2,j+1}) - y_{1j}$, if $\; j \geq 2$;
- $\widetilde{y}_{jj} = y_{j-1,j-1} + y_{j+1,j+1} - y_{jj}$, if $\; 2 \leq j \leq n-1$;
- $\widetilde{y}_{11} = y_{22} - y_{11}$.

Let λ be a partition and $\mu = (\mu_1, \dots, \mu_n)$ be a composition of the same size. Consider a Young tableau T of shape λ and weight μ. Remove the entry, say t_1 from the upper left corner of T, and apply the Schützenberger "jeu de taquin" algorithm [11,38] to the skew tableau that is left. This gives a tableau that we will denote by ΔT, whose diagram has one box removed from the diagram of T. Put the number $n + 1 - t_1$ in this box. Repeat the algorithm on ΔT, getting a smaller tableau $\Delta^2 T$, and putting $n + 1 - t_2$ in the box removed, where t_2 is the number in the upper left corner of ΔT. Continue until all the entries of T have been removed. As a result we will obtain a new Young tableau T^* of the same shape λ and weight $\overleftarrow{\mu} = (\mu_n, \mu_{n-1}, \dots, \mu_2, \mu_1)$. The map

$$\begin{array}{rcl} STY(\lambda, \mu) & \rightarrow & STY(\lambda, \overleftarrow{\mu}) \\ T & \rightarrow & T^* \end{array}$$

is called by *Schützenberger transformation* or evacuation map, and denoted by $\mathbb{S}$. The map $\mathbb{S}$ defines the map

$$\mathbb{S}_n : STY(\lambda, \leq n) \rightarrow STY(\lambda, \leq n), \tag{2.8}$$

which is called by Schützenberger involution on the set of semistandard Young tableaux $STY(\lambda, \leq n)$.

Example 2.8 Consider semistandard Young tableau $T = \begin{array}{llll} 1 & 1 & 2 & 3 \\ 2 & 2 & 3 & \\ 4 & & & \end{array}$, and apply the algorithm for construction of the Schützenberger transform T^* of the tableau T as it was described above. We display the construction of T^* step by step:

$$\begin{array}{llll} \textcircled{1} & 1 & 2 & 3 \\ 2 & 2 & 3 & \\ 4 & & & \end{array} \rightarrow \begin{array}{llll} \textcircled{1} & 2 & 2 & 3 \\ 2 & 3 & \boxed{4} & \\ 4 & & & \end{array} \rightarrow \begin{array}{llll} \textcircled{2} & 2 & 2 & 3 \\ 3 & \boxed{4} & \boxed{4} & \\ 4 & & & \end{array} \rightarrow \begin{array}{llll} \textcircled{2} & 2 & 3 & \boxed{3} \\ 3 & \boxed{4} & \boxed{4} & \\ 4 & & & \end{array} \rightarrow \begin{array}{llll} \textcircled{2} & 3 & \boxed{3} & \boxed{3} \\ 3 & \boxed{4} & \boxed{4} & \\ 4 & & & \end{array}$$

$$\rightarrow \begin{array}{llll} \textcircled{3} & 3 & \boxed{3} & \boxed{3} \\ 4 & \boxed{4} & \boxed{4} & \\ \boxed{3} & & & \end{array} \rightarrow \begin{array}{llll} \textcircled{3} & \boxed{2} & \boxed{3} & \boxed{3} \\ 4 & \boxed{4} & \boxed{4} & \\ \boxed{3} & & & \end{array} \rightarrow \begin{array}{llll} \textcircled{4} & \boxed{2} & \boxed{3} & \boxed{3} \\ \boxed{2} & \boxed{4} & \boxed{4} & \\ \boxed{3} & & & \end{array} \rightarrow \begin{array}{llll} \boxed{1} & \boxed{2} & \boxed{3} & \boxed{3} \\ \boxed{2} & \boxed{4} & \boxed{4} & \\ \boxed{3} & & & \end{array} .$$

Hence, $T^* = \mathbb{S}(T) = \begin{matrix} 1\,2\,3\,3 \\ 2\,4\,4 \\ 4 \end{matrix}$. It is not difficult to check that $T^* = q_3(T)$, where $q_3 = t_1t_2t_1t_3t_2t_1$.

Theorem 2.9 ([13,24]) *The involution q_{n-1} given on the space of triangles X_n being restricted to the set of semistandard Young tableaux $STY(\lambda, \le n)$, coincides with the Schützenberger involution $\mathbb{S}_n$ as it was defined above, see (2.8).*

2.7 Action of the symmetric group on the set of Young tableaux

In this Section we recall the definition of an action of the symmetric group S_n on the set of semistandard Young tableaux $STY(\lambda, \le n)$ due to A. Lascoux and M.-P. Schützenberger [30]. Given a positive integer j, $1 \le j \le n-1$, and a tableau $T \in STY(\lambda, \beta)$, $l(\beta) \le n$, consider the word $w(T)$ that corresponds to the tableau T. Recall, that by definition, the word $w(T)$ can be obtained from the tableau T by reading the elements of T from the right to the left in consecutive rows, starting from the top one, see e.g. [34], Chapter I, Section 9, p.142. For convenience, we will replace j by a and $j+1$ by b. As a first step we extract from the word $w(T)$ a subword w' that contains the letters a and b only. As a second step, we consequently remove from the word w' all pairs ab. As a result, we obtain a subword of type $b^m a^n$. We replace it by the word $b^n a^m$ and, after this, recover all the removed pairs ab and all the letters which are different from a and b. Finally, we obtain a new word $\widetilde{w}$. This word corresponds to some semistandard Young tableau $\widetilde{T} \in STY(\lambda, (j, j+1)\beta)$. We will denote the latter tableau by $\tau_j(T) := \widetilde{T}$.

Proposition 2.10 (A. Lascoux and M.-P. Schützenberger [30]) *Transformations τ_j, $1 \le j \le n-1$, satisfy the defining relations of the symmetric group S_n:*

1) $\tau_j^2 = 1, \quad 1 \le j \le n-1;$
2) $(\tau_j\tau_{j+1})^3 = 1, \quad 1 \le j \le n-2;$
3) $(\tau_i\tau_j)^2 = 1, \quad 1 \le i, j \le n-1, \quad \text{and} \quad |i-j| \ge 2.$

In other words, the involutions τ_j, $1 \le j \le n-1$, define a representation of the symmetric group on the set $STY(\lambda, \le n)$.

Theorem 2.11 (A.D. Berenstein and A.N. Kirillov [24]) *On the set of semistandard Young tableaux $STY(\lambda, \le n)$ one has*

$$\tau_j = q_j t_1 q_j.$$

Remark 2.12 The fact that involutions τ_j, $1 \le j \le n-1$, generate the symmetric group S_n has been proved by A. Lascoux and M.-P. Schützenberger [30] by subtle analysis of the properties of the *plactic* monoid and the Robinson–Schensted correspondence. The same action was rediscovered by M. Kashiwara [20,21] in the context of the theory of crystal bases, as an action of the Weyl group W (in our case $W \simeq S_n$) on the crystal basis corresponding to the highest weight λ irreducible representation V_λ of the Lie algebra $\mathfrak{gl}_n$.

Example 2.13 Let us illustrate on example the difference between the Bender–Knuth transformations t_j and those of Lascoux and Schützenberger s_j.

Take $T = \begin{matrix} 1\,1\,2\,2\,3 \\ 2\,2\,3\,3 \\ 3\,4\,4 \\ 4 \end{matrix}$, then $t_3(T) = \begin{matrix} 1\,1\,2\,2\,4 \\ 2\,2\,3\,4 \\ 3\,3\,4 \\ 4 \end{matrix}$, but

$$s_3(T) = t_1t_2t_1t_3t_2t_1t_2t_3t_1t_2t_1(T) = \begin{matrix} 1\,1\,2\,2\,4 \\ 2\,2\,3\,3 \\ 3\,4\,4 \\ 4 \end{matrix}.$$

Indeed,

$$w(T)_{3,4} = 3(3(34)4)(34) \xrightarrow{LS} 4(3(34)4)(34),$$

but

$$w(T)_{3,4} = 33(34)4(34) \xrightarrow{BK} 44(34)3(34).$$

2.8 Plane partitions

Let λ be a partition, a reverse plane partition of shape λ is an array $\pi = (x_{ij})$ of positive integers of shape λ (i.e. $1 \le i \le l(\lambda)$, $1 \le j \le \lambda_i$) that is weakly increasing in every row and every column. In the sequel, we will assume that λ has a rectangular shape (m^n). If $\pi = (x_{ij})$ is a reverse plain partition of shape (m^n), then $\widetilde{\pi} = (\widetilde{x}_{ij} := x_{n-i+1,m-j+1})$ is a plane partition of shape (m^n), i.e. an array of positive integers of shape (m^n) that is weakly decreasing in every row and in every column. With any plane partition $\widetilde{\pi}$ of square shape (n^n) one can associate a pair of GT–patterns (P,Q) such that P and Q have the same highest weights. Namely, starting from a square shape plane partition $\widetilde{\pi}$, let us define two triangles $P = \{\widetilde{x}_{ij} | n \ge i \ge j \ge 1\}$ and $Q = \{\widetilde{x}_{ij} | 1 \le i \le j \le n\}$. It is easy to see that P and Q are in fact GT–patterns and $\lambda(P) = \lambda(Q) = (\widetilde{x}_{11}, \widetilde{x}_{22}, \ldots, \widetilde{x}_{nn})$.

Using natural bijection between integral GT–patterns and semistandard Young tableaux described in Section 2.2, we see that the set of square shape plane partitions is in one–to–one correspondence with pairs of semistandard Young tableaux of the same shape. It is convenient to display a square shape plane partition $\widetilde{\pi} = (\widetilde{x}_{ij})$, $1 \le i, j \le n$, as a diamond

$$\Diamond\,\widetilde{x}_{ij} \quad = \quad \begin{array}{ccccccccc}
 & & & & \widetilde{x}_{1n} & & & & \\
 & & & \widetilde{x}_{1n-1} & & \widetilde{x}_{2n} & & & \\
 & & & & \vdots & & & & \\
 & \widetilde{x}_{12} & & & \cdots & & & \widetilde{x}_{n-1n} & \\
\widetilde{x}_{11} & & \widetilde{x}_{22} & & \cdots & \widetilde{x}_{n-1n-1} & & & \widetilde{x}_{nn} \\
 & \widetilde{x}_{21} & & & \cdots & & & \widetilde{x}_{nn-1} & \\
 & & & & \vdots & & & & \\
 & & & \widetilde{x}_{n-11} & & \widetilde{x}_{n2} & & & \\
 & & & & \widetilde{x}_{n1} & & & &
\end{array}$$

or by $\begin{smallmatrix} Q \\ P \end{smallmatrix}$, where P and Q are the GT–patterns defined above.

2.9 Charge and cocharge

Statistics *charge* was introduced by A. Lascoux and M.-P. Schützenberger [29] in their proof of Foulkes' conjecture. More precisely, let λ and μ be partitions of the same size, denote by $K_{\lambda\mu}(q)$ the corresponding Kostka–Foulkes polynomial. Recall that by definition, the Kostka–Foulkes polynomials $K_{\lambda\mu}(q)$ are matrix elements of the transition matrix between the Schur functions $s_\lambda(x)$ and the Hall–Littlewood ones:

$$s_\lambda(x) = \sum_\mu K_{\lambda\mu}(q) P_\mu(x;q). \tag{2.9}$$

We refer the reader to [34], Chapter III, Sections 2 and 6 for more details.

Proposition 2.14 (Lascoux–Schützenberger [29]) *Let λ and μ be partitions, then*

$$K_{\lambda\mu}(q) = \sum_{T \in STY(\lambda,\mu)} q^{c(T)}, \tag{2.10}$$

summed over the set $STY(\lambda,\mu)$ of semistandard Young tableaux of shape λ and weight μ, and $c(T)$ denotes the charge of tableau T.

We refer the reader to [34], Chapter III, Section 6, for a definition of the statistic *charge*. Follow [28], see also [8], we define two statistics plus and minus *charges* $c_n^{\pm}(\bullet)$ on the set of triangles X_n, namely, if $x \in X_n$, then

$$c_n^{\pm}(x) := \begin{matrix}\max\\ \min\end{matrix} \left\{ \sum_{j=1}^{n-1} (n-j)\kappa_j(s(x)) | s \in S_n := \langle s_1, \ldots, s_{n-1} \rangle \right\}.$$

On the set of dominant weight tableaux $STY_+(\leq n)$ the statistics c_n^-, coincides with statistics *charge* introduced by A. Lascoux and M.-P. Schützenberger [29], see [28] for the proof and further details. One can show that on the set $STY_+(\leq n)$ both statistics c_n^- and c_n^+ are equidistributed.

2.10 Transportation matrices and transportation polytope

Let λ and μ be compositions of the same size such that $l(\lambda) \leq n$, $l(\mu) \leq n$. Denote by $M(\lambda, \mu)$ the set of n by n matrices with non–negative real entries which have prescribed sums of rows λ and those of columns μ. This is a convex integral polytope which is called by *transportation polytope* of type $(\lambda; \mu)$. We denote by

$$M_{\mathbb{Z}}(\lambda, \mu) = M(\lambda, \mu) \cap \mathbb{Z}^{n^2}$$

the set of integer points of the polytope $M(\lambda, \mu)$, and call matrices from the set $M_{\mathbb{Z}}(\lambda, \mu)$ by *transportation matrices* of type $(\lambda; \mu)$.

Let $d = (d_{ij}) \in M_{n \times n}(\mathbb{R})$ be a matrix, follow [7,42] denote by

$$h_{a,\pm}^{(l)}(d) = \begin{matrix}\max\\ \min\end{matrix} \left\{ \sum_{j=k}^{n} d_{a+1,j+l} - \sum_{j=k+1}^{n} d_{a,j+l} \Big| \; 1 \leq k \leq n \right\}, 1 \leq a \leq n, \quad (2.11)$$

$$\widetilde{h}_{b,\pm}^{(l)}(d) = \begin{matrix}\max\\ \min\end{matrix} \left\{ \sum_{i=k}^{n} d_{i+l,b+1} - \sum_{i=k+1}^{n} d_{i+l,b} \Big| \; 1 \leq k \leq n \right\}, 1 \leq b \leq n, \quad (2.12)$$

the plus and minus row and column *exponents* of matrix d. Here and further we assume that the sum $i + l$ is considered by modulo n. In other words, we assume that $d_{i,j} = d_{i',j'}$ if either $i \equiv i' (\text{mod } n)$, or $j \equiv j' (\text{mod } n)$. We define also the plus and minus indexes of matrix $d \in M_{n \times n}(\mathbb{R})$, denoted by $ind_{\pm}(d)$, to be

$$ind_{\pm}(d) = \sum_{k=1}^{n-1} (n-k) h_{k,\pm}^{(0)}(d).$$

Finally, for two compositions λ and μ, and two partitions ν and η (all of lengths $\leq n$) denote by $M_{\lambda\mu}^{\nu\eta}$ the rational convex polytope

$$\left\{ d \in M(\lambda - \nu, \mu - \eta) \;\middle|\; \begin{array}{l} h_{i,+}^{(0)}(d) \leq \nu_i - \nu_{i+1}, \quad 1 \leq i \leq n-1; \\ \widetilde{h}_{j,+}^{(0)}(d) \leq \eta_j - \eta_{j+1}, \quad 1 \leq j \leq n-1 \end{array} \right\}.$$

In the sequel we will display a matrix $d = (d_{ij}) \in M_{n \times n}(\mathbb{R})$ by a square $\boxed{d_{ij}}$, or by a square with the main diagonal $(d_{ii})_{1 \leq i \leq n}$ 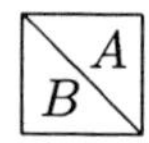

, or by a square with the secondary diagonal $(d_{j,n+1-j})_{1 \leq j \leq n}$ 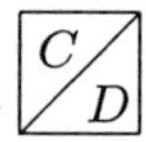

, where A, B, C, D are the corresponding parts of the matrix d.

2.11 Action of the extended affine symmetric group $\widehat{W}(A_{n-1}^{(1)})$ on the space of triangles X_n and on the set of GT–patterns

Let $t_1, t_2, \ldots, t_{n-1}$ be the Bender–Knuth involutions acting on the space of triangles, see Section 2.5. Follow [24], define transformations

$$s_i := q_i t_1 q_i, \quad 1 \leq i \leq n-1, \tag{2.13}$$

which act on the space of triangles X_n. It is easy to see that

$$s_i^2 = 1, \quad 1 \leq i \leq n-1,$$

and it follows from Theorem 2.6 that involutions s_i, $1 \leq i \leq n-1$, satisfy the defining relations of the symmetric group S_n.

Theorem 2.15 (A.D. Berenstein and A.N. Kirillov [24]) *Involutions s_i, $1 \leq i \leq n-1$, satisfy the following relations*

1) $s_i^2 = 1$;
2) $(s_i s_{i+1})^3 = 1, \quad 1 \leq i \leq n-1$;
3) $(s_i s_j)^2 = 1, \quad \text{if} \quad |i-j| \geq 2$.

Therefore, the involutions s_i, $1 \leq i \leq n-1$, define a continuous piece–wise linear representation of the symmetric group on the space of triangles X_n. It is clear from the very definition that the cone of GT–patterns K^λ is invariant under this action of S_n.

To continue, let us define for any $\epsilon \in \mathbb{R}$ a translation operator T_ϵ acting on the space of triangles: if $x \in X_n$, then

$$T_\epsilon(x) = \widetilde{x},$$

where $(\widetilde{x})_{ij} = x_{ij}$ if $(i,j) \neq (1,1)$, and $(\widetilde{x})_{11} = x_{11} - \epsilon$.

Finally, follow [24] let us define transformations

$$\begin{aligned} s_i^{(\epsilon)} &= q_i T_{-\epsilon} q_i, \quad 1 \le i \le n-1, \\ \pi &= s_{n-1} s_{n-2} \cdots s_2 s_1, \quad \text{and} \\ s_0 := s_0^{(\epsilon)} &= \pi T_\epsilon s_1 \pi^{-1}. \end{aligned}$$

Theorem 2.16 (A.D. Berenstein and A.N. Kirillov [24]) *Transformations t_i, $1 \le i \le n-1$, and T_ϵ, $\epsilon \in \mathbb{R}$, satisfy the following relations*

- $(t_1 T_\epsilon)^2 = 1, \quad t_j T_\epsilon = T_\epsilon t_j \;\text{ if }\; j \ge 3, \quad T_\epsilon T_\delta = T_{\epsilon+\delta};$
- (Yang – Baxter relation)
  $t_2 T_\epsilon t_2 T_{-\epsilon+\delta} t_2 T_\delta = T_\delta t_2 T_{-\delta+\epsilon} t_2 T_{-\epsilon} t_2, \quad \forall\ \epsilon, \delta \in \mathbb{R};$
- $(T_\epsilon t_2 t_1 T_\delta t_2 t_1)^3 = 1, \quad \forall\ \epsilon, \delta \in \mathbb{R};$
- $(t_1 T_\epsilon q_j t_1 T_\delta q_j)^2 = 1 \;\text{ and }\; T_\epsilon q_j T_\delta q_j = q_j T_\delta q_j T_\epsilon \;\text{ if }\; j \ge 3 \;\text{ and }\; \epsilon, \delta \in \mathbb{R}.$

Corollary 2.17 *Transformations $s_i^{(\epsilon)}$ satisfy the following relations*

- $s_i^{(\epsilon)} s_i^{(\eta)} = s_i^{(\epsilon+\eta)}, \quad 1 \le i \le n-1, \quad \epsilon, \eta \in \mathbb{R};$
- $s_i^{(\epsilon)} s_j^{(\eta)} = s_j^{(\eta)} s_i^{(\epsilon)} \;\text{ if }\; |i-j| \ge 2, \quad \epsilon, \eta \in \mathbb{R};$
- *(Yang – Baxter equation)*

$$s_i^{(\epsilon)} s_{i+1}^{(\epsilon+\eta)} s_i^{(\eta)} = s_{i+1}^{(\eta)} s_i^{(\epsilon+\eta)} s_{i+1}^{(\epsilon)}, \quad 1 \le i \le n-1, \quad \epsilon, \eta \in \mathbb{R}. \tag{2.14}$$

Note that if $x \in X_n$, and $\beta(x) = (\beta_1, \ldots, \beta_n)$, then

$$\beta(s_i(x)) = s_i(\beta(x)) = (\beta_1, \ldots, \beta_{i+1}, \beta_i, \ldots, \beta_n)$$

and

$$\beta(s_0^{(\epsilon)}(x)) = s_0^{(\epsilon)}(\beta(x)) = (\beta_n + \epsilon, \beta_2, \ldots, \beta_{n-1}, \beta_1 - \epsilon).$$

Corollary 2.18 *Transformations s_i, $0 \le i \le n$, satisfy the following relations*

- $s_i^2 = 1, \quad i = 0, 1, \ldots, n-1;$
- $(s_i s_{i+1})^3 = 1, \quad 0 \le i \le n-1,$ and by definition we set $s_n := s_0;$
- $(s_i s_j)^2 = 1 \;\text{ if }\; |i-j| \ge 2, \quad 0 \le i, j \le n-1.$

In other words, involutions s_i, $0 \le i \le n-1$, satisfy the relations of the affine symmetric group $W(A_{n-1}^{(1)})$. Therefore, the involutions s_i, $0 \le i \le n-1$, define a piece–wise linear representation of the affine symmetric group $W(A_{n-1}^{(1)})$ on the space of triangles X_n.

Remark 2.19 Follow [24] let us define transformations

$$\widetilde{s}_i^{(\epsilon)} := s_i s_i^{(\epsilon)}, \quad 1 \le i \le n-1,$$
$$\pi^{(\epsilon)} := \pi_n^{(\epsilon)} = \widetilde{s}_{n-1}^{(\epsilon)} \widetilde{s}_{n-2}^{(\epsilon)} \cdots \widetilde{s}_2^{(\epsilon)} \widetilde{s}_1^{(\epsilon)},$$
$$\widetilde{s}_0^{(\epsilon)} := \widetilde{s}_0^{(\epsilon,\delta)} = \pi^{(\epsilon)} T_\delta \widetilde{s}_1^{(\epsilon)} (\pi^{(\epsilon)})^{-1}, \quad \text{where} \quad \epsilon, \delta \in \mathbb{R}.$$

Using Theorem 2.16 one can show [24] that transformations $\widetilde{s}_i^{(\epsilon)}$, $0 \le i \le n-1$, satisfy the following relations:

- $(\widetilde{s}_i^{(\epsilon)})^2 = 1, \quad 0 \le i \le n-1, \quad$ in other words $\quad s_i s_i^{(\epsilon)} = s_i^{(-\epsilon)} s_i$;
- $(\widetilde{s}_i^{(\epsilon)} \widetilde{s}_{i+1}^{(\eta)})^3 = 1$ for any $\epsilon, \eta \in \mathbb{R}, \quad 0 \le i \le n-2$;
- $\widetilde{s}_i^{(\epsilon)} \widetilde{s}_j^{(\eta)} = \widetilde{s}_j^{(\eta)} \widetilde{s}_i^{(\epsilon)}$ if $|i-j| \ge 2, \quad \forall\, \epsilon, \eta \in \mathbb{R}$.

It follows from these relations that for any $\epsilon, \delta \in \mathbb{R}$ the involutions $\widetilde{s}_i^{(\epsilon)}$, $0 \le i \le n-1$, define a representation of the affine symmetric group $W(A_{n-1}^{(1)})$ on the space of triangles X_n. Note that if $\beta(x) = (\beta_1, \ldots, \beta_n)$ then

$$\beta(\widetilde{s}_0^{(\epsilon,\delta)}(x)) = (\beta_n + n\delta + \epsilon, \beta_2, \ldots, \beta_{n-1}, \beta_1 - n\delta - \epsilon).$$

We conclude this Section by a construction of the action of the extended affine symmetric group $\widehat{W}(A_{n-1}^{(1)})$ on the space of triangles X_n. To this end, let us introduce the following transformations:

$$g_i := g_i^{(\epsilon)} = \pi^{i-1} T_\epsilon t_1 \pi^{1-i}, \quad 1 \le i \le n,$$

and define $g_0 := g_n$.

Theorem 2.20 *Transformations $g_0, g_1, \ldots, g_{n-1}$ and π satisfy the relations of the extended affine symmetric group $\widehat{W}(A_{n-1}^{(1)})$. In other words,*

- $g_i^2 = 1, \quad i = 0, 1, \ldots, n-1$;
- $(g_i g_j)^2 = 1, \quad$ if $\quad |i-j| \ge 2, \quad 0 \le i, j \le n-1$;
- $(g_i g_{i+1})^3 = 1, \quad$ if $\quad 0 \le i \le n-1$;
- $\pi g_i \pi^{-1} = g_{i+1}$ if $\quad 0 \le i \le n-1$. In particular, $\pi^n = 1$.

Corollary 2.21 *The elements*

$$T_1 = \pi g_{n-1} g_{n-2} \cdots g_1,$$
$$T_k = g_{k-1} g_{k-2} \cdots g_1 \pi g_{n-1} \cdots g_k \quad \text{if} \quad 2 \le k \le n-1,$$
$$T_n = g_{n-1} g_{n-2} \cdots g_2 g_1 \pi$$

generate a commutative subgroup in the group $\widehat{W}(A_{n-1}^{(1)})$. Furthermore, $T_1 \cdots T_n = 1$.

2.12 Action of the symmetric group S_n on the space of matrices

In this Section we define an action of the extended affine symmetric group $\widehat{W}(A_{n-1}^{(1)})$ on the space of matrices $M_n = \mathbb{R}^{n^2}$.

Let $d = (d_{ij}) \in M_{n\times n}(\mathbb{R})$. In the sequel it is convenient to impose the periodicity conditions for matrix elements

$$d_{j+n,j} = d_{ij} = d_{i,j+n}, \quad 1 \le i \le n, \quad j \in \mathbb{Z}.$$

Let us introduce transformations σ_i, $1 \le i \le n$, acting on the space of matrices M_n:

$$\begin{aligned} \sigma_i(d_{jk}) &= d_{jk}, \quad \text{if} \quad j \ne i, i+1; \\ \sigma_i(d_{ij}) &= d_{i+1,j} + h_{i,+}^{(j)}(d) - h_{i,+}^{(j+1)}(d); \\ \sigma_i(d_{i+1,j}) &= d_{ij} + h_{i,+}^{(j+1)}(d) - h_{i,+}^{(j)}(d). \end{aligned}$$

See Section 2.10 for the definition of functions $h_{i,\pm}^{(j)}(d)$.

Proposition 2.22 *If $n \ge 3$, the ransformations σ_i, $1 \le i \le n$, satisfy the defining relations of the affine symmetric group $\widehat{S}_n = W(A_{n-1}^{(1)})$:*

- $\sigma_i^2 = 1, \quad 1 \le i \le n;$
- $(\sigma_i\sigma_{i+1})^3 = 1, \quad 1 \le i \le n-1, \quad (\sigma_1\sigma_n)^3 = 1;$
- $(\sigma_i\sigma_j)^2 = 1, \quad \text{if} \quad |i-j| \ge 2, \quad \text{and either} \quad (i,j) \ne (1,n), \quad \text{or} \quad (n,1).$

Therefore the involutions σ_i, $1 \le i \le n$, define a piece-wise linear representation of the affine symmetric group $W(A_{n-1}^{(1)})$ on the space of matrices M_n.

Remark 2.23 It is clear from the definition of transformation σ_i that

$$\sigma_i(d_{ij}) + \sigma_i(d_{i+1,j}) = d_{ij} + d_{i+1,j}, \quad 1 \le i \le n, \quad j \in \mathbb{Z}; \tag{2.15}$$

$$\sum_{j=1}^{n} \sigma_i(d_{ij}) = \sum_{j=1}^{n} d_{i+1,j}.$$

It is less obvious that one has the following splitting of the relations (2.15):

$$\begin{aligned} \max(\sigma_i(d_{ij}), \sigma_i(d_{i+1,j})) &= \max(d_{ij}, d_{i+1,j}), \\ \min(\sigma_i(d_{ij}), \sigma_i(d_{i+1,j})) &= \min(d_{ij}, d_{i+1,j}). \end{aligned}$$

Let us define also transformations $\widetilde{\sigma}_i$, $1 \le i \le n$, acting on the space of matrices M_n by the rule

$$\widetilde{\sigma}_i(d) = \sigma_i(d^t),$$

where d^t **denotes** the matrix which is transposed to d with respect to the main diagonal. It is clear that the transformations $\widetilde{\sigma}_i$, $1 \le i \le n$, also define a piece–wise linear representation of the affine symmetric group $\widehat{S}_n$ on the space of matrices M_n.

In order to define an action of the extended affine Weyl group on the space of matrices M_n, let us introduce the shift operator $\pi : M_n \to M_n$ by the following rule

$$\pi(d_{ij}) = d_{i+1,j}.$$

It is easy to see that

$$\pi(h_i^{(l)}(d)) = h_{i+1}^{(l)}(\pi(d)),$$

and therefore

$$\pi\sigma_a = \sigma_{a+1}\pi.$$

In other words we came to the following
Theorem 2.24 *If $n \ge 3$, the involutions σ_a, $1 \le a \le n$, and the shift operator π give rise to a piece–wise linear representation of the extended affine symmetric group $\widehat{W}(A_{n-1}^{(1)})$ on the space of matrices M_n.*

Finally, for any matrix $d \in M_{n\times n}(\mathbb{R})$ we define its left plus and minus $c_n^{\pm,L}(d)$ and right plus and minus $c_n^{\pm,R}(d)$ *charges* to be

$$c_n^{\pm,L}(d) = \begin{matrix}\max\\ \min\end{matrix}\Big\{ind(\sigma(d)) \;\Big|\; \sigma \in S_n := \langle\sigma_1, \sigma_2, \ldots, \sigma_{n-1}\rangle\Big\},$$

$$c_n^{\pm,R}(d) = \begin{matrix}\max\\ \min\end{matrix}\Big\{ind(\widetilde{\sigma}(d)) \;\Big|\; \widetilde{\sigma} \in S_n := \langle\widetilde{\sigma}_1, \widetilde{\sigma}_2, \ldots, \widetilde{\sigma}_{n-1}\rangle\Big\}.$$

2.13 Robinson–Schensted–Knuth correspondence

The Robinson–Schensted–Knuth correspondence is a bijection between the set of n by n matrices with non–negative integer entries, and the set of reverse plane partitions of square shape (n^n). We will display this correspondence as follows

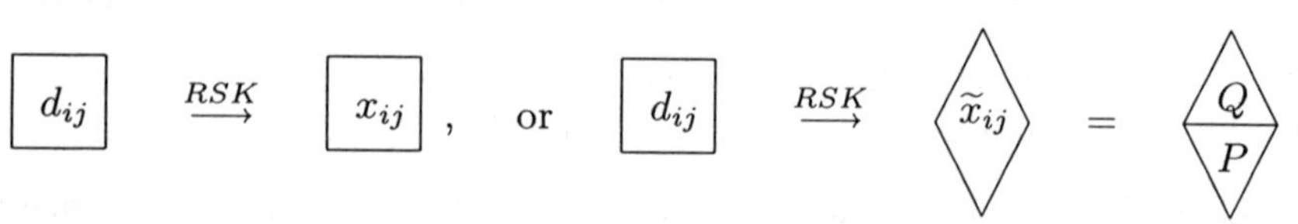

We refer the reader to the original paper by D. Knuth [26], or to [11], for the definition and basic properties of the Robinson–Schensted–Knuth correspondence. Note that in the literature there are several variants of the Robinson–Schensted–Knuth correspondence. In the sequel we will use the original construction due to Knuth [26]. We are not going to describe this construction explicitly, but illustrate it by example.

Example 2.25 Take a transportation matrix $m = \begin{pmatrix} 4 & 1 & 3 \\ 2 & 2 & 1 \\ 3 & 5 & 0 \end{pmatrix}$. Starting from the matrix m, define the corresponding two–rowed array, or multi–permutation,

$$w(m) = \begin{pmatrix} 11111111222222333333333 \\ 11112333112231112222 \end{pmatrix}.$$

Using the row bumping procedure, see e.g. [11], Section 4, or [26], we will come to the pair of semistandard Young tableaux of the same shape:

$$P = \begin{matrix} 1111111122222 \\ 22233 \\ 33 \end{matrix}, \qquad Q = \begin{matrix} 11111111233333 \\ 222223 \\ 33 \end{matrix}.$$

This pair of Young tableaux defines the corresponding diamond , reverse plane partition π and plane partition $\widetilde{\pi}$:

$$= \begin{matrix} & & 8 & & \\ & 9 & & 4 & \\ 14 & & 5 & & 2 \\ & 14 & & 3 & \\ & & 9 & & \end{matrix}, \qquad \pi = \begin{matrix} 2 & 4 & 8 \\ 3 & 5 & 9 \\ 9 & 14 & 14 \end{matrix}, \qquad \widetilde{\pi} = \begin{matrix} 14 & 9 & 8 \\ 14 & 5 & 4 \\ 9 & 3 & 2 \end{matrix}.$$

Proposition 2.26 *The Robinson–Schensted–Knuth correspondence (RSK for short) has the following properties:*

1. *if* 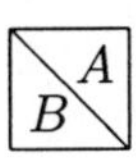

$\xrightarrow{RSK}$ 

, *then* 

$\xrightarrow{RSK}$ 

;

2. *if* 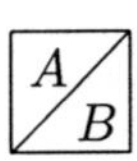

$\xrightarrow{RSK}$ 

, *then* 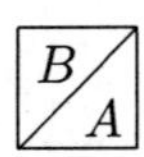

$\xrightarrow{RSK}$ 

,

where $\widehat{P} = \mathbb{S}P$ *and* $\widehat{Q} = \mathbb{S}Q$, *and* $\mathbb{S} = q_{n-1}$ *denotes the Schützenberger transformation;*

3. let 

, *then*

$$\kappa_j(P) = \widetilde{h}^{(0)}_{j,+}(d) := \max\left\{ \sum_{i=k}^{n} d_{i,j+1} - \sum_{i=k+1}^{n} d_{ij} | 1 \le k \le n \right\}, \ 1 \le j \le n-1,$$

$$\kappa_i(Q) = h^{(0)}_{i,+}(d) := \max\left\{ \sum_{j=k}^{n} d_{i+1,j} - \sum_{j=k+1}^{n} d_{ij} | 1 \le k \le n \right\}, \ 1 \le i \le n-1;$$

4. ([42]) *let μ, μ', ν and ν' be partitions, then the Robinson–Schensted–Knuth correspondence establishes a bijection*

$$M^{\nu\nu'}_{\mu\mu'} \xrightarrow{RSK} \coprod_{\lambda} STY(\lambda, \mu \parallel \nu) \times STY(\lambda, \mu' \parallel \nu').$$

Corollary 2.27 *i)* $\#\{m \in M(\mu,\mu) | m^t = m\} = \sum_\lambda |STY(\lambda,\mu)|$.

ii) $\#\{m \in M(\mu,\mu) | m^t = m, m^T = m\} = \sum_\lambda |\mathrm{Tab}^{(2)}(\lambda,\mu)|$,

where m^T denotes the matrix which is transposed to m with respect to the secondary diagonal, and $Tab^{(2)}(\lambda,\mu)$ denotes the set of semistandard domino tableaux of shape λ and content μ.

Theorem 2.28 *For any a such that $1 \le a \le n-1$, the following diagrams of maps are commutative*

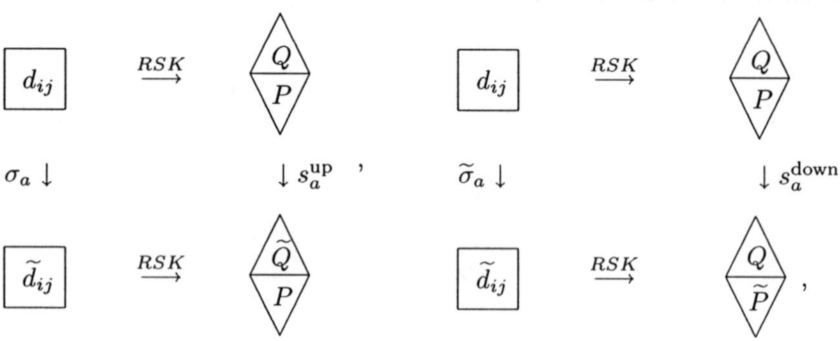

where $\widetilde{P}$ (resp. $\widetilde{Q}$) denotes the image $s_a(P)$ of the GT–pattern P (resp. Q) under the action of involution s_a.

Corollary 2.29 *The involutions* σ_a *and* $\widetilde{\sigma}_a$ *commute for all integer* a, $1 \le a \le n$.

Exercises.

1. Show that

$$i)\ s_i = t_{i-1} t_i s_{i-1} t_i t_{i-1}; \tag{2.16}$$

$$ii)\ q_i = q_{i-1} p_i = p_i^{-1} q_{i-1}, \quad \text{where } p_i = t_i t_{i-1} \cdots t_2 t_1; \tag{2.17}$$

$$iii)\ q_{n-1} s_i q_{n-1} = s_{n-i}, \quad q_{n-1} s_i^{(\epsilon)} q_{n-1} = s_{n-i}^{(\epsilon)} \quad \text{if } 1 \le i \le n-1; \tag{2.18}$$

$$iv)\ s_i = \sigma^{i-1} t_1 \sigma^{1-i}, \quad s_i^{(\epsilon)} = \sigma^{i-1} T_\epsilon \sigma^{1-i}, \tag{2.19}$$

where $\sigma = t_1 t_2 \cdots t_{n-1} := p_{n-1}^{-1}$.

Proofs of these statements may be found in [24], Proposition 1.4.

2. Define transformations $v_i = t_i t_{i+1} \cdots t_{n-1}$ and $u_i = v_{n-1} v_{n-2} \cdots v_{n-i}$.

Show that

- $u_i^2 = 1$;
- $\sigma^n = q_{n-1} u_{n-1}$;
- on the set of semistandard Young tableaux $STY(\lambda, \le n)$ the involution u_{n-1} coincides with the so–called *dual* Schützenberger involution [38].

3. a. Show that if λ is a rectangular diagram, $l(\lambda) \le n$, then on the set of semistandard Young tableaux $STY(\lambda, \le n)$ the Schützenberger and dual Schützenberger involutions coincide:

$$q_{n-1} = u_{n-1}.$$

b. Show that if λ is the staircase diagram $\lambda = (n-1, n-2, \ldots, 2, 1)$, and T is a *standard* Young tableau of shape λ, then

$$u_{n-1} q_{n-1}(T) = q_{n-1} u_{n-1}(T) = \sigma_{n-1}^n(T) = T',$$

where T' denotes the standard Young tableau which is conjugated to T.

The statement **3a** is due to M.-P. Schützenberger [38], and the statement **3b** is due to P. Edelman and C. Greene (Adv. Math. **63** (1987), 42-99).

4. a. Let m be an odd non–negative integer and $(\epsilon_1, \ldots, \epsilon_m) \in \mathbb{R}^m$ be a vector such that $\epsilon_1 + \cdots + \epsilon_m = 0$.

Show that

$$(T_{\epsilon_1} t_2 T_{\epsilon_2} t_2 \cdots T_{\epsilon_m} t_2)^2 = 1.$$

b. Let m be a non–negative integer such that $m = 6k$, $k \equiv 1 (\mathrm{mod}\ 2)$, and $\epsilon = (\epsilon_1, \ldots, \epsilon_m) \in \mathbb{R}^m$ be a vector such that for all l, $1 \le l \le 3k+3$, the sum $f_l := \epsilon_l + \epsilon_{l+3} + \cdots + \epsilon_{l+3k-3}$ depends only on the parity of l, i.e. $f_1 = f_3 = \cdots = f_{3k+2}$ and $f_2 = f_4 = \cdots = f_{3k+3}$.

Show that

$$T_{\epsilon_1}t_1t_2T_{\epsilon_2}t_1t_2\cdots T_{\epsilon_m}t_1t_2 = 1.$$

5. Action of the transformations $\sigma = t_1t_2\cdots t_{n-1}$ and $p_{n-1} = \sigma^{-1} = t_{n-1}t_{n-2}\cdots t_2t_1$ on the set of Young tableaux $STY(\lambda, \leq n)$.

a. Let λ be a partition and $\mu = (\mu_1, \ldots, \mu_n)$ be a composition of the same size. We assume that $\mu_1 > 0$. Given a Young tableau $T \in STY(\lambda, \mu)$, remove the entry 1 from the upper left corner of T and apply Schützenberger's "jeu de taquin" algorithm [11,38] to the skew tableau that is left. This gives a tableau, say ΔT, whose diagram has one box removed from the diagram of T. If the upper left corner of ΔT still contains 1, repeat the algorithm on ΔT, getting a smaller tableau $\Delta^2 T$, whose diagram has one box removed from the diagram of ΔT. Continue until all the entries of T which are equal to 1, have been removed. As a result we obtain a new tableau T" of content $(0, \mu_2, \ldots, \mu_n)$ and a horizontal strip θ_n, $|\theta_n| = \mu_1$. Let us subtract 1 from all the entries of tableau T". We obtain a tableau $\widetilde{T}$ of content $(\mu_2, \ldots, \mu_n)$. Finally, let us fill in the horizontal strip θ_n with the numbers n. The tableau $\widetilde{T}$ and the filled horizontal strip θ_n determine a tableau $\overline{T} \in STY(\lambda, \overline{\mu})$, where $\overline{\mu} = (\mu_2, \ldots, \mu_n, \mu_1)$.

Show that

$$\sigma^{-1}(T) = p_{n-1}(T) = \overline{T}.$$

b. Let λ be a partition and $\mu = (\mu_1, \ldots, \mu_n)$ be a composition of the same size. We assume that $\mu_n > 0$. Given a Young tableau $T \in STY(\lambda, \mu)$, remove the entry n from the bottom right corner of T, and apply Schützenberger's "jeu de taquin" algorithm [11,38] to the tableau that is left. This gives a skew tableau, say, ∇T, with the same right border strip, and whose left border strip has one box less. Repeating this algorithm, we remove all the entries equal to n from the tableau T. As a result we obtain a skew tableau T" of content $(0, \mu_2, \ldots, \mu_{n-1})$ and a horizontal strip θ_1, $|\theta_1| = \mu_n$. Let us fill in the strip θ_1 with 1's, and then add 1 to all the entries of tableau T". Let $\widetilde{T}$ be the resulting tableau. The tableau $\widetilde{T}$ and the filled horizontal strip θ_1 determine a tableau $\overline{T} \in STY(\lambda, \widehat{\mu})$, where $\widehat{\mu} = (\mu_n, \mu_1, \mu_2, \ldots, \mu_{n-1})$.

Show that

$$\sigma(T) = p_{n-1}^{-1}(T) = \overline{T}.$$

c. Show that

$$\mathbb{S} = p_1p_2\cdots p_{n-1},$$

where

$$\mathbb{S} \;:\; STY(\lambda,\mu) \to STY(\lambda, \overleftarrow{\mu})$$

denotes the Schützenberger transformation, see Section 2.6.

All these results are due to E. Gansner [13].

6. Lusztig's transformations.

For each triple of integers (ijk), $1 \le i < j < k \le n$, let us define the transformation

$$R_{ijk} \;:\; X_n \to X_n$$

of the space of triangles by the following rule: if $R_{ijk}(x) = \widetilde{x}$, then

$$\begin{aligned}
\widetilde{x}_{ij} &= x_{ij} + x_{ik} - x_{i,k-1} - \min(x_{jk} - x_{j,k-1}, x_{ij} - x_{i,j-1}),\\
\widetilde{x}_{jj} &= x_{jj} - x_{ik} + x_{i,k-1} + \min(x_{jk} - x_{j,k-1}, x_{ij} - x_{i,j-1}),\\
\widetilde{x}_{\alpha\beta} &= x_{\alpha\beta} \;\text{ if either }\; (\alpha,\beta) \ne (i,j) \;\text{ or }\; (\alpha,\beta) \ne (j,j).
\end{aligned}$$

Show that transformations R_{ijk} satisfy the following relations

i) $(R_{ijk})^2 = 1, \quad \beta(R_{ijk}(x)) = \beta(x)$;

ii) if $|(ijk) \cap (i'j'k')| \ne 2$, then $R_{ijk}R_{i'j'k'} = R_{i'j'k'}R_{ijk}$;

iii) (tetrahedral relations)

$$R_{ijk}R_{ijl}R_{ikl}R_{jkl} = R_{jkl}R_{ikl}R_{ijl}R_{ijk} \text{ for all } 1 \le i < j < k < l \le n;$$

iv) $R_{ijl}R_{ikl}R_{ijl} = R_{ikl}R_{ijl}R_{ikl}$ for all $1 \le i < j < k < l \le n$.

The statements i)–iii) are essentially due to G. Lusztig [33].

7. Factorization formula for involution s_j.

Define transformation θ_j of the space of triangles X_n by the following rule: if $x \in X_n$, then

$$\begin{aligned}
(\theta_j(x))_{jj} &= x_{jj} + \beta_{j+1}(x) - \beta_j(x),\\
(\theta_j(x))_{\alpha\beta} &= x_{\alpha\beta}, \;\text{ if }\; (\alpha,\beta) \ne (i,j).
\end{aligned}$$

Show that

$$s_j = R_{j-1,j}R_{j-2,j}\cdots R_{1j}\theta_j R_{1j}\cdots R_{j-2,j}R_{j-1,j}, \;\text{ for all }\; 1 \le j \le n-1,$$

where $R_{jk} := R_{jkk+1}$, $1 \le j < k \le n-1$.

Proof of this statement can be found in [24], Appendix.

3 Explicit formulae: piece–wise linear version

3.1 Piece–wise linear action of involutions s_k on the space of triangles

Follow [24] let us introduce the following linear functions on the space of triangles X_n:

$$\varphi_{ij} := x_{i-1,j-1} + x_{i,j+1} - x_{i-1,j} - x_{i,j}, \quad 2 \le i \le j \le n-1;$$
$$\varphi_{11} := 0, \quad \varphi_{1j} := x_{1,j+1} + x_{j+1,j+1} - x_{1,j} - x_{j,j}, \quad 2 \le j \le n-1.$$

Let us introduce also a piece-wise linear functions Q_n^k, $1 \le k \le n$, given on the space $\mathbb{R}^n$:

$$\begin{aligned} Q_n^1(a_1,\dots,a_n) &= a_1 + \max(0, a_2, a_2+a_3, \dots, a_2+\dots+a_n) \\ &\quad - \max(0, a_1, a_1+a_2, \dots, a_1+\dots+a_{n-1}); \\ Q_n^k(a_1,\dots,a_n) &= Q_n^1(a_k,\dots,a_n,a_1,a_2,\dots,a_{k-1}), \quad 2 \le k \le n. \end{aligned}$$

Theorem 3.1 ([24]) *For a positive integer k, $2 \le k \le n-1$, and a triangular $x \in X_n$, assume that $s_k(x) = \widetilde{x}$, $s_k^{(\epsilon)}(x) = \widetilde{\widetilde{x}}$. Then*

$$i)\ \widetilde{x}_{ik} = x_{ik} + Q_k^i(\varphi_{2,k}(x),\dots,\varphi_{k,k}(x),\varphi_{1,k}(x)); \tag{3.1}$$
$$ii)\ \widetilde{\widetilde{x}}_{ik} = x_{ik} + Q_k^i(\varphi_{2,k}(x),\dots,\varphi_{k,k}(x),\varphi_{1,k}(x)+\beta_{k+1}(x)-\beta_k(x)+\epsilon), \tag{3.2}$$

where $\beta(x) = (\beta_1(x),\dots,\beta_n(x))$ denotes the weight of the triangle x.

Examples 3.2 1) If $n = 2$, then

$$\begin{aligned} \widetilde{x}_{12} &= x_{11} - x_{22} + x_{23} + \max(x_{12}+x_{22}, x_{13}+x_{33}) \\ &\quad - \max(x_{12}+x_{22}, x_{11}+x_{23}) \\ &= y_{12} + y_{11} - y_{22} + \max(y_{13}+y_{33}, y_{22}+y_{23}) \\ &\quad - \max(y_{11}+y_{23}, y_{13}+y_{22}). \end{aligned}$$

2) If $n = 3$, then

$$\widetilde{x}_{13} = x_{13} + \max(\varphi_{13}, -\varphi_{33}, 0) + \min(\varphi_{33}, \varphi_{23}+\varphi_{33}, 0),$$

where

$$\begin{aligned} \varphi_{13} &= x_{14} + x_{44} - x_{13} - x_{33}, \\ \varphi_{23} &= x_{12} + x_{24} - x_{13} - x_{23}, \\ \varphi_{33} &= x_{22} + x_{34} - x_{23} - x_{33}. \end{aligned}$$

Exercises.

1. Show that functions $Q_n^1(a_1, \ldots, a_n)$ satisfy the following recurrence relations

- $Q_1^1(a_1) = a_1$;
- $$Q_n^1(a_1, \ldots, a_n) = \min(0, Q_{n-1}^1(a_1, \ldots, a_{n-1})) + \max(0, Q_{n-1}^1(a_2, a_3, \ldots, a_n)). \quad (3.3)$$

2. Let $s_k : X_n \to X_n$, $1 \le k \le n-1$, be piece–wise linear transformations of the space of triangles given by the formula (3.1).

Show that recurrence relations (3.3) imply that the transformation s_k satisfy the following relation (cf. (2.15)):

$$s_k = t_{k-1} t_k s_{k-1} t_k t_{k-1}.$$

3. Let j be an integer, $2 \le j \le n-1$. Show that the following piece–wise linear functions on the space of triangles X_n are s_j–invariant:

$$\psi_{1j}(x) = x_{jj} - x_{j+1,j+1} + \min(\varphi_{1j}(x), 0),$$
$$\psi_{ij}(x) = x_{i-1,j} - x_{i-1,j-1} + \min(\varphi_{ij}(x), 0),$$
$$\overline{\psi}_{1j}(x) = \max(\min(\varphi_{1j}(x), -\varphi_{jj}(x)), 0),$$
$$\overline{\psi}_{ij}(x) = \max(\min(\varphi_{ij}(x), -\varphi_{i-1,j}(x)), 0).$$

This result is due to A.D. Berenstein and A.N. Kirillov [24], Theorem 3.2.

3.2 Schützenberger involution q_{n-1}

We are going to describe the action of Schützenberger's involution q_{n-1} on the space of triangles X_n. It is convenient to introduce new parameters

$$d_{ij} = x_{ij} - x_{i,j-1}, \text{ if } 1 \le i < j \le n, \text{ and } d_{ii} = x_{ii}. \quad (3.4)$$

In order to state the main result of this Section, let us introduce a couple of notation. We will identify the index (i,j), $1 \le i,j \le n$, with the point on the plane $\mathbb{A}^2$ with coordinates (i,j). A lattice path p between two points (a,b) and (c,d) is a sequence of points

$$(a,b) = (i_1, j_1), (i_1, j_2), \ldots, (i_N, j_N) = (c,d)$$

such that for each k, $1 \le k \le N-1$, the point (i_{k+1}, j_{k+1}) is equal to either $(i_k + 1, j_k)$ or $(i_k, j_k + 1)$. Let us define the weight $wt(p)$ of a lattice path p to be

$$wt(p) = \sum_{k=1}^{N} d_{i_k, j_k}.$$

Two lattice paths p_1 and p_2 are called *non–crossing* (or *non–intersecting*) if they do not have common points. For given $1 \le j \le k \le n$ denote by $\mathcal{R}_{jk}^{(n)}$ the set of all j–tuples $\mathbf{p} = (p_1, \ldots, p_j)$ of non–crossing lattice paths between the points $(n-k+l, n-k+l)$ and (l, n), $1 \le l \le j$. Finally, let us denote by $x_{jk}^{(n)} := (q_{n-1}(x))_{jk}$, $1 \le j \le k \le n$.

Theorem 3.3 *Let* $1 \le j \le k \le n$, *then*

$$x_{jk}^{(n)} := x_{jk}^{(n)}(d) = \min \left\{ \sum_{s=1}^{j} wt(p_s) \mid \{p\} \in \mathcal{R}_{jk}^{(n)} \right\}, \tag{3.5}$$

where the minimum is taken over all sequences $\{p\} = \{p_1, \ldots, p_j\}$ *of non–crossing lattice paths from the set* $\mathcal{R}_{jk}^{(n)}$, *and parameters* $d = (d_{jk})_{1 \le j \le k \le n}$ *and* $(x_{jk})_{1 \le j \le k \le n}$ *are related by (3.4).*

Remark 3.4 In some different, but equivalent form, Theorem 3.3 has been proved for the first time by H. Knight and A. Zelevinsky [25]. In Section 4.4 we are going to present a sketch of pure algebraic proof of the *tropical* version of the formula (3.5), namely, the proof which is a corollary of Plücker's relations for the Macdonald 9-th variation of Schur functions.

3.3 Continuous analog of the Robinson–Schensted–Knuth correspondence

In what follows it is convenient to use new parameterization $t = (t_{ij})$, $1 \le i \le j \le n$, of the space of triangles X_n, and introduce new parameters

$$u_{ij} := u_{ij}(t) = \sum_{k=0}^{\min(i-1,j-1)} t_{i-k,j-k}.$$

Let us denote by $\mathcal{P}_{n,m}$ the rectangular

$$[1;n] \times [1;m] = \{(i,j) \in \mathbb{Z}^2 \mid 1 \le i \le n,\ 1 \le j \le m\}.$$

For any integer k, $1 \le k \le \min(n,m)$, denote by $\mathcal{R}_{n,m}^{(k)}$ the set consisting of all k–tuples $\mathbf{p} = (p_1, \ldots, p_r, \ldots, p_k)$ of non–crossing lattice paths, where p_r is a lattice path between the points $(1, r)$ and $(n, m-k+r)$, $1 \le r \le k$. For any k–tuples of lattice paths $\mathbf{p} = (p_1, \ldots, p_k) \in \mathcal{R}_{n,m}^{(k)}$, denote by

$$\mathrm{Supp}(\mathbf{p}) = \{(i,j) \in \mathcal{P}_{n,m} | (i,j) \in p_l \text{ for some } 1 \le l \le k\}$$

the support of $\mathbf{p}$. Finally, let us define the weight $\widetilde{wt}(\mathbf{p})$ of $\mathbf{p} \in \mathcal{R}_{n,m}^{(k)}$ to be

$$\widetilde{wt}(\mathbf{p}) = \sum_{(a,b) \in \mathcal{P}_{n,m} \setminus \mathrm{Supp}(\mathbf{p})} d_{a,b}.$$

Theorem 3.5 (Berenstein A.D. and Kirillov A.N. [4]) *Let* $\mathbf{d} = (d_{ij})$ *be a* n *by* n *matrix with non–negative integer coefficients, and* $t := t(\mathbf{d}) = (t_{ij})$ *be the reverse plane partition that corresponds to the matrix* $\mathbf{d}$ *under the Robinson–Schensted–Knuth correspondence.*

- *If* $1 \le i \le j < n$, *then*

$$u_{ij}(t) = \min\{\sum_{\mathbf{p}} \widetilde{w}t(\mathbf{p}) \mid \mathbf{p} \in \mathcal{R}^{(n-j)}_{n+i-j,n}\}, \tag{3.6}$$

- $t_{jn} = \prod_{1\le i\le j,\ 1\le k\le n} d_{ik}$.

- *If* $1 \le i \le j \le n$, *then* $u_{ji}(t(\mathbf{d})) = u_{ij}(t(\mathbf{d}^t))$, *where* $\mathbf{d}^t$ *denotes the matrix which is transposed to* $\mathbf{d}$ *with respect to the main diagonal.*

4 Explicit formulae: tropification

The main goal of this Section is to describe the tropical images of transformations defined in Sections 2 and 3.

4.1 Tropification and ultradiscretization

A function $\varphi(x)$ given on the space of triangles X_{n+1} is called continuous piece–wise linear function (cpl–function for shot) if there exist a nonnegative integer N and collections of $\mathbb{Z}$–linear forms

$$l_k(x) = \sum_{1\le i\le j\le n+1} a_{ij}^{(k)} x_{ij},\quad a_{ij}^{(k)} \in \mathbb{Z} \ \text{ for all } \ 1 \le k \le n+1,$$

and

$$m_k(x) = \sum_{1\le i\le j\le n+1} b_{ij}^{(k)} x_{ij},\quad b_{ij}^{(k)} \in \mathbb{Z} \ \text{ for all } \ 1 \le k \le n+1$$

such that

$$\varphi(x) = \max(l_1(x), \ldots, l_N(x)) - \max(m_1(x), \ldots, m_N(x)).$$

We denote by $CPL(n+1)$ the set of all cpl–functions on the set X_{n+1}.

Follow M.-P. Schützenberger (private communication) and G. Lusztig [31,32,33], define a map

$$\theta : CPL(n+1) \to \mathbb{Q}(x_{ij} | 1 \le i \le j \le n+1)$$

from the set of cpl–functions on X_{n+1} to the field of rational functions in the variables $\{x_{ij}\}_{1\le i\le j\le n+1}$ with rational coefficients, which sends any

$\mathbb{Z}$–linear form $l(x) = \sum a_{ij}x_{ij}$ to the product $\theta(l(x)) = \prod x_{ij}^{a_{ij}}$, and function $\max(l(x), \ldots, l_N(x))$ to the sum

$$\theta(\max(l(x), \ldots, l_N(x)) = \sum_{j=1}^{N} \theta(l_j(x)).$$

Follow M.-P. Schützenberger (private communication) we are calling the map θ by *tropification map*, and the image $\theta(\varphi)$ of any cpl–function φ under the tropification map θ by its *tropical image.*

As far as I am aware, the name *ultradiscretization* was invented by B. Grammaticos for certain limiting procedure (see below) which allows to construct Cellular Automata (CAs) from continuous equations (cf. the 9-th problem posed by S. Wolfram [41]). Recall that a cellular automaton is a discrete dynamical system in which dependent variables take on a finite set of discrete values, see e.g. [41,39,40] for details. One of the most well–developed examples of CAs is the soliton cellular automaton (SCA) proposed by D. Takahashi and J. Satsuma [39]. Now this SCA and its numerous generalizations are known in literature as Ball–Box Systems (BBS), see e.g. [40]. Recently it was discovered by several groups of people, see e.g. [17] and the literature quoted therein, that BBS are closely related to Combinatorics of transportation matrices, Young tableaux, RSK, crystal bases and so on.

Finally, *ultradiscretization* of a certain object $\Phi(A, B, C, \ldots)$ depending on variables $A, B, C, \ldots$ (say, function, differential equation, ...) is the following procedure:

first of all let us put $A = e^{a/\epsilon}$, $B = e^{b/\epsilon}$, $C = e^{c/\epsilon}$, ..., and then take the limit (if exists)

$$\lim_{\epsilon \to +0} \Phi(e^{a/\epsilon}, e^{b/\epsilon}, e^{c/\epsilon}, \ldots).$$

The basic rules are

$$\lim_{\epsilon \to +0} \epsilon(A + B + C + \cdots) = \max(a, b, c, \ldots),$$

$$\lim_{\epsilon \to +0} \epsilon e^{a/\epsilon} e^{b/\epsilon} e^{c/\epsilon} \cdots = a + b + c + \cdots.$$

Thus, *ultradiscretization* (if exists) is the inverse procedure to *tropification.*

4.2 Tropical image of involutions s_i and $\widetilde{s}_i^{(\epsilon)}$

It is convenient to change slightly notation and use variables $\{y_{ij}\}_{1 \le i \le j \le n+1}$ instead of those $\{x_{ij}\}_{1 \le i \le j \le n+1}$. Thus let

$$\mathcal{F}_{n+1} = \mathbb{Q}(y_{ij} | 1 \le i \le j \le n+1)$$

be the field of rational functions in the variables y_{ij}, $1 \le i \le j \le n+1$, with rational coefficients. Define "elementary" birational transformations t_j of the field $\mathcal{F}_{n+1}$:

- $(t_j(y))_{ik} = y_{ik}$ if $k \neq j$;

- $$(t_j(y))_{ij} = \begin{cases} (y_{2,j+1} + y_{1,j-1}y_{1,j+1})/y_{1j}, & \text{if } i = 1; \\ (y_{i,j-1}y_{i,j+1} + y_{i-1,j-1}y_{i+1,j+1})/y_{ij}, & \text{if } 1 < i < j; \\ y_{i-1,j-1}y_{i+1,j+1}/y_{ij}, & \text{if } i = j. \end{cases} \quad (4.1)$$

In the latter formulae we assume that $y_{0j} = 1$ if $j \ge 0$, and $y_{i0} = \delta_{i,0}$.

Let ϵ be a new variable, and consider the field

$$\mathcal{F}_{n+1}(\epsilon) = \mathcal{F}_{n+1} \otimes \mathbb{Q}(\epsilon),$$

where $\mathbb{Q}(\epsilon)$ denotes the field of rational functions in the variable ϵ. Follow Section 2.10, (2.13) let us define "*translation* operator" T_ϵ which is a birational automorphism of the field $\mathcal{F}_n(\epsilon)$:

- $(T_\epsilon(y))_{ij} = y_{ij}$ if $(i,j) \neq 1$; $(T_\epsilon(y))_{11} = y_{11}\epsilon^{-1}$,

and birational automorphisms q_i, s_i and s_i^ϵ of the field $\mathcal{F}_{n+1}(\epsilon)$:

- $q_i = t_1 \underbrace{t_2 t_1} \cdots \underbrace{t_i t_{i-1} \cdots t_1}$, $\quad s_i = q_i t_1 q_i$, $\quad s_i^{(\epsilon)} = q_i T_\epsilon q_i$, $1 \le i \le n$.

In order to state our main theorem of this Section, let us introduce the Laurent polynomials

$$Q_{k,n}^{(\epsilon)}(y) = \sum_{j=1}^{k} \frac{y_{j-1,n-1}y_{j,n+1}}{y_{j-1,n}y_{j,n}} + \epsilon \sum_{j=k+1}^{n} \frac{y_{j-1,n-1}y_{j,n+1}}{y_{j-1,n}y_{j,n}}, \quad (4.2)$$

where $1 \le k \le n$, and as before we assume that $y_{0j} = 1$, if $j \ge 1$. Note that Laurent's polynomial $Q_{n,n}^{(\epsilon)}(y)$ doesn't depend on ϵ, so we will denote it by $Q_{n,n}(y)$.

Remark 4.1 It is not difficult to see that if $x \in X_n$ is a triangle and j is an integer between 1 and $n-1$, then Laurent's polynomial $Q_{jj}(y)$ coincides with the *tropical image* of the j-th exponents $\kappa_j(x)$ of the triangle x, see Section 2.4, formula (2.6), when the former is written using the y–coordinates. For example, if $x \in X_4$, then

$$\begin{aligned} \kappa_3(x) &= \max(x_{14} - x_{13}, x_{14} - 2x_{13} + x_{12} + x_{24} - x_{23}, \\ &\qquad x_{14} - 2x_{13} + x_{12} + x_{24} - 2x_{23} + x_{22} + x_{34} - x_{33}) \\ &= \max(y_{14} - y_{13}, y_{12} + y_{24} - y_{13} - y_{23}, y_{22} + y_{23} - y_{33} - y_{34}). \end{aligned}$$

The tropical image of the latter max(...) is exactly Laurent's polynomial

$$Q_{33}(y) = \frac{y_{14}}{y_{13}} + \frac{y_{12}y_{24}}{y_{13}y_{23}} + \frac{y_{22}y_{23}}{y_{33}y_{34}}.$$

Theorem 4.2 (Tropical image of transformation $s_n^{(\epsilon)}$)

$$(s_n^{(\epsilon)}(y))_{kn} = \epsilon^{-1} y_{kn} \frac{Q_{k,n}^{(\epsilon)}(y)}{Q_{n,n}(y)}, \quad 1 \le k \le n. \tag{4.3}$$

Corollary 4.3

$$(s_n(y))_{kn} = y_{kn} \frac{Q_{k,n}(y)}{Q_{n,n}(y)} \cdot \frac{y_{n-1,n-1}y_{n+1,n+1}}{y_{n,n}^2}, \tag{4.4}$$

where $Q_{k,n}(y) = Q_{k,n}^{(\eta)}(y)$ *with* $\eta = \dfrac{y_{nn}^2}{y_{n-1,n-1}y_{n+1,n+1}}$.

Corollary 4.4

$$(\widetilde{s}_n^{(\epsilon)}(y))_{kn} = y_{kn} \frac{Q_{k,n}^{(\delta)}(y)}{\delta Q_{n,n}(y)}, \tag{4.5}$$

where $\delta = \dfrac{\epsilon y_{nn}^2}{y_{n-1,n-1}y_{n+1,n+1}}$. *Recall that* $\widetilde{s}_n^{(\epsilon)} := s_n s_n^{(\epsilon)} = q_n T_\epsilon t_1 q_n$.

Our proof of Theorem 4.2 is based on some remarkable properties of the Laurent polynomials $Q_{kn}^{(\epsilon)}(y)$ and $Q_{kn}(y)$ which we will state in Theorem 4.5 below.

Theorem 4.5 *Let* s_n, $s_n^{(\epsilon)}$ *and* $s_n^{(\eta)}$ *be birational transformations of the field* $\mathcal{F}_{n+1}(\epsilon, \eta)$ *given by formulae (4.3) and (4.4). Then*

- $Q_{nn}(s_n^{(\epsilon)}(y)) = \epsilon Q_{nn}(y);$ (4.6)
- $Q_{kn}^{(\epsilon)}(s_n^{(\eta)}(y)) = \eta Q_{kn}^{(\epsilon * \eta)}(y) Q_{nn}(y)/Q_{kn}^{(\eta)}(y),$ (4.7)

where we put by definition $\epsilon * \eta := \epsilon s^{(\epsilon)}(\eta)$;

- *if* $n \pm 1 \ge 1$, *then for any integer* k *between 1 and* n *the Laurent polynomials*

$$y_{kn} Q_{kn}(s_{n\pm1}(y)) Q_{n\pm1,n\pm1}(y) \frac{y_{n\pm1,n\pm1}^2}{y_{n\pm2,n\pm2}y_{n,n}} \tag{4.8}$$

are invariant with respect to the action of involution s_n.

For example, if $\epsilon = y_{nn}^2/y_{n-1,n-1}y_{n+1,n+1}$, then $s^{(\epsilon)}(\epsilon) = \epsilon^{-1}$, and $\epsilon * \epsilon = 1$. In particular, it follows from (4.7) that

- $$Q_{kn}(s_n(y)) = \frac{Q_{nn}^2(y)}{Q_{kn}(y)} \frac{y_{nn}^2}{y_{n-1,n-1}y_{n+1,n+1}}. \tag{4.9}$$

Corollary 4.6 *1) For any integer $j \geq 1$ the Laurent polynomial $y_{jj}Q_{jj}(y)$ is a $s_j^{(\epsilon)}$–invariant. In particular, the latter Laurent polynomial is invariant with respect to the action of involution s_j.*

2) The following Laurent's polynomials

$$y_{ij}^2 Q_{ij}(y)/y_{jj}, \quad 1 \leq i \leq j,$$

are s_j–invariants.

Remark 4.7 1) It is clear from the very definition (4.2) that Laurent's polynomials $Q_{kn}(y)$, $1 \leq k \leq n$, belong to the ring $\mathbb{Z}[y_{a,n-1}^{\pm 1}, y_{b,n}^{\pm 1}, y_{c,n+1}^{\pm 1}]$, and therefore they are s_j–invariants if $1 \leq j \leq n-2$.

2) It follows from (4.4) that if $y \in X_N$, where $N \geq n+1$, then $s_n(y) = y$ if and only if the following condition holds:

$$y_{n+1,n+1}y_{n-1,n-1} = y_{n,n}^2.$$

Therefore, for a triangle $y \in X_{n+1}$ the condition $s_j(y) = y$ for all j, $1 \leq j \leq n$, is equivalent to the following one: $y_{kk} = y_{11}^k$, $1 \leq k \leq n+1$. The latter condition means that the triangle y has *zero* weight. Summarizing, the action of the symmetric group S_n on the set of Young tableaux $STY(< n+1)$ constructed by A. Lascoux and M.-P. Schützenberger [30], as well as that of its tropical version constructed in this Section, see (4.4), acts trivially on the *zero* weight subspace, as well as on the zero weight set

$$X_{n+1}(0) = \{y \in X_{n+1} | y_{kk} = y_{11}^k, \ 1 \leq k \leq n+1\}.$$

3) It is easy to see that

$$Q_{k,n}^{(\epsilon)}(y) = Q_{n,n}(y^{\{k\}}), \quad 1 \leq k \leq n,$$

where for any triangle $y \in X_{n+1}$, the triangle $y^{\{k\}}$ is defined as follows:

$$(y^{\{k\}})_{ij} = y_{ij}, \quad \text{if} \quad j \neq n+1,$$
$$(y^{\{k\}})_{i,n+1} = \begin{cases} y_{i,n+1}, & \text{if } i \leq k, \\ \epsilon y_{i,n+1}, & \text{if } k < i \leq n. \end{cases}$$

In particular,

$$Q_{kn}(y) = Q_{nn}(y^{\{\eta\}}), \quad 1 \leq k \leq n,$$

where $\eta = y_{nn}^2 / y_{n-1,n-1}y_{n+1,n+1}$

Finally, let us define the following birational automorphisms of the field $\mathcal{F}_n(\epsilon) = \mathbb{Q}(\epsilon)(y_{ij}|1 \le i \le j \le n)$:

$$\pi := \pi_n = s_{n-1}s_{n-2}\cdots s_2s_1;$$
$$s_0 := s_0^{(\epsilon)} = \pi T_\epsilon t_1 \pi^{-1};$$
$$g_i := g_i^{(\epsilon)} = \pi^{i-1}T_\epsilon t_1 \pi^{1-i}, \quad 1 \le i \le n, \quad g_0 := g_n$$
$$T_k = g_{k-1}g_{k-2}\cdots g_1 \pi g_{n-1}\cdots g_k, \text{ if } 2 \le k \le n-1, \text{ and}$$
$$T_1 = \pi g_{n-1}g_{n-2}\cdots g_2g_1, \; T_n = g_{n-1}g_{n-2}\cdots g_2g_1\pi.$$

Corollary 4.8 *1) Birational transformations s_i, $i = 0, 1, \ldots, n-1$, define a birational representation of the affine symmetric group $W(A_{n-1}^{(1)})$ on the field $\mathcal{F}_n(\epsilon)$;*

2) birational transformations g_i, $i = 0, 1, \ldots, n-1$, and π define a birational representation of the extended symmetric group $\widehat{W}(A_{n-1}^{(1)})$ on the field $\mathcal{F}_n(\epsilon)$;

3) birational transformations $T_1, \ldots, T_n$ generate a commutative subgroup in $\widehat{W}(A_{n-1}^{(1)})$, and $T_1T_2\cdots T_n = 1$;

4) (Birational solution to the Yang–Baxter equation)
let i be an integer between 0 and $n-1$, and $\epsilon, \eta \in \mathbb{R}$, then

$$s_i^{(\epsilon)}(y)s_{i+1}^{(\epsilon+\eta)}(y)s_i^{(\eta)}(y) = s_{i+1}^{(\eta)}(y)s_i^{(\epsilon+\eta)}(y)s_{i+1}^{(\epsilon)}(y),$$

where matrices $s_i^{(\epsilon)}(y)$ and $s_{i+1}^{(\eta)}(y)$ are given by (4.3) with $n = i$ and $n = i+1$ correspondingly.

We conclude this Section by describing a certain set of simple invariants of the involution $s_n : \mathcal{F}_{n+1} \to \mathcal{F}_{n+1}$.

Theorem 4.9 (cf. [24], Theorem 3.2, (3.19))
The following Laurent polynomials

- $\psi_{1,n}(y) := \dfrac{y_{1n}}{y_{1,n+1}} + \dfrac{y_{n-1,n}y_{n+1,n+1}}{y_{nn}y_{n,n+1}}$,

- $\psi_{i,n}(y) := \dfrac{y_{i-2,n}y_{i-1,n-1}}{y_{i-2,n-1}y_{i-1,n}} + \dfrac{y_{i-1,n+1}y_{in}}{y_{i-1,n}y_{i,n+1}}, \quad 2 \le i \le n,$

are invariant with respect to the action of the involution s_n.

Other examples of s_n–invariants which are given by Laurent's polynomials, one can find in Theorem 4.5, formula (4.8), and Corollary 4.8, as well as in Remark 4.10 and Theorem 4.32. It looks a challenging task to describe a system of *fundamental* s_n–invariants for the birational action of the symmetric group S_{n+1} on the field $\mathcal{F}_{n+1}$. In other words to find a minimal set of

s_n–invariants such that any s_n–invariant can be written as a rational function of the former.

Remark 4.10 Let $n \geq 2$ be an integer. It is easy to see that

$$s_j(y_{jj}/y_{j-1,j-1}) = y_{j+1,j+1}/y_{j,j}, \quad 1 \leq j \leq n-1, \quad y_{00} := 1,$$

and therefore the coefficients $\theta_k(y)$ of the expansion of the product

$$\prod_{1 \leq i<j \leq n} \left(1 + t\left(\frac{y_{i,i}}{y_{i-1,i-1}} + \frac{y_{j,j}}{y_{j-1,j-1}}\right)\right) = \sum_{k=0}^{n(n-1)/2} \theta_k(y) t^k$$

are invariants with respect to the birational action of the symmetric group S_n constructed in this Section.

Note that the product

$$\prod_{1 \leq i<j \leq n} (y_{i,i}/y_{i-1,i-1} + y_{j,j}/y_{j-1,j-1}) := \theta_{\frac{n(n-1)}{2}}(y)$$

is equal to the tropification of the function

$$n(\beta(y)) = \sum_{1 \leq i<j \leq n} \min(\beta_i(y), \beta_j(y))$$

given on the space of triangles X_n, where for any triangle $y \in X_n$, $\beta(y) := (\beta_1(y), \ldots, \beta_n(y)) \in \mathbb{R}^n$ denotes its weight, cf [34], Chapter I, formulae (1.5) and (1.6).

4.3 Tropical image of involution σ_a

Let us denote by $\mathcal{M}_n = \mathbb{Q}(d_{ij})_{1 \leq i,j \leq n}$ the field of rational functions in the variables $\mathbf{d} = (d_{ij})_{1 \leq i,j \leq n}$ with rational coefficients. In the sequel it is convenient to impose the periodicity condition

$$d_{i+n,j} = d_{ij} = d_{i,j+n}, \quad 1 \leq i \leq n, \quad j \in \mathbb{Z}.$$

To continue, let us consider the following Laurent polynomials

$$h_a^{(j)}(\mathbf{d}) := h_{a,n}^{(j)}(\mathbf{d}) = \sum_{k=1}^{n} \prod_{i=k}^{n} d_{a+1,i+j} \prod_{i=k+1}^{n} d_{a,i+j}^{-1}, \tag{4.10}$$

where $0 \leq j \leq n-1$ and $1 \leq a \leq n$.

Remark 4.11 i) It is clear from the very definition (4.10) of Laurent's polynomial $h_a^{(l)}$ that the former depends only on 2 by n matrix

$$\begin{pmatrix} d_{a,1}, \ldots, d_{a,n} \\ d_{a+1,1}, \ldots, d_{a+1,n} \end{pmatrix}.$$

It is natural therefore to define for any 2 by n matrix $\begin{pmatrix} a_1, \dots, a_n \\ b_1, \dots, b_n \end{pmatrix}$ its exponents $h^{(l)}(a,b)$, $0 \le l \le n$, to be

$$h^{(l)}(a,b) = \sum_{j=1}^{n} \prod_{i=j}^{n} a_{i+l} \prod_{i=j+1}^{n} b_{i+l}^{-1}.$$

Note that Laurent's polynomial $h^{(0)}(a,b)$ coincides with *tropical image* of the "local energy" $H(\overleftarrow{\mathbf{a}}, \overleftarrow{\mathbf{b}})$ of states $\overleftarrow{\mathbf{a}}$ and $\overleftarrow{\mathbf{b}}$, see Section 4.9. Here for any sequence $\mathbf{a} = (a_1, \dots, a_n)$ we denote by $\overleftarrow{\mathbf{a}} := (a_n, a_{n-1}, \dots, a_1)$ the reverse sequence.

ii) In the paper [17] the combinatorial R–matrix for fundamental representation of algebra $U_q(\widehat{sl}(n))$ had been computed. The *tropical image* R of the latter can be described as follows:

$$R : \mathbb{C}^n \otimes \mathbb{C}^n \to \mathbb{C}^n \otimes \mathbb{C}^n$$
$$(a_1, \dots, a_n) \otimes (b_1, \dots, b_n) \to (a'_1, \dots, a'_n) \otimes (b'_1, \dots, b'_n),$$

where

$$a'_k = b_k L^{(k-1)}(a,b)/L^{(k)}(a,b),$$
$$b'_k = a_k L^{(k)}(a,b)/L^{(k-1)}(a,b), \quad 1 \le k \le n,$$

and

$$L^{(k)}(a,b) = h^{(k)}(a^{-1}, b^{-1}) \prod_{j=1}^{n} a_j = \sum_{j=1}^{n} \prod_{i=1}^{j-1} a_{k+i} \prod_{i=j+1}^{n} b_{k+i}.$$

Recall that

$$h^{(l)}(a,b) := \sum_{j=1}^{n} \prod_{i=j}^{n} a_{i+l} \prod_{i=j+1}^{n} b_{i+l}^{-1}, \quad 1 \le l \le n,$$

denotes the *exponents* of the following 2 by n matrix $\begin{pmatrix} b_1, \dots, b_n \\ a_1, \dots, a_n \end{pmatrix}$.

Final remark, the interpretation of *tropical* version of the combinatorial R–matrix presented above, I had learned from the lecture given by Prof. Y. Yamada during the International Workshop "Physics and Combinatorics", August 21–26, 2000, Graduate School of Mathematics, Nagoya University.

Now let us define birational transformations $\sigma_a = \sigma_{a,n}$, $1 \le a \le n$, of the field $\mathcal{M}_n$ (or affine space $\mathbb{A}^{n^2}$):

- $(\sigma_a(\mathbf{d}))_{ij} = d_{ij}$ if $i \ne a, a+1$;

- $(\sigma_a(\mathbf{d}))_{aj} = d_{a+1,j} h_a^{(j-1)}(\mathbf{d})/h_a^{(j)}(\mathbf{d});$ (4.11)
- $(\sigma_a(\mathbf{d}))_{a+1,j} = d_{a,j} h_a^{(j)}(\mathbf{d})/h_a^{(j-1)}(\mathbf{d}).$

Theorem 4.12 (Tropical image of transformations σ_a)

i) If $n \geq 3$, the transformations σ_a, $1 \leq a \leq n$, given by formulae (4.11) define a birational representation of the affine symmetric group $\widehat{S}_n = W(A_{n-1}^{(1)})$ on the field $\mathcal{M}_n$ (or affine space $\mathbb{A}^{n^2}$).

ii) The ultradiscretization of this action coincides with the cpl–action of the affine symmetric group $\widehat{S}_n$ on the space of matrices M_n introduced in Section 2.11.

Proposition 4.13 *Involutions σ_a, $1 \leq a \leq n$, have the following properties:*

- $\sigma_a(d_{aj}d_{a+1,j}) = d_{aj}d_{a+1,j};$
- $\sigma_a(d_{a,j+1}^{-1} + d_{a+1,j}^{-1}) = d_{a,j+1}^{-1} + d_{a+1,j}^{-1}, \quad j \in \mathbb{Z}.$

In other words, for any integer a, $1 \leq a \leq n$, the "linear" forms $d_{a,j+1}^{-1} + d_{a+1,j}^{-1}$, and the products $d_{aj}d_{a+1,j}$, $j \in \mathbb{Z}$, are invariant with respect to the action of involution σ_a.

Corollary 4.14 *For any integer a, $1 \leq a \leq n$, the "linear" forms*

$$H_a := H_a(\mathbf{d}) = \sum_{k=0}^{n-1} d_{n-k,a+k}^{-1}$$

are invariants with respect to the action of all involutions σ_j, $1 \leq j \leq n$.

Our proof of Theorem 4.12 is based on the following remarkable properties of Laurent's polynomials $h_a^{(j)}(\mathbf{d})$:

Theorem 4.15 *For any integers a and j between 1 and n the Laurent polynomials $h_a^{(j)}(\mathbf{d})$ and $h_a^{(j)}(\sigma_{a\pm1,n}(\mathbf{d}))h_{a\pm1}^{(j)}(\mathbf{d})$ are invariant with respect to the action of the involution σ_a:*

i) $h_a^{(j)}(\sigma_a(\mathbf{d})) = h_a^{(j)}(\mathbf{d});$

ii) $h_a^{(j)}(\sigma_a\sigma_{a\pm1}(\mathbf{d}))h_{a\pm1}^{(j)}(\sigma_a(\mathbf{d})) = h_a^{(j)}(\sigma_{a\pm1}(\mathbf{d}))h_{a\pm1}^{(j)}(\mathbf{d}).$

Remark 4.16 i) The Laurent polynomials $h_a^{(j)}(\mathbf{d})$, given by formulae (4.10), coincide with the tropical images of the *row exponents* $h_{a,+}^{(j)}(d)$of matrix d, see definition in Section 2.10. It is convenient also to introduce polynomial analog $L_{j,n}^{(a)}(\mathbf{d})$ of functions $h_{j,n}^{(a)}(\mathbf{d})$, see Remark 4.11. Namely,

$$L_{j,n}^{(a)}(\mathbf{d}) := L_j^{(a)}(\mathbf{d}) = h_{a,n}^{(j)}(\mathbf{d}^{-1}) \prod_{j=1}^{n} d_{a+1,j} = \sum_{k=1}^{n} \prod_{j=1}^{k-1} d_{a+1,j+l} \prod_{i=k+1}^{n} d_{a,i+j},$$

where by definition we set $(\mathbf{d}^{-1})_{ij} = d_{ij}^{-1}$, $1 \le i,j \le n$. It follows from Theorem 4.12 that birational transformations $\sigma_a^* = \sigma_{a,n}^*$, $1 \le a \le n$, of the affine space $\mathbb{A}^{n^2}$ given by the following formulae

$$\begin{aligned} &\bullet\ (\sigma_a^*(\mathbf{d}))_{ij} = d_{ij} \ \text{ if } \ i \neq a, a+1; \\ &\bullet\ (\sigma_a^*(\mathbf{d}))_{aj} = d_{a+1,j} L_j^{(a)}(\mathbf{d}) / L_{j-1}^{(a)}(\mathbf{d}); \\ &\bullet\ (\sigma_a^*(\mathbf{d}))_{a+1,j} = d_{a,j} L_{j-1}^{(a)}(\mathbf{d}) / L_j^{(a)}(\mathbf{d}), \end{aligned} \tag{4.12}$$

also define a birational representation of the affine symmetric group $W(A_{n-1}^{(1)})$ on the field $\mathcal{M}_n$.

ii) It follows from the definition (4.10) of polynomials $h_a^{(j)}(\mathbf{d})$ that if for some k, $1 \le k \le n$, one has

$$d_{k,1} d_{k,2} \cdots d_{k,n} = d_{k+1,1} d_{k+1,2} \cdots d_{k+1,n},$$

then $\sigma_k(\mathbf{d}) = \mathbf{d}$; in other words in the case under consideration the action of involution σ_k on the field $\mathcal{M}_n$ is trivial. Therefore, in the case when the products $\prod_{j=1}^n d_{a,j}$ are the same for all a, $1 \le a \le n$, the action of affine Weyl group $W(A_{n-1}^{(1)}) = \langle \sigma_1, \ldots, \sigma_n \rangle$ on the field $\mathcal{M}_n := \mathbb{Q}(d_{i,j})_{1 \le i,j \le n}$, constructed in this Section, is trivial.

iii) Let us define *shift* operator π acting on the field $\mathcal{M}_n = \mathbb{Q}(d_{ij})_{1 \le i,j \le n}$ by the following rule:

$$\pi(d_{i,j}) = d_{i+1,j}, \quad 1 \le i,j \le n.$$

For example, $\pi(d_{n,j}) = d_{1,j}$, $1 \le j \le n$. It is easy to see that

$$\pi(h_{a,n}^{(j)}(\mathbf{d})) = h_{a+1,n}^{(j)}(\pi(\mathbf{d})),$$

and therefore $\pi\sigma_a = \sigma_{a+1}\pi$, $1 \le a \le n$. In other words we came to the following

Theorem 4.17 *If $n \ge 3$, the involutions σ_a, $1 \le a \le n$, and the shift operator π give rise to a birational representation of the extended affine Weyl group $\widehat{W}(A_{n-1}^{(1)})$ on the field of rational functions $\mathcal{M}_n$.*

4.4 *Tropical image of Schützenberger involution q_{n-1}*

Let $\{y_{ij} | 1 \le i \le j \le n\}$ be the set of independent variables, consider the field of rational functions $\mathcal{F}_n = \mathbb{Q}(y_{ij})$. Let us introduce also the set of variables $\{d_{ij} | 1 \le i \le j \le n\}$ which are related to the variables $\{y_{ij} | 1 \le i \le j \le n\}$ by means of the following birational transformation:

$$d_{ij} = \frac{y_{ij} y_{i-1,j-1}}{y_{i-1,j} y_{i,j-1}} \ \text{ if } \ i < j; \ \ d_{ii} = \frac{y_{ii}}{y_{i-1,i}}; \ \ y_{ij} = \prod_{k \le i,\ l \le j} d_{k,l}.$$

We assume in the above formulae that $y_{0j} = 1$, if $j \geq 1$.

Now consider the set $\mathcal{R}_{jk}^{(n)}$ of all collections $\mathbf{p} = (p_1, \ldots, p_j)$ of j–tuples of non–crossing lattice paths $p_1, \ldots, p_l, \ldots, p_j$ between the points $(n-k+l, n-k+l)$ and (l, n), $1 \leq l \leq j$, correspondingly. For any such collection of lattice paths $\mathbf{p} = (p_1, \ldots, p_j)$ define its weight $wt(\mathbf{p})$ to be

$$wt(\mathbf{p}) = \prod_{(i_a, j_a) \in \mathrm{Supp}(\mathbf{p})} d_{i_a, j_a},$$

where $\mathrm{Supp}(\mathbf{p}) = \{(a,b) \in \mathbb{Z}^2 | (a,b) \in p_l \text{ for some } 1 \leq l \leq j\}$ denotes the support of the family of non–crossing lattice paths $\mathbf{p}$.

Theorem 4.18 (Tropical image of Schützenberger involution q_{n-1}) *Assume that* $q_{n-1}(y) = (y_{jk}^{(n)})$, *then*

$$y_{jk}^{(n)} = \sum_{\mathbf{p} \in \mathcal{R}_{jk}^{(n)}} wt(\mathbf{p}). \tag{4.13}$$

Remark 4.19 The number of terms in the RHS (4.13) is equal to the number of plane partitions contained in the box $j \times (k-j) \times (n-k)$, and is equal also to the dimension of highest weight $\lambda = ((k-j)^j)$ irreducible representation of the general linear group $\mathfrak{gl}(n-k+j)$.

Remark 4.20 Formula (4.13) for $y_{jk}^{(n)}$ can be written as a determinant. To this end, let $x = (a,b)$ and $y = (c,d)$ be two points with integer coordinates in the plane $\mathbb{A}^2$. Denote by $\mathcal{P}(x;y)$ the set of lattice paths from the point x to that y, and denote by $E(x;y)$ the generating function

$$E(x;y) := E((a,b);(c,d)) = \sum_{p \in \mathcal{P}(x;y)} wt(p).$$

Theorem 4.21 (Determinantal formula)

$$y_{jk}^{(n)}(d) = \det |E((n-k+a, n-k+a);(b,n))|_{1 \leq a,b \leq j}. \tag{4.14}$$

Finally we are going to prove that polynomials

$$y_{ik}^{(n)} := y_{ik}^{(n)}(d) \in \mathbb{Z}[d_{ij}|_{1 \leq i \leq j \leq n}]$$

satisfy the Hirota type discrete bilinear equations.

Theorem 4.22 *Polynomials* $y_{ik}^{(n)} := y_{ik}^{(n)}(d) = (q_{n-1}(y))_{ik}$, $y \in X_{n+1}$, *satisfy the following quadratic relations*

$$y_{ik}^{(n)} y_{ik}^{(n+1)} = y_{i-1,k-1}^{(n)} y_{i+1,k+1}^{(n+1)} + y_{i,k-1}^{(n)} y_{i,k+1}^{(n+1)}, \tag{4.15}$$

and "boundary conditions"

- $y_{ik}^{(k)} = y_{ik} := \prod_{1\le a\le i,\ 1\le b\le k} d_{a,b},\quad 1 \le i \le k \le n+1,$ (4.16)

- $y_{0k}^{(n)} = 1,\quad k \le n.$ (4.17)

Proof. First of all let us recall, see e.g. [24], Proposition 1.4, or the present paper, formula (2.15), that $q_n = p_n^{-1} q_{n-1}$, where $p_n^{-1} = t_1 t_2 \cdots t_n$. Secondly, let us observe that the Bender–Knuth involution t_j, $1 \le j \le n$, changes entries in the j-th row of a triangle $y \in X_{n+1}$ only. Finally, applying the formula (4.1) for the action of involution t_k to the triangle $\widetilde{y} = (\widetilde{y}_{ij})$, where $\widetilde{y}_{ij} = y_{ij}^{(n+1)}$, if $j > k$, and $\widetilde{y}_{ij} = y_{ij}^{(n)}$ if $j \le k$, we are coming to the recurrence relation (4.15). ∎

It is easy to see that the recurrence relations (4.15) together with boundary conditions (4.16) and (4.17) define $y_{ik}^{(n)}$ *uniquely* as a function of y_{ik}, $1 \le i \le k \le n+1$. On the other hand one can show that the determinant (4.14) also satisfies the recurrence relations (4.15). A proof of the latter statement is a standard application of the non–crossing lattice paths technique, see e.g. [27], and will be given in a separate publication. We treat the relation (4.15) as the Plücker relation for the Macdonald 9-th variation of Schur functions, see e.g. [34], Chapter I, Section 3, Exercise 21, corresponding to a rectangular shape.

Exercise.

1. Show that if $n \ge 2$, then

$$y_{1,2}^{(n)} = \frac{y_{n-1,n-1}}{y_{n-2,n-2}} Q_{n-1,n-1}(y).$$

4.5 Tropical image of the Robinson–Schensted–Knuth correspondence

The tropical image of the Robinson–Schensted–Knuth correspondence is a certain birational transformation of the affine space $\mathbb{A}^{n^2}$; in other words it is a birational automorphism of the field of rational functions $\mathcal{M}_n = \mathbb{Q}(\mathbf{d})$, where $\mathbf{d} = (d_{ij})$, $1 \le i,j \le n$, denotes the set of variables.

In order to describe this birational automorphism, it is convenient to introduce new sets of variables (x_{ij}), (z_{ij}), and (v_{ij}), $1 \le i,j \le n$, which are connected by birational transformations

$$z_{ij} := z_{ij}(x) = \prod_{k=0}^{\min(n-i,n-j)} x_{i+k,j+k},$$

$$v_{ij} := v_{ij}(x) = \prod_{k=0}^{\min(i-1,j-1)} x_{i-k,j-k},$$

$$x_{ij} = x_{ij}(z) = z_{ij}/z_{i+1,j+1},$$
$$x_{ij} = x_{ij}(v) = v_{ij}/v_{i-1,j-1}.$$

Here and further we assume that $x_{ij} = 1$, if $(i,j) \notin [1;n] \times [1;n]$.

To continue, let us denote by $\mathcal{P}_{n,m}$ the set of points $(i,j) \in \mathbb{Z}^2$ such that $1 \le i \le n-m+1$ and $1 \le j \le n$. For each integer k such that $1 \le k \le n-m+1$, denote by $\mathcal{P}_{n,m}^{(k)}$ the set of all k–tuples of non–crossing lattice paths $\mathbf{p} = (p_1, \ldots, p_k)$, where the path p_r, $1 \le r \le k$, is a lattice path between the points $(1,r)$ and $(n-m+1, n-k+r)$.

For any path p between two points in the square $[1;n] \times [1;n]$ we define its *weight* $wt(p)$ to be

$$wt(p) = \prod_{(i_k,j_k)\in p} d_{i_k j_k}.$$

Theorem 4.23 *Tropical image of the Robinson–Schensted–Knuth correspondence*

$$\boxed{d_{ij}} \quad \overset{RSK}{\longrightarrow} \quad \boxed{x_{ij}}$$

is given by the following formulae:

- *assume that* $x := x(\mathbf{d}) = RSK(\mathbf{d})$, *then*

$$z_{i,j}(x) = \sum_{\mathbf{p}\in\mathcal{P}_{n,j-i+1}^{(n-j+1)}} \prod_{p_r\in\mathbf{p}} wt(p_r), \quad \text{if} \quad j \ge i \ge 2, \tag{4.18}$$

$$z_{1,k}(x) = \prod_{\substack{1\le i\le n-k+1\\ 1\le j\le n}} d_{ij}, \qquad z_{k,1}(x) = \prod_{\substack{1\le i\le n\\ 1\le j\le n-k+1}} d_{ij};$$

- *if* $i \le j$, *then*

$$z_{ji}(x(\mathbf{d})) = z_{ij}(x(\mathbf{d}^t)). \tag{4.19}$$

Remark 4.24 The cardinality of the set $\mathcal{P}_{n,m}^{(k)}$ is equal to

$$\prod_{a=1}^{k} \binom{2n-m-k}{n-a} \Big/ \binom{2n-m-k}{k-a}.$$

The latter product is equal to the dimension of irreducible representation V_λ of the Lie algebra $\mathfrak{gl}(n)$ corresponding to the highest weight $\lambda = ((n-m-k+1)^k)$.

Corollary 4.25 (Determinantal formula)
Assume that $x := x(\mathbf{d}) = RSK(\mathbf{d})$. *If* $j \geq i \geq 2$, *then*

$$z_{ij}(x) = \det |E((1,a);(n-j+i-1, b+j-1))|_{1 \leq a,b \leq j-i+1}.$$

See Remark 4.20 for definition of function $E((a,b);(c,d))$.

Remark 4.26 For any diagram λ denote by $STY(\lambda, \leq n)$ the set of semistandard Young tableaux of shape λ filled by integer numbers $1, 2, \ldots, n$. Using well–known bijection between the set of certain non–crossing paths and that of semistandard (i.e. column strict) Young tableaux, see e.g. [16], one can write the tropical image of the formula (3.6) in the following form:

if $i \leq j$, then

$$\prod_{k=0}^{i-1} x_{i-k,j-k} = \sum_{T \in STY((i^j), \leq n)} \prod_{x \in (i^j)} d_{T(x)+c(x), n+1-T(x)}, \tag{4.20}$$

where the sum is taken over the set of semistandard Young tableaux T of rectangular shape $(\underbrace{i, \ldots, i}_{j})$, filled by integer numbers from the interval $[1; n]$; the product is taken over all cells in the rectangular (i^j); $T(x)$ denotes the entry of tableau T located in the cell x, and if $x = (i,j)$, then $c(x) = j - i$ denotes the *content* of the cell x. Note that if $i \leq j$, then

$$v_{ji}(x(\mathbf{d})) = v_{ij}(x(\mathbf{d}^t)). \tag{4.21}$$

To state our next result which describes interrelations between RSK and certain birational actions of the symmetric group S_n on the affine space $\mathbb{A}^{n^2}$, we need a few additional definitions.

First of all, let us define a birational transformation $\widetilde{\sigma}_a : \mathbb{A}^{n^2} \to \mathbb{A}^{n^2}$ by the following rule:

$$\widetilde{\sigma}_a(\mathbf{d}) = \sigma_a(\mathbf{d}^t),$$

where $\mathbf{d}^t$ denotes the matrix which is transposed to $\mathbf{d}$ with respect to the main diagonal. Next, let us define a birational map s_a^{up} (corr. s_a^{down}) as follows: consider a matrix of size n by n $\boxed{x_{ij}}$, where x_{ij}, $1 \leq i,j \leq n$, are the set of variables. Let 

be the diamond corresponding to the matrix $\boxed{x_{ij}}$, see Section 2.8. Recall that P and Q stand for the following triangles

$$Q = \begin{array}{ccccc} x_{nn}, & x_{n-1,n-1}, & \ldots, & x_{22}, & x_{11} \\ & x_{n-1,n}, & \ldots, & x_{12} & \\ & & \vdots & & \\ & & x_{2n}, & x_{1,n-1} & \\ & & x_{1,n} & & \end{array}$$

$$P = \begin{array}{ccccc} x_{nn}, & x_{n-1,n-1}, & \ldots, & x_{22}, & x_{11} \\ & x_{n,n-1}, & \ldots, & x_{21} & \\ & & \vdots & & \\ & & x_{n2}, & x_{n-1,1} & \\ & & x_{n1} & & \end{array}$$

Then, by definition,

$$s_a^{up}\left(\boxed{x_{ij}}\right) = \genfrac{}{}{}{}{\widetilde{Q}}{P} = \boxed{\widetilde{x}_{ij}},$$

$$s_a^{down}\left(\boxed{x_{ij}}\right) = \genfrac{}{}{}{}{Q}{\widetilde{P}} = \boxed{\widetilde{\widetilde{x}}_{ij}},$$

where $\widetilde{P} = s_aP$ and $\widetilde{Q} = s_aQ$ stand for the images of triangles P and Q under the involution s_a, $1 \le a \le n-1$.

Now we are ready to state our result:

Theorem 4.27 (Tropical version of Theorem 2.28) *For any a, $1 \le a \le n-1$, the following diagrams of birational maps are commutative*

$$\begin{array}{ccc} \mathbb{A}^{n^2} & \xrightarrow{RSK} & \mathbb{A}^{n^2} \\ \sigma_a \downarrow & & \downarrow s_a^{up} \\ \mathbb{A}^{n^2} & \xrightarrow{RSK} & \mathbb{A}^{n^2} \end{array}, \qquad \begin{array}{ccc} \mathbb{A}^{n^2} & \xrightarrow{RSK} & \mathbb{A}^{n^2} \\ \widetilde{\sigma}_a \downarrow & & \downarrow s_a^{down} \\ \mathbb{A}^{n^2} & \xrightarrow{RSK} & \mathbb{A}^{n^2} \end{array}.$$

Corollary 4.28 (Tropical version of "conserved quantities for Ball–Box systems") *If* $1 \le i \le j \le n$, *then Laurent's polynomials* $z_{ji}(x(\mathbf{d}))$, *given by formulae (4.18) and (4.19), are invariant with respect to the action of symmetric group* $S_n = \langle \sigma_1, \ldots, \sigma_{n-1} \rangle$.

We conclude this Section by stating the result which is a tropical version of Proposition 2.26, item 3. To this end, let $n \ge 2$ be an integer and $\mathbf{d} = (d_{ij})$ be n by n matrix, where we treat the entries d_{ij} of matrix d, $1 \le i, j \le n$, as the set of independent variables. For these data in Section 4.3 we introduced Laurent's polynomials

$$h_a^{(l)}(\mathbf{d}) = \sum_{k=1}^{n} \prod_{j=k}^{n} d_{a+1,j+l} \prod_{k+1}^{n} d_{a,j+l}^{-1}, \quad 1 \le a \le n. \tag{4.22}$$

Note that the latter Laurent polynomial coincides with the tropification of the row exponent $h_{a,+}^{(l)}(d)$ of matrix $d \in M_{n \times n}(\mathbb{R})$. To state our result we need a couple of notation. First of all, for a given n by n matrix $x = (x_{ij})$ we denote by x^+ the following triangle:

$$x^+ = \begin{array}{ccccccccc} x_{nn} & & x_{n-1,n-1} & & \cdots & x_{22} & & x_{11} \\ & x_{n-1,n} & & & \cdots & & x_{12} & \\ & & & & \vdots & & & \\ & & & x_{2,n} & & x_{1,n-1} & & \\ & & & & x_{1,n} & & & \end{array}$$

Note that the triangle x^+ coincides with that Q from Theorem 4.27.

Finally, let us consider the birational map $RSK : \mathbb{A}^{n^2} \to \mathbb{A}^{n^2}$ defined by formulae (4.18) and (4.19), see Theorem 4.23.

Theorem 4.29 *Let* $\mathbf{d} = (d_{ij})$ *and* $x = (x_{ij})$, $1 \le i, j \le n$, *be two sets of variables connected by the birational map RSK:* $x = RSK(\mathbf{d})$. *Then for any integer* a, $1 \le a \le n-1$, *we have*

$$h_{a,+}^{(0)}(\mathbf{d}) = Q_{aa}(y(x^+)). \tag{4.23}$$

Example 4.30 Let us take $n = 3$, and $\mathbf{d} = (d_{ij})$ and $x = (x_{ij})$ be 3 by 3 matrices. Then

$$h_{1,+}^{(0)}(\mathbf{d}) = d_{23} + d_{22}d_{23}/d_{13} + d_{21}d_{22}d_{23}/d_{12}d_{13},$$

$$h_{2,+}^{(0)}(\mathbf{d}) = d_{33} + d_{32}d_{33}/d_{23} + d_{31}d_{32}d_{33}/d_{22}d_{23},$$

$$h_{3,+}^{(0)}(\mathbf{d}) = d_{13} + d_{12}d_{13}/d_{33} + d_{11}d_{12}d_{13}/d_{32}d_{33},$$

$$x^+ = \begin{array}{ccccc} x_{33} & & x_{22} & & x_{11} \\ & x_{23} & & x_{12} & \\ & & x_{13} & & \end{array}, \quad y^+ = \begin{array}{ccccc} x_{33} & & x_{33}x_{22} & & x_{11}x_{22}x_{33} \\ & x_{23} & & x_{12}x_{23} & \\ & & x_{13} & & \end{array},$$

$$Q_{11}(y^+) = y^+_{23}/y^+_{13} = x_{23}/x_{13},$$
$$Q_{22}(y^+) = y^+_{33}/y^+_{23} + y^+_{13}y^+_{22}/y^+_{23}y^+_{12} = x_{33}/x_{23} + x_{13}x_{22}x_{33}/x_{12}x^2_{23}.$$

Now we are going to use our assumption that $x = RSK(\mathbf{d})$. In the case $n = 3$ the corresponding formulae can be found in Appendix 4, Subsection 2, item ii). First of all, it is easy to see from these formulae that

$$Q_{11}(y^+) = x_{23}(d)/x_{13}(d) = h^{(0)}_{1,+}(d).$$

The second equality, namely

$$Q_{22}(y^+) = x_{33}(d)/x_{23}(d) + x_{13}(d)x_{22}(d)x_{33}(d)/x_{12}(d)x^2_{23}(d) = h^{(0)}_{2,+}(\mathbf{d})$$

is more difficult to check, and we leave details to the reader.

4.6 Tropical image of the inverse Robinson–Schensted–Knuth transformation

Given natural numbers n, i and j such that $1 \leq i < j \leq n$, denote by $SS^{(n)}_{ij}$ the set of all tableau T of rectangular shape $(\underbrace{i, \dots, i}_{j})$ filled by integer numbers from the interval $[i + 1; 2n - j]$ such that

• part of the tableau T filled by numbers from the interval $[i + 1; n]$ is a columns strict one;

• part of the tableau T filled by numbers from the interval $[n + 1, 2n - j]$ is a rows strict one.

For any tableau $T \in SS^{(n)}_{\min(n-j,i),j}$ define its weight $wt(T)$ to be

$$wt(T) = \prod_{\substack{s = (k,l) \\ 1 \leq k \leq j,\ 1 \leq l \leq \min(i, n-j) \\ T(s) \leq n}} \frac{t_{T(s)-k, \max(n-i,j)+l-k+1}}{t_{T(s)-k, \max(n-i,j)+l-k}}$$

$$\cdot \prod_{\substack{s = (k,l);\ 1 \leq k \leq j, \\ 1 \leq l \leq \min(i, n-j) \\ T(s) \geq n+1}} \frac{t_{\max(n-i,j)+l-k+1, \max(n-i,j)+l-k+n+1-T(s)}}{t_{\max(n-i,j)+l-k, \max(n-i,j)+l-k+n+1-T(s)}}$$

$$\cdot \prod_{k \geq 0} t_{i+j-n-k, j-k}.$$

Here $t = (t_{ij})_{1\le i,j\le n}$ denotes the set of variables and we assume that $t_{ij} = 1$ if either $i \le 0$ or $j \le 0$, so that the index k in the third product in the above formula belongs to the interval $[0, \max(i+j-n-1, 0)]$, which is non–empty iff $n < i+j \le 2n$.

Theorem 4.31 (cf. [4]) (Tropical version of the inverse Robinson–Schensted–Knuth correspondence) *The inverse birational transformation $d = RSK^{-1}(t)$ to that given by formulae (4.20) and (4.21) can be described as follows:*

- *if* $1 \le i \le j \le n$, *then*

$$\prod_{1\le k\le i,\ 1\le l\le j} d_{kl}(t) = \sum_{T\in SS^{(n)}_{\min(n-j,i),j}} wt(T);$$

- *if* $1 \le i \le j \le n$, *then*

$$\prod_{1\le k\le j,\ 1\le l\le i} d_{kl}(t) = \prod_{1\le k\le i,\ 1\le l\le j} d_{kl}((t_{ij})^t).$$

4.7 Tropical image of cocharge

Follow [24] denote by $\mathbb{P}_n \subset K_n \cap \mathbb{Z}^{n(n+1)/2}$ the subset of the set of GT–patterns K_n consisting of all triangles $\mathbf{p} = (p_{ij}) \in X_n \cap \mathbb{Z}^{n(n+1)/2}_{\ge 0}$ such that

i) $p_{nn} = p_{n-1,n-1} = 0$;
ii) $p_{i,n} - 1 \le p_{ii} \le p_{i+1,n},\ \ 1 \le i \le n-1$;
iii) $p_{ij} - p_{i+1,j+1} \le 1,\ \ 1 \le i \le j \le n-1$;
iv) $p_{ij} - p_{i+1,j} \le 1,\ \ 1 \le i < j \le n$.

It is easy to see that for each i, $1 \le i \le n-1$, there exists a unique index k_i, $0 \le k_i \le n-i$ such that

$$p_{in} = \cdots = p_{i,n-k_i} > p_{i,n-k_i-1} = \cdots = p_{ii}.$$

We will denote this index k_i by $k_i(\mathbf{p})$. For example, $k_i(\mathbf{p}) = n-i$ if and only if $p_{in} = p_{i,n-1} = \cdots = p_{ii}$. It is easy to see that the correspondence

$$\mathbb{P} \to (k_1(p), \ldots, k_{n-1}(p))$$

defines a bijection between the set $\mathbb{P}_n$ and the set of all $(n-1)$–tuples of integers $(k_1, \ldots, k_{n-1})$ such that $0 \le k_i \le n-i$ for all $i = 1, \ldots, n-1$. In particular, $|\mathbb{P}_n| = n!$.

For any $\mathbf{p} \in \mathbb{P}_n$ and any triangle $y \in X_{n+1}$ let us define $\mathbf{p}$–*weight* of y, denoted by $wt_{\mathbf{p}}(y)$, as follows

$$wt_{\mathbf{p}}(y) = \prod_{2\le j\le n} \left(\frac{y_{1,j+1}y_{j-1,j}y_{j+1,j+1}}{y_{1,j}y_{j,j}y_{j,j+1}}\right)^{\overline{p}_{1j}} \prod_{2\le i\le j\le n} \left(\frac{y_{i-1,j-1}y_{i-2,j}y_{i,j+1}}{y_{i-2,j-1}y_{i,j}y_{i-1,j+1}}\right)^{\overline{p}_{ij}},$$

where $\overline{p}_{ij} := p_{n+1-j,n+1-i}$, and by definition $y_{0j} = 1$, if $j \geq 1$.

Finally, follow [24], formulae (3.5)–(3.7), for any triangle $y \in X_{n+1}$ we define Laurent polynomials $\widetilde{c}_{n+1}(y)$ and $\overline{c}_{n+1}(y)$:

$$\widetilde{c}_{n+1}(y) = \left\{ \sum_{\mathbf{p} \in \mathbb{P}_n} wt_{\mathbf{p}}(y) \right\} \prod_{j=1}^{n-1} \frac{y_{j,n}}{y_{j,n+1}}, \quad \text{if} \quad n \geq 2; \tag{4.24}$$

$$\widetilde{c}_1(y) = \widetilde{c}_2(y) = 1,$$

$$\overline{c}_{n+1}(y) = \prod_{j=2}^{n+1} \widetilde{c}_j(y), \quad \text{if} \quad n \geq 2.$$

Theorem 4.32 *Let $y \in X_{n+1}$ be a triangle, then*

- $$\widetilde{c}_{n+1}(y) = \sum_{k=0}^{n-2} \left\{ \sum_{\substack{\mathbf{p} \in \mathbb{P}_{n-1} \\ k_1(\mathbf{p}) = k}} \epsilon_n^{p_{1,n-1}} wt_{\mathbf{p}}(y) \right\} Q_{k+1,n}(y) \cdot \frac{y_{1,n} \cdots y_{n-1,n}}{y_{1,n+1} \cdots y_{n,n+1}} \frac{y_{n-1,n} y_{n+1,n+1}}{y_{n,n}}, \tag{4.25}$$

where $\epsilon_n = \dfrac{y_{n-1,n-1} y_{n+1,n+1}}{y_{n,n}^2}$*;*

- *the cocharge function $\overline{c}_{n+1}(y)$ is an invariant with respect to the action of the affine symmetric group $W(A_n^{(1)})$. In other words, for any i, $0 \leq i \leq n$,*

$$\overline{c}_{n+1}(s_i(y)) = \overline{c}_{n+1}(y);$$

- (stability) *assume that the first row $(y_{k,n+1})_{k=1}^{n+1}$ of a triangle $y \in X_{n+1}$ is such that $y_{k,n+1} = y_{k,n}$ if $1 \leq k \leq n$, and $y_{n+1,n+1} = y_{n,n}$, then*

$$CT[\widetilde{c}_{n+1}(y)] = 1;$$

- *the ultra–discretization of the function $\overline{c}_{n+1}$ being restricted to the set of semistandard Young tableaux $STY(\lambda, \leq n+1)$ coincides with the cocharge statistics $\overline{c}$ introduced by A. Lascoux and M.-P. Schützenberger.*

Here for any Laurent polynomial

$$f(x) \in \mathbb{C}[x_1^{\pm 1}, \ldots, x_N^{\pm 1}],$$

the symbol $CT[f(x)]$ denotes the constant term of $f(x)$.

Let us remark that Laurent's polynomial $\overline{c}_{n+1}(y)$, $y \in X_{n+1}$, contains exactly $n!? = 1! \cdot 2! \cdots n!$ non–zero monomials, if $n \geq 2$. Note also, that cardinality of the set $\{\mathbf{p} \in \mathbb{P}_n \mid k_1(\mathbf{p}) = k\}$ is equal to $(n-1)!$ for all integer k between 0 and $n-1$.

Our proof of Theorem 4.32 and, in particular, the proof of statement that the function cocharge $\overline{c}_{n+1}(y)$ is *invariant* under the action of the affine Weyl group $W(A_n^{(1)})$, is based on two results which we are going to state now. One result is an explicit expression for the sum in the braces of formula (4.25), and another one is *inductive* formulae for the function $\widetilde{c}_{\bullet}(y)$.

Theorem 4.33 *Given integers $n \geq 4$ and k, $0 \leq k \leq n-2$, and a triangle $y \in X_{n+1}$, then*

$$\sum_{\substack{\mathbf{p} \in \mathbb{P}_{n-1} \\ \kappa_1(\mathbf{p}) = k}} \epsilon_n^{p_{1,n-1}} wt_{\mathbf{p}}(y) = \widetilde{c}_{n-1}(y^{[k]}) \frac{\beta_{n-1}(y)}{\beta_{n-2}(y)} \frac{y_{k+1,n} y_{k,n-2}}{y_{n-1,n} y_{k,n-1} y_{k+1,n-1}} \cdot \left(\frac{y_{n-2,n-2} y_{n+1,n+1}}{y_{n-1,n-1} y_{n,n}} \right)^{n-2-k},$$

where

- *the triangle $y^{[k]} \in X_{n-1}$, $0 \leq k \leq n-2$, is defined as follows:*

$$(y^{[k]})_{ij} = y_{ij}, \quad \text{if } 1 \leq i \leq j < n-1;$$

$$(y^{[k]})_{i,n-1} = \begin{cases} y_{i,n-1}, & \text{if } i \leq k \\ y_{i,n-1} \frac{y_{n-2,n-2} y_{n+1,n+1}}{y_{n-1,n-1} y_{n,n}}, & \text{if } k < i \leq n-1; \end{cases}$$

- *for any integer k, $1 \leq k \leq n+1$, and a triangle $y \in X_{n+1}$ we set by definition $\beta_k(y) = y_{1k} y_{2k} \cdots y_{kk}$.*

Theorem 4.34 *Let $n \geq 2$ be an integer, $y = (y_{ij} | 1 \leq i \leq j \leq n+1)$ be the set of variables and $\widetilde{c}_{n+1}(y)$ be Laurent's polynomial defined by formula (4.24) Then*

$$\widetilde{c}_{n+1}(y) = \widetilde{c}_n(s_n(y)) \left\{ \frac{\widetilde{Q}_{n,n}(y)}{y_{n+1,n+1}} \prod_{j=1}^{n} \frac{\widetilde{Q}_{j,n}(y)}{\widetilde{Q}_{n,n}(y)} \prod_{j=1}^{n-1} \frac{y_{n-1,n-1}}{y_{j,n-1}} \prod_{j=1}^{n} \frac{y_{n+1,n+1}}{y_{j,n+1}} \right\}, \tag{4.26}$$

where $\widetilde{Q}_{j,n}(y) = \dfrac{y_{j,n}^2}{y_{n,n}} Q_{j,n}(y)$, $1 \leq j \leq n$.

Note that according to Corollary 4.6, the Laurent polynomials $\widetilde{Q}_{j,n}(y)$, $1 \leq j \leq n$, are invariant under the action of involution s_n. Therefore the product in the braces of the RHS(4.26), is s_n–invariant as well.

Corollary 4.35 (Inductive formula) *Let $n \geq 2$ be an integer, then*

$$i) \quad \widetilde{c}_{n+1}(y) = \widetilde{c}_n(s_n(y)) \left\{ \frac{y_{n+1,n+1} Q_{nn}(y) \beta_n(y)}{\beta_{n-1}(y) \beta_{n+1}(y)} \prod_{k=1}^{n-1} (s_n(y))_{kn} \right\}, \tag{4.27}$$

where as before $\beta_k(y)$ *denotes the product* $\beta_k(y) := y_{1k} \cdots y_{kk}$;

$$ii) \quad \widetilde{c}_{n+1}(y) = \left\{ \sum_{k=0}^{n-2} \widetilde{c}_{n-1}(y^{[k]}) \frac{y_{k+1,n} y_{k,n-2}}{y_{k,n-1} y_{k+1,n}} \xi_n^{n-2-k} Q_{k+1,n}(y) \right\} \tag{4.28}$$

$$\cdot \frac{\beta_{n-1}(y)\beta_n(y)}{\beta_{n-2}(y)\beta_{n+1}(y)} \left(\frac{y_{n+1,n+1}}{y_{n,n}} \right)^2,$$

where $\xi_n = \dfrac{y_{n-2,n-2} y_{n+1,n+1}}{y_{n-1,n-1} y_{n,n}} = s_{n-1}(s_n(y_{nn})/y_{nn})$;

$$iii) \quad \overline{c}_{n+1}(y) = \frac{\overline{c}_n(y)\overline{c}_n(s_n(y))}{\overline{c}_{n-1}(y)} \left\{ \frac{y_{n+1,n+1} Q_{n,n}(y)\beta_n(y)}{\beta_{n-1}(y)\beta_{n+1}(y)} \prod_{k=1}^{n-1} (s_n(y))_{k,n} \right\}. \tag{4.29}$$

Corollary 4.36 *Let* $n \geq 2$ *be an integer, then*

1) *Laurent's polynomial* $\widetilde{c}_{n+1}(y)\widetilde{c}_n(y)$ *is* s_n*–invariant;*

2) *Laurent's polynomial* $\widetilde{c}_n(y)$ *is* s_j*–invariant for all* j, $1 \leq j \leq n-2$.

At the end of this Section we are going to study in more details the function cocharge $\overline{c}_{n+1}(y)$ in the case when the triangle y has weight zero, i.e.

$$y \in X_{n+1}(0) = \{ y \in X_{n+1} | y_{kk} = y_{11}^k, \ 1 \leq k \leq n+1 \}.$$

First of all let us recall, see Remark 4.7, 2), that if the triangle $y \in X_{n+1}$ satisfies the following condition:

$$y_{k+1,k+1} y_{k-1,k-1} = y_{k,k}^2$$

for some k, $1 \leq k \leq n$, then the involution s_k acts on the triangle y trivially, i.e. $s_k(y) = y$. In this case the formula (4.26) takes the form:

$$\widetilde{c}_{k+1}(y) = \widetilde{c}_k(y) \frac{y_{k+1,k+1}}{y_{k,k}} \frac{Q_{k,k}(y)\beta_k^2(y)}{\beta_{k-1}(y)\beta_{k+1}(y)}. \tag{4.30}$$

Furthermore, if $y \in X_{n+1}(0)$, then the equality (4.30) holds for all k, $1 \leq k \leq n$, and therefore

$$\widetilde{c}_{n+1}(y) = \prod_{j=1}^{n} Q_{j,j}(y) \frac{y_{n+1,n+1}\beta_n(y)}{\beta_{n+1}(y)}.$$

From the latter equality one can deduce the following expression for the value of function cocharge $\overline{c}_{n+1}(y)$ on a zero weight triangle $y \in X_{n+1}(0)$:

$$\overline{c}_{n+1}(y) = \left\{ \prod_{j=1}^{n} (Q_{j,j}(y))^{n+1-j} \right\} \frac{y_{11}^{(n+1)(n+2)/2}}{\beta_{n+1}(y)}. \tag{4.31}$$

The formula (4.31) is the tropical version or *tropification*, of the following well–known formula for the cocharge of a zero–weight tableau $T \in STY(\lambda, (l^n))$:

$$\overline{c}(T) = \sum_{j=1}^{n-1} (\lambda_j - \kappa_j(T))(n-j) - l\binom{n}{2},$$

or in equivalent form for the value of statistics *charge*

$$c_n(T) = \sum_{j=1}^{n-1} (n-j)\kappa_j(T),$$

where $\lambda = (\lambda_1, \lambda_2, \ldots, \lambda_n)$ is the shape of tableau T, and $l = |\lambda|/n$, see e.g., A. Lascoux, B. Leclerc and J.-Y. Thibon [28], or A.N. Kirillov, Journ. Geom. and Phys. **5** (1988), 365-389.

4.8 Tropical image of charge

Let $n \geq 2$ be an integer. Recall that statistics *charge* $c_n(T)$ on the set of dominant weight semistandard Young tableaux $T \in STY_+(\leq n)$ was discovered and invented by A. Lascoux and M.P.-Schützenberger [29], who had used it to give a combinatorial description of the Kostka–Foulkes polynomials $K_{\lambda\mu}(q)$. On the set of tableaux $STY_+(\leq n)$ statistics *charge* c_n and *cocharge* $\overline{c}_n$ are connected by the following relation:

for partitions λ and μ, and tableau $T \in STY(\lambda, \mu)$ one has

$$c_n(T) + \overline{c}_n(T) = n(\mu) - n(\lambda),$$

where for any partition ν, $n(\nu) = \sum_{j \geq 1} (i-1)\nu_i$.

It seems natural to ask: how one can define a tropical version of statistics *charge* ? We know, see Theorem 4.27, that the tropical version of statistics *cocharge* $\overline{c}_n$ is $W(A_{n-1})$–invariant. We would like to keep this property for tropical version of statistics charge as well. Also we would like to have for the former an analog of formula (4.31). Keeping in mind these restrictions, we are coming to the following

Definition 4.37 *i) The function tropical index ind(y) on the space of triangles X_n is defined to be*

$$ind(y) = \prod_{k=1}^{n-1} (Q_{kk}(y))^{n-k}; \tag{4.32}$$

ii) the function tropical charge $c_n(y)$ on the space of triangles X_n is defined to be

$$c_n(y) = \sum_{w \in S_n} ind(w(y)). \tag{4.33}$$

Theorem 4.38 *The ultradiscritization of the function tropical charge c_n, see (4.33), being restricted to the set of dominant weight semistandard Young tableaux $STY_+(\leq n)$, coincides with statistics charge introduced by A. Lascoux and M.P.-Schützenberger* [29].

4.9 Energy function and charge

Let $\mathbf{a} = (a_1, \ldots, a_n)$ and $\mathbf{b} = (b_1, \ldots, b_n)$ be two n–tuples of real numbers, follow [19,10], (6.1), define "the local energy of states" **a** and **b** to be

$$H(\mathbf{a}, \mathbf{b}) = \max\{h_i(\mathbf{a}, \mathbf{b})\}_{i=1}^{n},$$

where $h_i = a_1 + \cdots + a_i - b_1 - \cdots - b_{i-1} = a_1 + \cdots + a_i + b_i + \cdots + b_n - |\mathbf{b}|$. We assume here that $b_0 = 0$, and by definition $|\mathbf{b}| = b_1 + \cdots + b_n$.

Now for any matrix $\mathbf{d} \in M_{n\times n}(\mathbb{R})$ we define its *energy* $E(\mathbf{d})$ to be [19]

$$E(\mathbf{d}) = \sum_{j=1}^{n-1} (n-j) H(\mathbf{d}^{(j)}, \mathbf{d}^{(j+1)}), \tag{4.34}$$

where $\mathbf{d}^{(j)} = (d_{j1}, \ldots, d_{jn})$, $1 \leq j \leq n$.

Theorem 4.39 *Assume that for all i, $1 \leq i \leq n$, we have $\sum_{j=1}^{n} d_{ij} = l$, and*

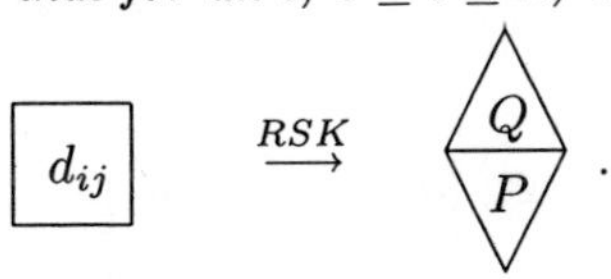

Then

i) $E(\mathbf{d}) = c_n(Q)$, where for any triangle $P \in X_n$, $c_n(Q)$ denotes its charge, i.e.

$$c_n(Q) = l\binom{n}{2} - \sum_i (n-i)wt_i(Q) - \overline{c}_n(Q),$$

and $\overline{c}_n(Q)$ stands for the cocharge of Q as it was defined in Section 2.9.

ii) More precisely, if $1 \leq j \leq n-1$, then

$$H(\mathbf{d}^{(j)}, \mathbf{d}^{(j+1)}) = \kappa_j(Q),$$

where $\kappa_j(Q)$ denotes the j-th exponent of triangle Q.

Let us remark that in the case under consideration if $1 \leq j \leq n-1$, then

$$H(d^{(j)}, d^{(j+1)}) = h_{j,+}^{(0)}(d),$$

where $h_{j,+}^{(0)}(d)$ denotes the j-th row exponent of matrix d, see Section 2.10.

To treat the general case, let us define for a matrix $d \in M_{n\times n}(\mathbb{R})$ and integer j between 1 and $n-1$, the "local energy" functions

$$\widehat{E}_j^{\pm}(d) = \begin{matrix}\max \\ \min\end{matrix} \left\{ H((\sigma d)^{(j)}, (\sigma d)^{(j+1)}) | \sigma \in S_n = \langle \sigma_1, \ldots, \sigma_{n-1} \rangle \right\},$$

and the energy functions

$$E_n^{\pm}(d) := \sum_{j=1}^{n-1} (n-j) \widehat{E}_j^{\pm}(d).$$

Denote also by

$$\mathcal{E}_n^{\pm}(d) = (E_1^{\pm}(d), \ldots, E_{n-1}^{\pm}(d))$$

the vectors of local energy functions, and

$$h_{\pm}^{(0)} = (h_{1,\pm}^{(0)}(d), \ldots, h_{n-1,\pm}^{(0)}(d))$$

the vectors of exponents.

Theorem 4.40 *For any matrix $d \in M_{n\times n}(\mathbb{R})$ there exist permutations*

$$\sigma^{\pm} \in S_n := \langle \sigma_1, \ldots, \sigma_{n-1} \rangle$$

such that

$$\mathcal{E}_n^{\pm}(d) = h_{\pm}^{(0)}(\sigma(d)).$$

Corollary 4.41 (cf. [35]) *Let $d \in M_{n\times n}(\mathbb{R})$, and* $\boxed{d_{ij}} \xrightarrow{RSK} \begin{smallmatrix} Q \\ P \end{smallmatrix}$, *then*

$$E_n^{\pm}(d) = c_n^{\pm}(Q),$$

in other words the energy of matrix $d \in M_n$ is equal to the charge of the corresponding "recording triangle" Q.

We conclude this Section by stating some interesting property of the energy function E. To this end, let us define an involution $I_n : X_n \to X_n$ by setting for any triangle $x \in X_n$

$$(I_n(x))_{ij} = x_{1n} - x_{j+1-i,j}, \quad 1 \leq i \leq j \leq n.$$

Proposition 4.42 *If $x \in X_n$, then*
i) $\overline{c}_n(I_n(x)) = \overline{c}_n(x)$, $\forall x \in X_n$;

ii) assume that $\boxed{d_{ij}} \xrightarrow{RSK} \genfrac{}{}{}{}{Q}{P}$, *and for all* i, $1 \le i \le n$, *we have* $\sum_{j=1}^{n} d_{ij} = l$, *then*

$$E(\mathbf{d}) = ind(I_n(Q)),$$

where by definition $ind(Q) = \sum_{j=1}^{n} (n-j)\kappa_j(Q) := ind(\mathbf{d})$.

Acknowledgments

This notes grew out of series of lectures delivered at the Workshop "Combinatorial methods in representation theory", RIMS, Kyoto University (July 1998), organized by S. Okada et al., at the Workshop "Representation Theory and Integrable Systems", Okinawa (December 1998), organized by M. Okado et al., at the Erwin Schrödinger International Institute for Mathematical Physics, Vienna (July 2000), and at the International Workshop "Physics and Combinatorics" (August 2000), Graduate School of Mathematics, Nagoya University.

I would like to thank those who attended for helpful comments, suggestions and support. I would like to thank Professors V.G. Kac and A.A. Kirillov for their hospitality during my visit to ESI, Vienna.

I wish to thank with much gratitude Dr. N.A. Liskova for inestimable help and support on all stages of the paper writing.

Appendix: *Explicit formulae for small n*

1. Tropical image of Bender–Knuth's involutions t_n, $n = 2, 3$:

$$t_2 \begin{pmatrix} y_{11} & y_{12} & y_{13} \\ 0 & y_{22} & y_{23} \\ 0 & 0 & y_{33} \end{pmatrix} = \begin{pmatrix} y_{11}, & \dfrac{y_{11}y_{13} + y_{23}}{y_{12}}, & y_{13} \\ 0, & \dfrac{y_{11}y_{33}}{y_{22}}, & y_{23} \\ 0, & 0, & y_{33} \end{pmatrix},$$

$$t_3\begin{pmatrix} y_{11} & y_{12} & y_{13} & y_{14} \\ 0 & y_{22} & y_{23} & y_{24} \\ 0 & 0 & y_{33} & y_{34} \\ 0 & 0 & 0 & y_{44} \end{pmatrix} = \begin{pmatrix} y_{11}, & y_{12}, & \dfrac{y_{12}y_{14}+y_{24}}{y_{13}}, & y_{14} \\ 0, & y_{22}, & \dfrac{y_{12}y_{34}+y_{22}y_{24}}{y_{23}}, & y_{24} \\ 0, & 0, & \dfrac{y_{22}y_{44}}{y_{33}}, & y_{34} \\ 0, & 0, & 0, & y_{44} \end{pmatrix}$$

2. Tropical image of involution s_n, $n = 2, 3$:

$$s_2\begin{pmatrix} y_{11} & y_{12} & y_{13} \\ 0 & y_{22} & y_{23} \\ 0 & 0 & y_{33} \end{pmatrix} = \begin{pmatrix} y_{11}, & \dfrac{y_{11}y_{12}(y_{13}y_{33}+y_{22}y_{23})}{y_{22}(y_{11}y_{23}+y_{13}y_{22})}, & y_{13} \\ 0, & \dfrac{y_{11}y_{33}}{y_{22}}, & y_{23} \\ 0, & 0, & y_{33} \end{pmatrix};$$

$$s_3\begin{pmatrix} y_{11} & y_{12} & y_{13} & y_{14} \\ 0 & y_{22} & y_{23} & y_{24} \\ 0 & 0 & y_{33} & y_{34} \\ 0 & 0 & 0 & y_{44} \end{pmatrix}$$

$$= \begin{pmatrix} y_{11}, & y_{12}, & \dfrac{y_{13}(y_{33}^2y_{24}y_{12}+y_{22}y_{14}y_{44}y_{23}+y_{33}y_{34}y_{22}y_{13})}{y_{33}(y_{33}y_{14}y_{23}+y_{33}y_{24}y_{12}+y_{34}y_{22}y_{13})}, & y_{14} \\ 0, & y_{22}, & \dfrac{y_{23}(y_{44}y_{14}y_{23}+y_{33}y_{13}y_{34}+y_{44}y_{24}y_{12})y_{22}}{y_{33}(y_{33}y_{14}y_{23}+y_{33}y_{24}y_{12}+y_{34}y_{22}y_{13})}, & y_{24} \\ 0, & 0, & \dfrac{y_{44}y_{22}}{y_{33}}, & y_{34} \\ 0, & 0, & 0, & y_{44} \end{pmatrix}$$

For the reader's convenience we list below explicit formulae for the Laurent polynomials $Q_{kn}(y)$, $1 \le k \le n$, $n = 2, 3$:

$$Q_{12}(y) = \frac{y_{13}}{y_{12}} + \frac{y_{22}y_{23}}{y_{12}y_{33}}, \quad Q_{22}(y) = \frac{y_{13}}{y_{12}} + \frac{y_{11}y_{23}}{y_{12}y_{22}},$$

$$Q_{13}(y) = \frac{y_{14}}{y_{13}} + \frac{y_{12}y_{24}y_{33}^2}{y_{13}y_{22}y_{23}y_{44}} + \frac{y_{33}y_{34}}{y_{23}y_{44}},$$

$$Q_{23}(y) = \frac{y_{14}}{y_{13}} + \frac{y_{12}y_{24}}{y_{13}y_{23}} + \frac{y_{33}y_{34}}{y_{23}y_{44}},$$

$$Q_{33}(y) = \frac{y_{14}}{y_{13}} + \frac{y_{12}y_{24}}{y_{13}y_{23}} + \frac{y_{22}y_{34}}{y_{23}y_{33}}.$$

3. Tropical image of Schützenberger involution q_n, $n = 2, 3$:

$$q_2 \begin{pmatrix} y_{11} & y_{12} & y_{13} \\ 0 & y_{22} & y_{23} \\ 0 & 0 & y_{33} \end{pmatrix} = \begin{pmatrix} \frac{y_{33}}{y_{22}}, & \frac{y_{11}y_{23} + y_{13}y_{22}}{y_{11}y_{12}}, & y_{13} \\ 0, & \frac{y_{33}}{y_{11}}, & y_{23} \\ 0, & 0, & y_{33} \end{pmatrix};$$

$$q_3 \begin{pmatrix} y_{11} & y_{12} & y_{13} & y_{14} \\ 0 & y_{22} & y_{23} & y_{24} \\ 0 & 0 & y_{33} & y_{34} \\ 0 & 0 & 0 & y_{44} \end{pmatrix} = \begin{pmatrix} \frac{y_{44}}{y_{33}}, & y_{12}^{(4)}, & y_{13}^{(4)}, & y_{14} \\ 0, & \frac{y_{44}}{y_{22}}, & y_{23}^{(4)}, & y_{24} \\ 0, & 0, & \frac{y_{44}}{y_{11}}, & y_{34} \\ 0, & 0, & 0, & y_{44} \end{pmatrix},$$

where

$$\begin{aligned} y_{12}^{(4)} &= \frac{y_{12}y_{24}y_{33} + y_{14}y_{23}y_{33} + y_{13}y_{22}y_{34}}{y_{13}y_{22}y_{23}} \\ &= d_{33}d_{23}d_{24}d_{14} + d_{33}d_{23}d_{13}d_{14} + d_{33}d_{34}d_{24}d_{14}; \end{aligned}$$

$$\begin{aligned} y_{13}^{(4)} &= \frac{y_{11}y_{14}y_{23} + y_{13}y_{14}y_{22} + y_{11}y_{12}y_{24}}{y_{11}y_{12}y_{13}} \\ &= d_{22}d_{23}d_{13}d_{14} + d_{22}d_{12}d_{13}d_{14} + d_{22}d_{23}d_{24}d_{14}; \end{aligned}$$

$$y_{23}^{(4)} = \frac{y_{11}y_{23}y_{34} + y_{13}y_{22}y_{34} + y_{12}y_{24}y_{33}}{y_{11}y_{12}y_{23}} = d_{22}d_{23}d_{13}d_{14}d_{33}d_{34}d_{24}$$

$$+\, d_{22}d_{12}d_{13}d_{14}d_{33}d_{34}d_{24} + d_{22}d_{12}d_{13}d_{14}d_{33}d_{23}d_{24}.$$

4. Tropical image of the Robinson–Schensted–Knuth correspondence and its inverse.

1) $n = 2$

i) (min, +)–version

$$t_{11} = d_{12} + d_{21}, \quad t_{12} = d_{11}d_{12}, \quad t_{21} = d_{11}d_{21}, \quad t_{11}t_{22} = d_{11}d_{12}d_{21}d_{22}.$$

• Inverse Robinson–Schensted–Knuth correspondence

$$d_{11} = \frac{t_{12}}{t_{11}} + \frac{t_{21}}{t_{11}}, \quad d_{11}d_{12} = t_{12}, \quad d_{11}d_{21} = t_{21}, \quad d_{22} = \frac{t_{22}}{t_{12}} + \frac{t_{22}}{t_{21}}.$$

ii) (max, +)–version

$$x_{11}x_{22} = d_{11}d_{12}d_{21}d_{22}, \quad x_{12} = d_{11}d_{12}, \quad x_{21} = d_{11}d_{21},$$
$$x_{22} = d_{11}d_{22}(d_{12} + d_{21});$$

• Inverse Robinson–Schensted–Knuth correspondence

$$d_{11}^{-1} = \frac{x_{11}}{x_{12}} + \frac{x_{11}}{x_{21}}, \quad d_{11}d_{12} = x_{12}, \quad d_{11}d_{21} = x_{21},$$
$$d_{12}d_{22} = \frac{x_{11}x_{22}}{x_{21}}, \quad d_{21}d_{22} = \frac{x_{11}x_{22}}{x_{12}}, \quad d_{22}^{-1} = \frac{x_{12}}{x_{22}} + \frac{x_{21}}{x_{22}}.$$

2) $n = 3$

i) (min, +)–version

$$t_{11} = d_{13} + d_{22} + d_{31},$$
$$t_{12} = d_{12}d_{13} + d_{13}d_{21} + d_{21}d_{22} = L_{3,3}^{(1)}(\mathbf{d}),$$
$$t_{21} = d_{21}d_{31} + d_{12}d_{31} + d_{12}d_{22} = L_{3,3}^{(1)}(\mathbf{d}^t),$$
$$t_{13} = d_{11}d_{12}d_{13}, \quad t_{31} = d_{11}d_{21}d_{31},$$
$$t_{11}t_{22} = d_{12}d_{13}d_{22}d_{23} + d_{12}d_{13}d_{23}d_{31} + d_{12}d_{13}d_{31}d_{32} + d_{13}d_{21}d_{23}d_{31}$$
$$+\, d_{13}d_{21}d_{31}d_{32} + d_{21}d_{22}d_{31}d_{32} = t_{12}d_{31}d_{32} + t_{21}d_{13}d_{23},$$
$$t_{12}t_{23} = d_{11}d_{12}d_{13}d_{21}d_{22}d_{23}, \quad t_{21}t_{32} = d_{11}d_{21}d_{31}d_{12}d_{22}d_{32},$$
$$t_{11}t_{22}t_{33} = d_{11}d_{12}d_{13}d_{21}d_{22}d_{23}d_{31}d_{32}d_{33};$$

• Inverse Robinson–Schensted–Knuth correspondence:

$$
\begin{aligned}
d_{11} &= \frac{t_{13}}{t_{12}} + \frac{t_{31}}{t_{21}} + \frac{t_{23}}{t_{22}} + \frac{t_{32}}{t_{22}}, \\
d_{11}d_{12} &= \frac{t_{13}}{t_{11}} + \frac{t_{13}t_{21}}{t_{11}t_{12}} + \frac{t_{21}t_{23}}{t_{11}t_{22}} + \frac{t_{21}t_{32}}{t_{11}t_{22}}, \\
d_{11}d_{21} &= \frac{t_{31}}{t_{11}} + \frac{t_{31}t_{12}}{t_{11}t_{21}} + \frac{t_{12}t_{32}}{t_{11}t_{22}} + \frac{t_{12}t_{23}}{t_{11}t_{22}}, \\
& d_{11}d_{12}d_{13} = t_{13}, \quad d_{11}d_{21}d_{31} = t_{31}, \\
d_{11}d_{12}d_{21}d_{22} &= \frac{t_{12}t_{21}t_{23}}{t_{11}t_{22}} + \frac{t_{12}t_{21}t_{32}}{t_{11}t_{22}}, \\
& d_{11}d_{12}d_{13}d_{21}d_{22}d_{23} = t_{12}t_{23}, \quad d_{11}d_{12}d_{21}d_{22}d_{31}d_{32} = t_{21}t_{32}, \\
& d_{11}d_{12}d_{13}d_{21}d_{22}d_{23}d_{31}d_{32}d_{33} = t_{11}t_{22}t_{33}.
\end{aligned}
$$

ii) (max, +)–version

$$
\begin{aligned}
x_{11}x_{22}x_{33} &= d_{11}d_{12}d_{13}d_{21}d_{22}d_{23}d_{31}d_{32}d_{33}, \\
x_{12}x_{23} &= d_{11}d_{12}d_{13}d_{21}d_{22}d_{23}, \\
x_{21}x_{32} &= d_{11}d_{21}d_{31}d_{12}d_{22}d_{32}, \\
x_{13} &= d_{11}d_{12}d_{13}, \\
x_{31} &= d_{11}d_{21}d_{31}, \\
x_{22}x_{33} &= d_{11}d_{12}d_{13}d_{21}d_{22}d_{23}d_{31}d_{32}d_{33}(1/d_{13} + 1/d_{22} + 1/d_{31}), \\
x_{23} &= d_{11}d_{12}d_{13}d_{23} + d_{11}d_{12}d_{22}d_{23} + d_{11}d_{21}d_{22}d_{23}, \\
x_{32} &= d_{11}d_{21}d_{31}d_{32} + d_{11}d_{21}d_{22}d_{32} + d_{11}d_{12}d_{22}d_{32}, \\
x_{33} &= d_{11}d_{21}d_{31}d_{32}d_{33} + d_{11}d_{21}d_{22}d_{32}d_{33} + d_{11}d_{21}d_{22}d_{23}d_{33} \\
&\quad + d_{11}d_{12}d_{22}d_{32}d_{33} + d_{11}d_{12}d_{22}d_{23}d_{33} + d_{11}d_{12}d_{13}d_{23}d_{33} \\
&= (x_{23} + x_{32})d_{33}.
\end{aligned}
$$

• Inverse Robinson–Schensted–Knuth correspondence

$$
\begin{aligned}
d_{11}^{-1} &= \frac{x_{12}}{x_{13}} + \frac{x_{21}}{x_{31}} + \frac{x_{22}}{x_{23}} + \frac{x_{22}}{x_{32}}, \\
d_{11}^{-1}d_{12}^{-1} &= \frac{x_{11}}{x_{13}} + \frac{x_{11}x_{12}}{x_{13}x_{21}} + \frac{x_{11}x_{22}}{x_{21}x_{23}} + \frac{x_{11}x_{22}}{x_{21}x_{32}}, \\
d_{11}^{-1}d_{21}^{-1} &= \frac{x_{11}}{x_{31}} + \frac{x_{11}x_{21}}{x_{31}x_{12}} + \frac{x_{11}x_{22}}{x_{12}x_{32}} + \frac{x_{11}x_{22}}{x_{12}x_{23}}, \\
& d_{11}d_{12}d_{13} = x_{13}, \quad d_{11}d_{21}d_{31} = x_{31}, \\
d_{11}^{-1}d_{12}^{-1}d_{21}^{-1}d_{22}^{-1} &= \frac{x_{11}x_{22}}{x_{12}x_{21}x_{23}} + \frac{x_{11}x_{22}}{x_{12}x_{21}x_{32}}, \\
& d_{11}d_{12}d_{13}d_{21}d_{22}d_{23} = x_{12}x_{23}, \quad d_{11}d_{21}d_{31}d_{12}d_{22}d_{32} = x_{21}x_{32},
\end{aligned}
$$

$$d_{11}d_{12}d_{13}d_{21}d_{22}d_{23}d_{31}d_{32}d_{33} = x_{11}x_{22}x_{33}.$$

It is not difficult to see that the (min, +)–version and the (max, +)–version of *tropical* image of the inverse Robinson–Schensted–Knuth correspondence are connected by the following change of variables: $d_{ij} \to d_{ij}^{-1}$, $t_{ij} \to x_{ij}^{-1}$ for all admissible i and j.

5. Tropical image of involutions $\sigma_a := \sigma_{a,n}$, $1 \le a \le n$, and $n = 2, 3$.
1) In the case $n = 2$ one has $h_a^{(l)} := h_a^{(l)}(\mathbf{d}) = d_{a+1,l}(d_{a+1,l+1} + d_{a,l})/d_{a,l}$, and

$$\sigma_1(d_{11}) = d_{21}h_1^0/h_1^1 = \frac{d_{11}d_{22}(d_{12}+d_{21})}{d_{12}(d_{11}+d_{22})},$$
$$\sigma_1(d_{12}) = d_{22}h_1^1/h_1^0 = \frac{d_{12}d_{21}(d_{11}+d_{22})}{d_{11}(d_{12}+d_{21})},$$
$$\sigma_1(d_{21}) = d_{11}h_1^1/h_1^0 = \frac{d_{12}d_{21}(d_{11}+d_{22})}{d_{22}(d_{12}+d_{21})},$$
$$\sigma_1(d_{22}) = d_{12}h_1^0/h_1^1 = \frac{d_{11}d_{22}(d_{12}+d_{21})}{d_{21}(d_{11}+d_{22})};$$
$$\sigma_2 = \sigma_1.$$

Note that

$$\sigma_1(d_{12}^{-1} + d_{21}^{-1}) = d_{12}^{-1} + d_{21}^{-1},$$
$$\sigma_1(d_{11}^{-1} + d_{22}^{-1}) = d_{11}^{-1} + d_{22}^{-1}.$$

2) In the case $n = 3$ one has

$$h_a^{(l)} := h_a^{(l)}(\mathbf{d}) = \frac{d_{a+1,l}(d_{a,l}d_{a,l+2} + d_{a,l+2}d_{a+1,l+2} + d_{a+1,l+1}d_{a+1,l+2})}{d_{a,l}d_{a,l+2}},$$

$$\sigma_a(d_{a,j}) = d_{a+1,j}h_a^{(j-1)}/h_a^{(j)},$$
$$\sigma_a(d_{a+1,j}) = d_{a,j}h_a^{(j)}/h_a^{(j-1)},$$

and

$$\sigma_a(d_{a,j}) = \frac{d_{aj}d_{a+1,j-1}(d_{a,j-1}d_{a,j+1} + d_{a,j+1}d_{a+1,j+1} + d_{a+1,j}d_{a+1,j+1})}{d_{a,j+1}(d_{a,j}d_{a,j+2} + d_{a,j+2}d_{a+1,j+2} + d_{a+1,j+1}d_{a+1,j+2})}.$$

Note that

$$\sigma_1(d_{13}^{-1} + d_{22}^{-1}) = d_{13}^{-1} + d_{22}^{-1},$$
$$\sigma_1(d_{12}^{-1} + d_{21}^{-1}) = d_{12}^{-1} + d_{21}^{-1},$$
$$\sigma_1(d_{11}^{-1} + d_{23}^{-1}) = d_{11}^{-1} + d_{23}^{-1}.$$

See Proposition 4.13 for general statement.

6. Tropical image of cocharge $\overline{c}_{n+1}$, $n = 2, 3$.

- $\overline{c}_3(y) = \dfrac{y_{12}}{y_{13}} + \dfrac{y_{12}y_{33}}{y_{22}y_{23}}$, if $y = \begin{pmatrix} y_{13} & & y_{23} & & y_{33} \\ & y_{12} & & y_{22} & \\ & & y_{11} & & \end{pmatrix}$.

In other words, $\overline{c}_3(x) = \min(x_{13} - x_{12}, x_{22} - x_{33})$, if $x \in X_3$.

- $$\overline{c}_4(y) = \frac{y_{13}y_{23}y_{44}}{y_{14}y_{33}y_{34}} \left(\frac{y_{12}}{y_{13}} + \frac{y_{12}y_{33}}{y_{22}y_{23}} \right) \left(\frac{y_{12}}{y_{13}} + \frac{y_{12}y_{33}}{y_{22}y_{23}} + \frac{y_{13}y_{34}}{y_{23}y_{24}} + \frac{y_{14}y_{23}}{y_{13}y_{24}} + \frac{y_{33}y_{34}}{y_{24}y_{44}} + \frac{y_{14}y_{44}}{y_{24}y_{33}} \right).$$

In other words, if $x \in X_4$, then [24]

$$\overline{c}_4(x) = \min(x_{13} - x_{12}, x_{22} - x_{33}) + x_{14} - x_{13} + x_{33} - x_{44}$$
$$+ \min(x_{23} - x_{34}, x_{24} - x_{23}, x_{13} - x_{12}, x_{22} - x_{33}, \beta_4(x) - x_{34}, x_{24} - \beta_4(x)),$$

where $(\beta_1(x), \beta_2(x), \beta_3(x), \beta_4(x)) = wt(x)$ denotes *weight* of triangle x.

As an illustration of the formula (4.25), let us compute the function $\widetilde{c}_5(y)$. First of all, the set $\mathbb{P}_3$ contains 6 integral GT–patterns:

$$\mathbb{P}_3 = \left\{ \begin{matrix} 1 & & 1 & & 0 \\ & 1 & & 0 & \\ & & 0 & & \end{matrix}, \begin{matrix} 1 & & 0 & & 0 \\ & 1 & & 0 & \\ & & 0 & & \end{matrix}, \begin{matrix} 1 & & 0 & & 0 \\ & 0 & & 0 & \\ & & 0 & & \end{matrix}, \begin{matrix} 2 & & 1 & & 0 \\ & 1 & & 0 & \\ & & 1 & & \end{matrix}, \begin{matrix} 1 & & 1 & & 0 \\ & 1 & & 0 & \\ & & 1 & & \end{matrix}, \begin{matrix} 0 & & 0 & & 0 \\ & 0 & & 0 & \\ & & 0 & & \end{matrix} \right\}.$$

$k_1(\mathbf{p})$	1	1	0	0	2	2

As the second step, let us compute the corresponding weights $wt_{\mathbf{p}}(y)$. By definition, if $\mathbf{p} \in \mathbb{P}_3$, is a GT–pattern, then

$$wt_{\mathbf{p}}(y) = \left(\frac{y_{13}y_{33}}{y_{22}y_{23}} \right)^{p_{23}} \left(\frac{y_{14}y_{23}y_{44}}{y_{13}y_{33}y_{34}} \right)^{p_{13}} \left(\frac{y_{11}y_{23}}{y_{13}y_{22}} \right)^{p_{22}}$$
$$\cdot \left(\frac{y_{12}y_{24}}{y_{14}y_{23}} \right)^{p_{12}} \left(\frac{y_{13}y_{22}y_{34}}{y_{12}y_{24}y_{33}} \right)^{p_{11}}.$$

In particular one has

$$wt_{\substack{1\,1\,0\\1\,0\\0}}(y)=\frac{y_{12}y_{24}y_{33}y_{44}}{y_{22}y_{23}y_{33}y_{34}};\qquad wt_{\substack{1\,0\,0\\1\,0\\0}}(y)=\frac{y_{12}y_{24}y_{44}}{y_{13}y_{33}y_{34}};$$

$$wt_{\substack{1\,0\,0\\0\,0\\0}}(y)=\frac{y_{14}y_{23}y_{44}}{y_{13}y_{33}y_{34}};\qquad wt_{\substack{2\,1\,0\\1\,0\\1}}(y)=\frac{y_{14}y_{44}^2}{y_{33}^2y_{34}};$$

$$wt_{\substack{1\,1\,0\\1\,0\\1}}(y)=\frac{y_{13}y_{44}}{y_{23}y_{33}};\qquad wt_{\substack{0\,0\,0\\0\,0\\0}}(y)=1.$$

As the third step, let us compute the sums

$$B_k:=\left\{\sum_{\mathbf{p}\in\mathbb{P}_3,\ k_1(\mathbf{p})=k}\epsilon^{p_{1,n-1}}wt_{\mathbf{p}}(y)\right\}\frac{y_{14}y_{24}y_{34}}{y_{15}y_{25}y_{35}y_{45}}\frac{y_{34}y_{55}}{y_{44}y_{k+1,4}},$$

for $k=0,1,2.$

After simple calculations one can find

$$B_0=\left(\frac{y_{23}}{y_{13}}+\frac{y_{55}}{y_{44}}\right)\frac{y_{14}y_{24}y_{34}y_{55}^2}{y_{15}y_{25}y_{35}y_{45}y_{44}^2};$$

$$B_1=\left(\frac{y_{12}}{y_{13}}+\frac{y_{12}y_{33}}{y_{22}y_{23}}\right)\frac{y_{14}y_{24}y_{34}y_{55}^2}{y_{15}y_{25}y_{35}y_{45}y_{44}^2};$$

$$B_2=\left(\frac{y_{44}}{y_{55}}+\frac{y_{13}}{y_{23}}\right)\frac{y_{14}y_{24}y_{34}y_{55}^2}{y_{15}y_{25}y_{35}y_{45}y_{44}^2}.$$

Finally, we obtain the following formula for the function $\widetilde{c}_5(y)$:

$$\widetilde{c}_5(y)=\frac{y_{14}y_{24}y_{34}y_{55}^2}{y_{15}y_{25}y_{35}y_{45}y_{44}^2}\left\{\left(\frac{y_{23}}{y_{13}}+\frac{y_{55}}{y_{44}}\right)(y_{14}Q_{14}(y))\right.$$
$$\left.+\left(\frac{y_{12}}{y_{13}}+\frac{y_{12}y_{33}}{y_{22}y_{23}}\right)(y_{24}Q_{24}(y))+\left(\frac{y_{44}}{y_{55}}+\frac{y_{13}}{y_{23}}\right)(y_{34}Q_{34}(y)\right\},$$

where

$$Q_{14}(y)=\frac{y_{15}}{y_{14}}+\frac{y_{13}y_{44}^2y_{25}}{y_{14}y_{24}y_{33}y_{55}}+\frac{y_{23}y_{35}y_{44}^2}{y_{24}y_{33}y_{34}y_{55}}+\frac{y_{44}y_{45}}{y_{34}y_{55}},$$

$$Q_{24}(y)=\frac{y_{15}}{y_{14}}+\frac{y_{13}y_{25}}{y_{14}y_{24}}+\frac{y_{23}y_{44}^2y_{35}}{y_{24}y_{33}y_{34}y_{55}}+\frac{y_{44}y_{45}}{y_{34}y_{55}},$$

$$Q_{34}(y)=\frac{y_{15}}{y_{14}}+\frac{y_{13}y_{25}}{y_{14}y_{24}}+\frac{y_{23}y_{35}}{y_{24}y_{34}}+\frac{y_{44}y_{45}}{y_{34}y_{55}},$$

$$Q_{44}(y)=\frac{y_{15}}{y_{14}}+\frac{y_{13}y_{25}}{y_{14}y_{24}}+\frac{y_{23}y_{35}}{y_{24}y_{34}}+\frac{y_{33}y_{45}}{y_{34}y_{44}}.$$

Recall, that Laurent's polynomials $Q_{kn}(y) = Q_{kn}^{(\eta)}(y)$, where $\eta = y_{nn}^2/y_{n-1,n-1}y_{n+1,n+1}$ and $1 \le k \le n$, describe the action of the symmetric group S_{n+1} on the space of triangles X_{n+1}:

$$(s_n(y))_{kn} = y_{kn}\frac{Q_{kn}(y)}{Q_{nn}(y)} \cdot \frac{y_{n-1,n-1}y_{n+1,n+1}}{y_{n,n}^2},$$

see Corollary 4.3.

For the reader's convenience and for references we give below the explicit formula for the cocharge $\overline{c}_5(x)$ of a triangle $x \in X_5$ which is obtained from the latter formula for $\widetilde{c}_5(y)$ by means of the ultradiscretization map:

$$\begin{aligned}
\overline{c}_5(x) = \overline{c}_4(x) &+ 2x_{15} - 2x_{14} + x_{25} - x_{24} + x_{34} - x_{45} + x_{44} - x_{55} \\
&+ \min\{x_{13} - x_{12}, x_{22} - x_{33}, x_{23} - x_{34}, x_{24} - x_{23}, \\
&\quad x_{22} - x_{34} + x_{45} - x_{44}, \\
&\quad x_{13} - x_{12} + x_{33} - x_{34} + x_{45} - x_{44}, \\
&\quad x_{13} - x_{12} + x_{14} - x_{15} + x_{44} - x_{55}, \\
&\quad x_{14} - x_{13} + x_{24} - x_{23} + x_{45} - x_{34}, \\
&\quad x_{14} - x_{13} + x_{35} - 2x_{34} + x_{45}, \\
&\quad x_{14} - x_{15} + x_{22} - x_{33} + x_{44} - x_{55}, \\
&\quad x_{14} - x_{15} + x_{23} - x_{34} + x_{44} - x_{55}, \\
&\quad x_{14} - x_{15} + x_{24} - x_{23} + x_{44} - x_{55}, \\
&\quad x_{24} - x_{23} + x_{33} - x_{34} + x_{45} - x_{44}, \\
&\quad x_{25} - x_{24} + x_{35} - 2x_{34} + x_{45}, \\
&\quad x_{33} - 2x_{34} + x_{35} - x_{44} + x_{45}, \\
&\quad 2x_{14} - x_{12} + x_{24} - x_{15} + x_{44} - x_{25} - x_{55}, \\
&\quad x_{14} - x_{15} + x_{23} - x_{25} + x_{24} - x_{34} + x_{33} - x_{55}, \\
&\quad x_{14} - x_{15} + 2x_{24} - x_{25} + x_{33} - x_{35} - x_{55}, \\
&\quad 2x_{14} - x_{13} + x_{22} - x_{15} + x_{24} - x_{25} + x_{44} - x_{33} - x_{55}, \\
&\quad 2x_{14} - x_{13} + x_{23} - x_{15} + x_{24} - x_{25} + x_{44} - x_{34} - x_{35}, \\
&\quad 2x_{14} - x_{13} + 2x_{24} - x_{15} + x_{44} - x_{25} - x_{35} - x_{55}, \\
&\quad x_{14} - x_{15} + 2x_{24} - x_{25} + x_{34} - x_{35} + x_{44} - x_{45} - x_{55}, \\
&\quad x_{15} - x_{14} + x_{25} - x_{24} + x_{35} - 2x_{34} + x_{45} - x_{44} + x_{55},
\end{aligned}$$

$$2x_{14} - 2x_{15} + 2x_{24} - x_{25} + x_{34} - x_{35} + 2x_{44} - x_{45} - 2x_{55}\}.$$

It is very interesting problem to describe explicitly the linearity domains ("fan") of the continuous piece–wise linear function $\bar{c}_n(x)$ on the space of triangles $X_n \simeq \mathbb{R}^{n(n+1)/2}$, and investigate the corresponding toric variety.

References

1. Baccelli F., Cohen G., Olsder G.J. and Quadrat S.-P., *Synchronization and linearity; an algebra for discrete event systems,* Wiley Series in Probability and Mathematical Statistics, John Wiley & Sons, Chichester, 1992;
2. Bender A. and Knuth D., *Enumeration of plain partitions,* Journ. Comb. Theory, ser. A **13** (1972), 40-54;
3. Berenstein A.D., Fomin S.V. and Zelevinsky A.V., *Parametrizations of canonical bases and totally positive matrices,* Adv. Math. **122** (1996), 49-149;
4. Berenstein A.D. and Kirillov A.N., *The Robinson–Schensted–Knuth bijection, quantum matrices and piece–wise linear combinatorics,* in preparation;
5. Berenstein A.D. and Zelevinsky A.V., *Tensor product multiplicities and convex polytopes in partition space,* J. Geom. Phys. **5** (1988), 453-472;
6. Berenstein A.D. and Zelevinsky A.V., *Triple multiplicities for $sl(r+1)$ and the spectrum of the exterior algebra of the adjoint representation,* Journ. Algebraic Comb. **1** (1992), 7-22.
7. Berenstein A.D. and Zelevinsky A.V., *Canonical bases for the quantum group of type A_r and piece–wise linear combinatorics,* Duke Math. Journ. **82** (1996), 473-502;
8. Donin I.F., *Multiplicities of S_n–modules and the index and charge of tableaux,* Funct. Anal. Appl. **27** (1993), no. 4, 280-282;
9. Eilenberg S., *Automata, languages and machines,* vol.A and vol.B, Academic Press, New York, 1974, 1976;
10. Frenkel E. and Szenes A., *Crystal bases, dilogarithm identities and torsion in algebraic K–groups,* Journ. Amer. Math. Soc. **8** (1995), no. 3, 629-664;
11. Fulton W., *Young Tableaux,* Cambridge Univ. Press, Cambridge, 1997.
12. Fulton W. and Harris G., *Representation theory,* Graduate Texts in Math., vol 129, Springer–Verlag, 1999;
13. Gansner E.R., *On the equality of two plane partition correspondences,* Discrete Math. **30** (1980), 121-132;

14. Gelfand I.M. and Zelevinsky A.V., *Polyhedra in a space of diagrams and the canonical basis in irreducible representations of* $\mathfrak{gl}_n$, Funkt. Analis i Prilozhen. **19** (1985), No. 2, 72-75 (in Russian);
15. Gelfand I.M. and Zelevinsky A.V., *Multiplicities and proper bases for* $\mathfrak{gl}_n$, Group Theoretical Methods in Physics (Yurmala, 1985), vol.2, Nauka, Moscow, 1986, pp.22-45; English translation: VNU Sci. Press, Utrecht, 1986, pp.127-159;
16. Gessel I. and Viennot G., *Binomial determinants, paths and hook length formulae,* Adv. Math. **58** (1985), 300-321;
17. Hatayama G., Hikami K., Inoue R., Kuniba A., Takagi T. and Tokihiro T., *The* $A_M^{(1)}$ *automata related to crystals of symmetric tensors,* Preprint math.QA/9912209, 45 p.;
18. Golan J.S., *Semirings and their applications,* Kluwer Academic Publ., Dordrecht, 1999;
19. Jimbo M., Misra K.C., Miwa T. and Okado M., *Combinatorics of representations of* $U_q(\widehat{sl}(n))$ *at* $q = 0$, Commun. Math. Phys. **136** (1991), 543-566;
20. Kashiwara M., *On crystal bases of the q–analogue of the universal enveloping algebra,* Duke Math. J. **63** (1991), 465-516;
21. Kashiwara M. and Nakashima T., *Crystal graphs for representations of the q–analog of classical Lie algebras,* J. Algebra **165** (1994), 295-345;
22. Kirillov A.N., *Ubiquity of Kostka polynomials,* Preprint math.QA/ 9912094, 104p.;
23. Kirillov A.N., *Discrete bilinear Hirota equations, Schützenberger involutions and rational representations of Symmetric group,* Lectures given in Okinawa, December 1998, in preparation, 1998, 20p.;
24. Kirillov A.N. and Berenstein A.D., *Groups generated by involutions, Gelfand–Tsetlin patterns, and combinatorics of Young tableaux,* St. Petersburg Math. J. **7** (1996), no.1, 77-127;
25. Knight H. and Zelevinsky A.V., *Representations of quivers of type A and the multisegment duality,* Adv. Math. **117** (1996), 273-293;
26. Knuth D., *Permutations, matrices, and generalized Young tableaux,* Pacific Journ. Math. **34** (1970), 709-727;
27. Krattenthaler C., *The major counting of nonintersecting lattice paths and generating functions for tableaux,* Memoirs of AMS **115** (1995);
28. Lascoux A., Leclerc B. and J.-Y. Thibon, *Polynômes de Kostka–Foulkes et graphes cristallins des groupes quantiques de type* A_n, C.R. Acad. Sci. Paris **320** (1995), 131-134;
29. Lascoux A. and Schützenberger M.-P., *Sur une conjecture de H.O. Foulkes,* C. R. Acad. Sci. Paris **286A** (1978), 323-324;

30. Lascoux A. and Schützenberger M.-P., *La monoid plaxique,* in "Non-commutative structures in Algebra and Geometric Combinatorics", ROMA, CNR, 1981, 129-156;
31. Lusztig G., *Total positivity in reductive groups,* in: Lie theory and geometry, Progr. Math. **123**, Birkhäuser, Boston, MA, 1994, pp.531-568;
32. Lusztig G., *Total positivity and canonical bases,* in: Algebraic groups and Lie groups, Austral. Math. Soc. Lect. Ser. **9**, Cambridge Univ. Press, Cambridge, 1997, pp.281-295;
33. Lusztig G., *Introduction to Quantum groups,* Progress in Math. **110**, Birkhäuser, Boston. MA, 1993;
34. Macdonald I.G., *Symmetric functions and Hall polynomials,* 2nd ed., Oxford, 1995;
35. Nakayashiki A. and Yamada Y., *Kostka polynomials and energy functions in solvable lattice models*, Selecta Math., New Ser. **3** (1997), 547-599;
36. Noumi M. and Yamada Y., *Affine Weyl groups, discrete dynamical systems and Painlevé equations,* Comm. Math. Phys. **199** (1998), 281-295;
37. Pin J.-E., *Tropical semirings,* Idempotency Analysis (J. Grunawardena ed.), Publ. Newton Inst. **11**, Cambridge Univ. Press, Cambridge, 1998, 50-69;
38. Schützenberger M.-P., *Promotion des morphisms d'ensembles ordonnés,* Discr. Math. **2** (1972), 73-94;
39. Takahashi D. and Satsuma J., *A soliton celluar automaton,* Journ. of the Physical Society of Japan **59** (1990), 3514-3519;
40. Tokihiro T., Takahashi D. and Matsukidaira J., *Box and ball system as a realization of ultradiscrete nonautonomous KP equation,* J. Phys. A **33** (2000), 607-619;
41. Wolfram S., ed. Theory and applications of cellular automata, World Scientific, Singapore, 1986;
42. Zelevinsky A.V., *A generalization of the Littlewood–Richardson rule and the Robinson–Schensted–Knuth correspondence,* Journ. of Algebra **69** (1981), 82-94.

BETHE'S STATES FOR GENERALIZED XXX AND XXZ MODELS

ANATOL N. KIRILLOV & NADEJDA LISKOVA
Graduate School of Mathematics, Nagoya University
Chikusa–ku, Nagoya 486–8602, Japan
and
Steklov Mathematical Institute at St.Petersburg
Fontanka 27, St.Petersburg 191011, Russia
E-mail: kirillov@math.nagoya-u.ac.jp

For any rational number $p_0 \geq 1$ we prove an identity of Rogers–Ramanujan-Gordon–Andrews' type. Bijection between the space of states for XXZ model and that of XXX model is constructed.

1 Introduction

The main goal of our paper is to study a combinatorial relationship between the space of states for generalized XXZ model and that for XXX one. In our previous paper [4] we gave a combinatorial description of states for generalized XXZ model in terms of the so–called $sl(2)$–XXZ rigged configurations. On the other hand it is well–known that when the anisotropy parameter p_0 of XXZ model goes to infinity then the XXZ model under consideration transforms to the XXX one. We are going to describe this transformation from combinatorial point of view in the case when p_0 is an integer.

A combinatorial completeness of Bethe's states for generalized XXX–model had been proved in [1] and appeared to be a starting point for numerous applications to combinatorics of Young tableaux and representation theory of symmetric and general linear groups, see e.g. [2]. Here we mention only a "fermionic" formula for the Kostka–Foulkes polynomials, see e.g. [2], and the relationship of the latter with $\widehat{sl}(2)$–branching functions $b_\lambda^{k\Lambda_0}(q)$, see e.g. [3]. We will show in Section 1, Theorem 2.3 and Remark 2.5, that q–counting of the number of XXZ states using Bethe's ansatz approach [5,6], gives rise to the Rogers–Ramanujan–Gordon–Andrews' type identity for any rational number $p_0 \geq 1$.

It seems an interesting problem to find a polynomial version of the Rogers–Ramanujan type identity (2.12) from our Theorem 2.3.

Another question which we are interested in is to understand a combinatorial nature of the limit

$$XXZ \underset{p_0 \to +\infty}{\longrightarrow} XXX.$$

In Section 3 we describe a combinatorial rule which shows how the XXZ–configurations fall to the XXX pieces. For simplicity we consider in our paper only the case $p_0 > \sum_m s_m$. General case will be considered elsewhere.

2 Rogers–Ramanujan's type identity

This paper is a continuation of our previous work [4]. Let us remind the main definitions, notation and results from [4].

For fixed $p_0 \in \mathbf{R}$, $p_0 \geq 1$ let us define (cf. [5]) a sequence of real numbers p_i and sequences of integer numbers ν_i, m_i, y_i, z_i:

$$p_0 := p_0,\ p_1 = 1,\ \nu_i = \left[\frac{p_i}{p_{i+1}}\right],\ p_{i+1} = p_{i-1} - \nu_{i-1}p_i,\ i = 1, 2, \ldots \quad (2.1)$$

$$y_{-1} = 0,\ y_0 = 1,\ \ y_1 = \nu_0,\ y_{i+1} = y_{i-1} + \nu_i y_i,\ i = 0, 1, 2, \ldots \quad (2.2)$$

$$z_{-1} = 0,\ z_0 = 1,\ z_1 = \nu_1,\ z_{i+1} = z_{i-1} + \nu_{i+1} z_i,\ i = 0, 1, 2, \ldots \quad (2.3)$$

$$m_0 = 0,\ m_1 = \nu_0,\ m_{i+1} = m_i + \nu_i,\ i = 0, 1, 2, \ldots \quad (2.4)$$

$$r(j) = i,\ \text{if } m_i \leq j < m_{i+1},\ j = 0, 1, 2, \ldots \quad (2.5)$$

It is clear that integer numbers ν_i define the decomposition of p_0 into continuous fraction

$$p_0 = [\nu_0, \nu_1, \nu_2, \ldots] = \nu_0 + \cfrac{1}{\nu_1 + \cfrac{1}{\nu_2 + \ldots}}.$$

Let us define (see Fig. 1) a piecewise linear function n_j, $j \geq 0$,

$$n_j := y_{i-1} + (j - m_i)y_i, \quad \text{if} \quad m_i \leq j < m_{i+1}. \quad (2.6)$$

It is clear that for any integer $n > 1$ there exists the unique rational number t such that $n = n_t$.

Let us introduce additionally the following functions (see [4])

$$q_j = (-1)^i(p_i - (j - m_i)p_{i+1}), \quad \text{if} \quad m_i \leq j < m_{i+1}, \quad (2.7)$$

$$\Phi_{k,2s} = \begin{cases} \dfrac{1}{2p_0}(q_k - q_k n_\chi), & \text{if } n_k > 2s, \\ \dfrac{1}{2p_0}(q_k - q_\chi n_k) + \dfrac{(-1)^{r(k)-1}}{2}, & \text{if } n_k \leq 2s, \end{cases}$$

where $2s = n_\chi - 1$.

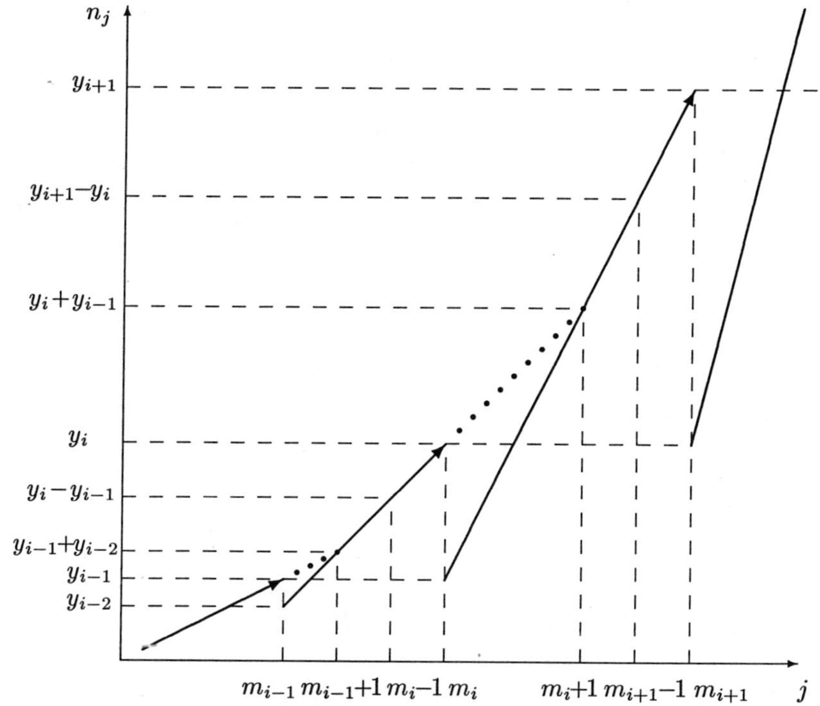

Fig.1. Image of piecewise linear function n_j in the interval $[m_{i-1}, m_{i+1}]$

In what follows we assume that the anisotropy parameter $p_0 \geq 1$ is a rational number, and $p_0 = [\nu_0, \nu_1, \ldots, \nu_\alpha]$ denotes its decomposition into continuous fraction. It is not difficult to see that

$$p_0 = [\nu_0, \nu_1, \ldots, \nu_\alpha] = \frac{y_{\alpha+1}}{z_\alpha},$$
$$\bar{p}_0 := [\nu_0, \nu_1, \ldots, \nu_{\alpha-1}] = \frac{y_\alpha}{z_{\alpha-1}},$$

where the numbers $\{y_j\}_{j=0}^{\alpha+1}$ and $\{z_j\}_{j=0}^{\alpha}$ are defined by (2.2) and (2.3) correspondingly. We assume that if $\alpha > 0$, then $\nu_\alpha \geq 2$.

In order to formulate our main result of the paper [4] about the number of Bethe's states for generalized XXZ model, let us consider the following symmetric matrix $\Theta^{-1} = (c_{ij})_{1 \leq i,j \leq m_{\alpha+1}}$:

$$i)\ c_{ij} = c_{ji} \text{ and } c_{ij} = 0, \text{ if } |i - j| \geq 2.$$

$$ii) \quad c_{j-1,j} = (-1)^{i-1}, \quad \text{if } \ m_i \le j < m_{i+1}.$$

$$iii) \quad c_{jj} = \begin{cases} 2(-1)^i, & \text{if } m_i \le j < m_{i+1} - 1, \ \ i \le \alpha, \\ (-1)^i, & \text{if } j = m_{i+1} - 1, \ \ i \le \alpha, \\ (-1)^{\alpha+1}, & \text{if } j = m_{\alpha+1}. \end{cases}$$

Example 1 Let us take $p_0 = \dfrac{16}{7}$, then $p_0 = [2,3,2]$, $\alpha = 2$, $\bar{p} = [2,3] = \dfrac{7}{3}$,

$$\begin{aligned} &\nu_0 = 2, \ \ \nu_1 = 3, \ \ \nu_2 = 2; \\ &m_0 = 0, \ \ m_1 = 2, \ \ m_2 = 5, \ \ m_3 = 7; \\ &y_0 = 1, \ \ y_1 = 2, \ \ y_2 = 7, \ \ y_3 = 16; \\ &z_0 = 1, \ \ z_1 = 3, \ \ z_2 = 7. \end{aligned}$$

Therefore, $p_0 = \dfrac{y_{\alpha+1}}{z_\alpha}$, $\bar{p}_0 = \dfrac{y_\alpha}{z_{\alpha-1}}$, and

$$n_j = \begin{cases} j, & \text{if } 0 \le j < 2, \\ 1 + 2(j-2), & \text{if } 2 \le j < 5, \\ 2 + 7(j-5), & \text{if } 5 \le j < 7, \\ 7 + 16(j-7), & \text{if } 7 \le j. \end{cases}$$

Finally,

$$\Theta = \frac{1}{16}\begin{pmatrix} 9 & 7 & 5 & 3 & 2 & 1 & 1 \\ 7 & -7 & -5 & -3 & -2 & -1 & -1 \\ 5 & -5 & -15 & -9 & -6 & -3 & -3 \\ 3 & -3 & -9 & -15 & -10 & -5 & -5 \\ 2 & -2 & -6 & -10 & 4 & 2 & 2 \\ 1 & -1 & -3 & -5 & 2 & 9 & 9 \\ 1 & -1 & -3 & -5 & 2 & 9 & -7 \end{pmatrix}, \quad \Theta^{-1} = \begin{pmatrix} 1 & 1 & & & & & \\ 1 & -2 & 1 & & & & \\ & 1 & -2 & 1 & & & \\ & & 1 & -1 & -1 & & \\ & & & -1 & 2 & -1 & \\ & & & & -1 & 1 & 1 \\ & & & & & 1 & -1 \end{pmatrix},$$

and $\det|\Theta^{-1}| = 16$.

Note, see [4], Theorem 4.7, that the absolute value of the determinant $\det(\Theta^{-1})$ is equal to $y_{\alpha+1}$, i.e. the numerator of p_0.

To continue, let us consider the matrix $E = (e_{jk})_{1 \le j,k \le m_{\alpha+1}}$, where

$$e_{jk} = (-1)^{r(k)}(\delta_{j,k} - \delta_{j,m_{\alpha+1}-1} \cdot \delta_{k,m_{\alpha+1}} + \delta_{j,m_{\alpha+1}} \cdot \delta_{k,m_{\alpha+1}-1}).$$

One can check that the vacancy numbers $P_j(\lambda)$, see [4], (3.9), can be computed as follows

$$P_j(\lambda)+\lambda_j=((E-2\Theta)\widetilde{\lambda}^t+b^t)_j,\quad 1\le j\le m_{\alpha+1},$$

where the j-th component of vector $b=(b_1,\dots,b_{m_{\alpha+1}})$ is defined by the following formula

$$b_j=(-1)^{r(j)}\left(n_j\left\{\frac{\sum 2s_mN_m-2l}{p_0}\right\}-\sum_m 2\Phi_{j,2s_m}\cdot N_m\right),$$

and for any sequence of integer numbers $\lambda=(\lambda_1,\dots,\lambda_{m_{\alpha+1}})$ we denote by $\widetilde{\lambda}$ the sequence $(\widetilde{\lambda}_1,\dots,\widetilde{\lambda}_{m_{\alpha+1}})$, where $\widetilde{\lambda}_j=(-1)^{r(j)}\lambda_j$, $1\le j\le m_{\alpha+1}$.

Theorem 2.1 ([4]) *The number of Bethe's states $Z^{XXZ}(N,s\mid l)$ of the generalized XXZ model is equal to*

$$\sum_\lambda\prod_j\binom{((E-B)\widetilde{\lambda}^t+b^t)_j}{\lambda_j},\tag{2.8}$$

where summation is taken over all sequences of non–negative integer numbers $\lambda=\{\lambda_k\}_{k=1}^{m_{\alpha+1}}$ such that

$$\sum_{k=1}^{m_{\alpha+1}} n_k\lambda_k=l,\quad \lambda_k\ge 0;$$

$$\widetilde{\lambda}=(\widetilde{\lambda}_1,\dots\widetilde{\lambda}_{m_{\alpha+1}}),\quad \widetilde{\lambda}_j=(-1)^{r(j)}\lambda_j,\quad B=2\Theta.$$

Recall, see [4], or Section 3, that N and s in the above formula for the number of states $Z^{XXZ}(N,s|l)$ denote vectors $N=(N_1,\dots,N_k)$ and

$$s=(\underbrace{s_1,\dots,s_1}_{N_1},\dots,\underbrace{s_k,\dots,s_k}_{N_k}),$$

i.e. N_m is equal to the number of spins in the XXZ–chain which are equal to s_m.

One of the main goal of the present paper is to consider a natural q–analog for (2.8). Namely, let us define the following q–analog of the sum (2.8)

$$\sum_\lambda q^{\frac12\widetilde{\lambda}B\widetilde{\lambda}^t}\prod_j\begin{bmatrix}((E-B)\widetilde{\lambda}^t+b^t)_j\\ \lambda_j\end{bmatrix}_{q^{\epsilon_j}},\tag{2.9}$$

where $\quad\epsilon_j=(-1)^{r(j)}$.

Let us recall that $\begin{bmatrix} M \\ N \end{bmatrix}_q$ is the Gaussian q–binomial coefficient:

$$\begin{bmatrix} M \\ N \end{bmatrix}_q = \begin{cases} \dfrac{(q;q)_M}{(q;q)_N(q;q)_{M-N}}, & \text{if } 0 \le N \le M, \\ 0, & \text{otherwise.} \end{cases}$$

Remark 2.2 In our previous paper [4], see (5.1) and (5.2), we had considered another q–analog of (2.8). It turned out however that the q–analog (5.1) from [4], probably, does not possess good combinatorial properties.

One of the main results of the present paper is the following:

Theorem 2.3 *Assume that $p_0 \ge 1$ be a rational number, and consider a rational function*

$$V_l(q) := V_l^{(p_0)}(q) = \sum_\lambda \frac{q^{\frac{1}{2}\widetilde{\lambda} B \widetilde{\lambda}^t}}{\prod_j (q^{\epsilon(j)}; q^{\epsilon(j)})_{\lambda_j}}, \tag{2.10}$$

summation in (2.10) is taken over all sequences of non–negative integer numbers $\lambda = \{\lambda_k\}_{k=1}^{m_\alpha+1}$ such that

$$l = \sum_{k \ge 1} n_k \lambda_k, \quad \lambda_k \ge 0.$$

Then we have

$$\sum_{l \ge 0} q^{\frac{l^2}{p_0}} V_l^{(p_0)}(q) = \tag{2.11}$$

$$1 + \sum_{\substack{k \ge m \ge 0 \\ (k,m) \ne (0,0)}} (-1)^{\frac{1+(-1)^\alpha}{2}k+m} q^{(ky_{\alpha+1}+my_\alpha)(kz_\alpha+mz_{\alpha-1})+\Delta_\alpha(k,m)} \cdot \frac{Q_{k,m}^{((-1)^\alpha)}(q)}{(q;q)_k},$$

where

$$Q_{k,m}^{(\pm 1)}(q) := Q_{k,m}^{(\pm)}(q) = q^{\frac{1\mp 1}{2}(k+m)} \begin{bmatrix} k-1 \\ m \end{bmatrix}_q + q^{\frac{1\pm 1}{2}(2k-m)} \begin{bmatrix} k-1 \\ m-1 \end{bmatrix}_q,$$

$$\Delta_\alpha(k,m) = \frac{1+(-1)^\alpha}{2}\binom{k-m}{2} + \frac{1-(-1)^\alpha}{2}\binom{m}{2}.$$

Let us emphasize that polynomials $Q^{(\pm)}_{k,m}(q)$ are the *same* for all rational numbers $p_0 \geq 1$.

Taking the sum with respect to the index m in the RHS(2.11), we obtain **Corollary 2.4** (Rogers–Ramanujan–Gordon–Andrews' type identity)

$$\sum_{l\geq 0} q^{\frac{l^2}{p_0}} V_l^{(p_0)}(q) = 1+\sum_{k>0}(-1)^{\frac{1-(-1)^{\alpha}}{2}k} q^{k^2(y_{\alpha+1}+y_{\alpha})(z_{\alpha}+z_{\alpha-1})}\cdot\frac{L_k^{(p_0)}(q)}{(q;q)_k}, \quad (2.12)$$

where $L_k^{(p_0)}(q) =$

$$\sum_{m\geq 0}(-1)^m q^{m^2 y_{\alpha}z_{\alpha-1}-km(y_{\alpha+1}z_{\alpha-1}+2y_{\alpha}z_{\alpha-1}+y_{\alpha}z_{\alpha})+\Delta_{\alpha}(k,k-m)} Q^{((-1)^{\alpha})}_{k,k-m}(q).$$

A proof of identity (2.11) is a "q–version" of that given in [4], Theorem 4.1.

Remark 2.5 (Gordon–Andrews' type identity) Let $p_0 \geq 1$ be an integer, then $\alpha = 0$, $y_1 = p_0$, $y_0 = 1$, $z_0 = 1$, $z_{-1} = 0$, and the RHS of (2.11) takes the following form

$$1+\sum_{\substack{k\geq m\geq 0\\(k,m)\neq(0,0)}}(-1)^{k+m}q^{(kp_0+m)k+\frac{(k-m)(k-m-1)}{2}}\cdot\frac{Q^{(+)}_{k,m}(q)}{(q;q)_k}$$

$$=1+\sum_{k>0}(-1)^k\frac{q^{k^2p_0+\frac{k(k-1)}{2}}}{(q;q)_k}\left\{\sum_{m=0}^{k}(-1)^m q^{\frac{m(m+1)}{2}}Q^{(+)}_{k,m}(q)\right\}. \quad (2.13)$$

It is not difficult to see that if $k > 0$, the sum in the brackets (2.13) is equal to $(1+q^k)(q;q)_k$. Hence, if $p_0 \geq 1$ is an integer, then we come to the following identity:

$$\sum_{l\geq 0} q^{\frac{l^2}{p_0}} V_l^{(p_0)}(q) = 1+\sum_{k>0}(-1)^k q^{k^2p_0+\frac{k(k-1)}{2}}(1+q^k). \quad (2.14)$$

Using the Jacobi triple identity, one can rewrite (2.14) in the following forms

$$\sum_{l\geq 0} q^{\frac{l^2}{p_0}} V_l^{(p_0)}(q) \quad (2.15)$$

$$=\prod_{n\geq 1}(1-q^{(2p_0+1)n})(1-q^{(2p_0+1)n-p_0-1})(1-q^{(2p_0+1)n-p_0}),$$

$$\frac{1}{(q;q)_\infty}\sum_{l\geq 0} q^{\frac{l^2}{p_0}} V_l^{(p_0)}(q) = \prod_{n\not\equiv 0,p_0,p_0+1(\mathrm{mod}\ 2p_0+1)}(1-q^n)^{-1}. \quad (2.16)$$

It looks very challenging task to find for any rational number $p_0 \geq 1$ an explicit product formula (Weyl's denominator identity) for the right hand side of identity (2.11).

We consider the identity (2.12) as an identity between *bosonic* and *fermionic* formulae for the character of "vacuum representation" of the generalized Kac–Moody algebra corresponding to the matrix Θ^{-1}.

3 $XXZ \to XXX$ bijection

In this section we are going to describe a bijection between the space of states for XXZ–model and that of XXX–model. Let us formulate the corresponding combinatorial problem more explicitly. First of all as it follows from the results of our previous paper, the combinatorial completeness of Bethe's states for the XXZ model is equivalent to the following identity

$$\prod_m (2s_m+1)^{N_m} = \sum_{l=0}^{N} Z^{XXZ}(N, s \mid l), \tag{3.1}$$

where $N = \sum_m 2s_m N_m$ and the numbers $Z^{XXZ}(N, s \mid l)$ are given by (2.8). On the other hand it follows from the combinatorial completeness of Bethe's states for XXX model (see [1]) that

$$\prod_m (2s_m+1)^{N_m} = \sum_{l=0}^{\frac{1}{2}N} (N-2l+1) Z^{XXX}(N, s \mid l), \tag{3.2}$$

where the number $Z^{XXX}(N, s \mid l)$ stands for the multiplicity of $(\frac{N}{2} - l)$-spin irreducible representation of $sl(2)$ in the tensor product

$$V_{s_1}^{\otimes N_1} \otimes \cdots \otimes V_{s_m}^{\otimes N_m}.$$

Let us remark that both numbers $Z^{XXZ}(N, s \mid l)$ and $Z^{XXX}(N, s \mid l)$ admit a combinatorial interpretation in terms of rigged configurations. The difference between the space of states of XXX model and that of XXZ model is the availability of the so–called 1^-–configurations (or 1^- string) in the space of states for the latter model. The presence of 1^-–strings in the space of states for XXZ–model is a consequence of broken $sl(2)$–symmetry of the XXZ–model. Our goal in this section is to understand from a combinatorial point of view how the anisotropy of XXZ model breaks the symmetry of the XXX chain. More exactly, we suppose to describe a bijection between XXZ–rigged configurations and XXX–rigged configurations. Let us start with recalling a definition of rigged configurations.

We consider at first the case of $sl(2)$ XXX–magnet. Given a composition $\mu = (\mu_1, \mu_2, \ldots)$ and a natural integer l, by definition a $sl(2)$–configuration of type (l, μ) is a partition $\nu \vdash l$ such that all vacancy numbers

$$P_n(\nu; \mu) := \sum_k \min(n, \mu_k) - 2 \sum_{k \le n} \nu'_k \tag{3.3}$$

are nonnegative. Here ν' denotes the conjugate partition to that ν. A rigged configuration of type (l, μ) is a configuration ν of type (l, μ) together with the collection of integer numbers $\{J_\alpha\}_{\alpha=1}^{m_n(\nu)}$ which satisfy the following inequalities

$$0 \le J_1 \le J_2 \le \cdots \le J_{m_n(\nu)} \le P_n(\nu; \mu).$$

Here $m_n(\nu)$ is equal to the number of parts of the partition ν which are equal to n. It is clear that the total number of rigged configurations of type (l, μ) is equal to the following number

$$Z(l \mid \mu) := \sum_{\nu \vdash l} \prod_{n \ge 1} \binom{P_n(\nu; \mu) + m_n(\nu)}{m_n(\nu)}.$$

The following result had been proved in [1].

Theorem 3.1 *Multiplicity of $(N - 2l + 1)$ dimensional irreducible representation of $sl(2)$ in the tensor product*

$$V_{s_1}^{\otimes N_1} \otimes \cdots \otimes V_{s_m}^{\otimes N_m}$$

is equal to the number $$Z\left(l \mid \underbrace{2s_1, \ldots, 2s_1}_{N_1}, \ldots, \underbrace{2s_m, \ldots, 2s_m}_{N_m}\right).$$

Example 2 One can check that

$$V_1^{\otimes 5} = 6V_0 + 15V_1 + 15V_2 + 10V_3 + 4V_4 + V_5.$$

In our case we have $\mu = (2^5)$. Let us consider $l = 5$. It turns out that there exist three configurations of type $(3, (2^5))$, namely

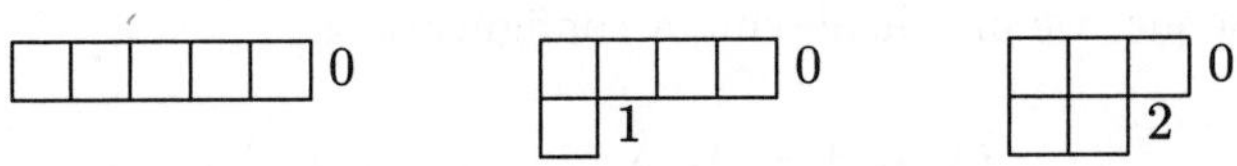

Hence, $Z(3 \mid (2^5)) = 1 + 2 + 3 = 6 = \mathrm{Mult}_{V_0}\left(V_1^{\otimes 5}\right)$.

Now let us give a definition of $sl(2)$–XXZ configuration. We consider in this Section only the case when the anisotropy parameter p_0 is an integer,

$p_0 \in \mathbf{Z}_{\geq 2}$. Under this assumption the formulae (2.6) and (2.7) take the following form:

$$
\begin{aligned}
& n_j = j, \text{ if } 1 \leq j < p_0, \ v_j = +1; \\
& n_{p_0} = 1, \ v_{p_0} = -1; \\
& 2\Phi_{k,2s} = \frac{2sk}{p_0} - \min(k, 2s), \text{ if } 1 \leq k < p_0, \ 2s+1 < p_0; \\
& 2\Phi_{p_0,2s} = \frac{2s}{p_0}, \text{ if } 2s+1 < p_0; \\
& b_{kj} = k - j, \text{ if } 1 \leq j \leq k < p_0; \\
& b_{kp_0} = 1, \text{ if } 1 \leq k < p_0; \\
& a_j := a_j(l \mid \mu) = \sum_m \min(j, \mu_m) - 2l - j \left[\frac{\sum_m \mu_m - 2l}{p_0} \right], \text{ if } 1 \leq j < p_0; \\
& a_{p_0}(l \mid \mu) = \left[\frac{\sum_m \mu_m - 2l}{p_0} \right].
\end{aligned}
$$

Definition 3.2 *A $sl(2)$–XXZ–configuration of type (l, μ) is a pair (λ, λ_{p_0}), where λ is a composition with all parts strictly less than p_0, $\sum_{j<p_0} j\lambda_j + \lambda_{p_0} = l$, and such that all vacancy numbers $P_j(\lambda \mid \mu)$ are nonnegative.*

Let us recall [4] that if the anisotropy parameter $p_0 \geq 2$ is an integer, then

$$
P_j(\lambda|\mu) := a_j(l \mid \mu) + 2 \sum_{j<k<p_0} (k-j)\lambda_k + \lambda_{p_0}, \text{ if } j < p_0 - 1; \quad (3.4)
$$

$$
P_{p_0-1}(\lambda \mid \mu) := a_{p_0-1}(l \mid \mu) + \lambda_{p_0};
$$

$$
P_{p_0}(\lambda \mid \mu) := a_{p_0}(l \mid \mu) + \lambda_{p_0-1}.
$$

In the sequel we are displaying a configuration

$$
(\lambda, \lambda_{p_0}) = (\lambda_1, \lambda_2, \ldots, \lambda_{p_0-1}, \lambda_{p_0})
$$

as the diagram of the following partition $(1^{\lambda_1+\lambda_{p_0}}, 2^{\lambda_2}, \ldots, (p_0-1)^{\lambda_{p_0-1}})$.

Example 3 Let us consider $p_0 = 6$, $s = \frac{3}{2}$, $N = 5$, $l = 5$. The total number of type $(5, (3^5))$ $sl(2)$–XXZ configurations is equal to 12.

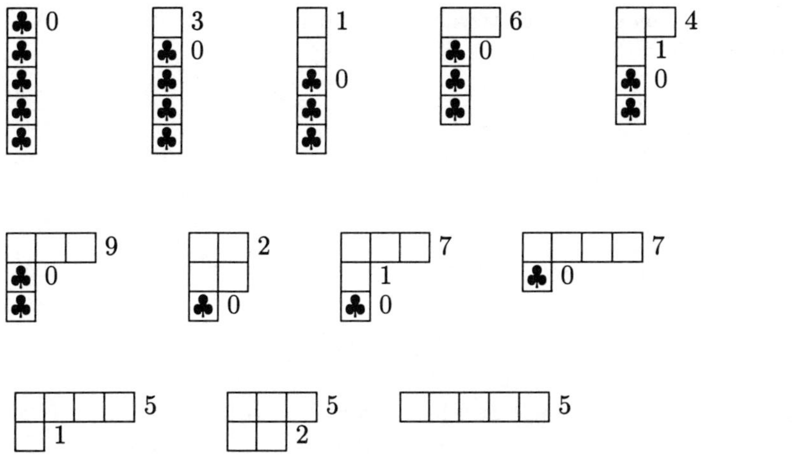

The total number of type $(5,(3^5))$ rigged configurations is equal to

$$Z^{XXZ}(5 \mid (3^5)) = 101 = 1+4+3+7+10+10+6+16+8+12+18+6.$$

Here we have used the symbol ♣ to mark the 1^-–strings.

Now we are ready to describe a map Π from the space of states for XXZ model to that of XXX one. More exactly we are going to describe a rule how a XXZ–configuration falls to the XXX–pieces. At first we describe this rule schematically:

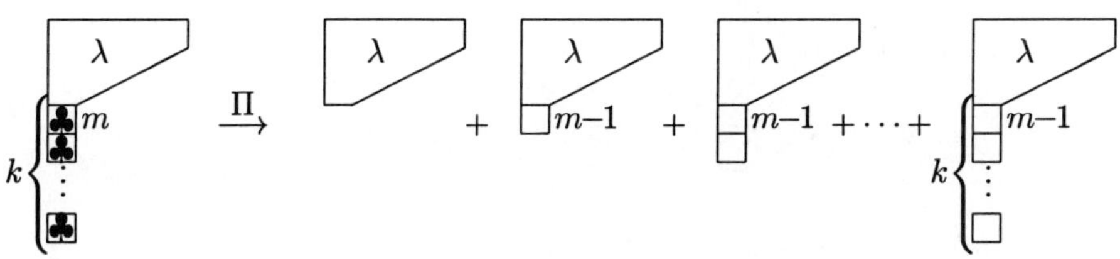

This decomposition corresponds to the well–known identity

$$\begin{bmatrix} m+k \\ k \end{bmatrix}_q = \sum_{j=0}^{k} q^j \begin{bmatrix} m+j-1 \\ j \end{bmatrix}_q .$$

In what follows we will assume that $p_0 > \sum_m s_m$.

Theorem 3.3 *The map* Π *is well–defined and gives rise to a bijection between the space of states of* XXZ*–model and that of* XXX *one.*

Proof. Let us start with rewriting the formulae (3.4) for the XXZ–vacancy numbers in more convenient form, namely,

$$P_j^{XXZ}(\widetilde{\nu} \mid \mu) = \sum_m \min(j, 2s_m) - 2\sum_{k\le j} \nu'_k - j\left[\frac{\sum_m 2s_m - 2l}{p_0}\right],$$

if $1 \le j < p_0 - 1$;

$$P_{p_0-1}^{XXZ}(\widetilde{\nu} \mid \mu) = p_0\left\{\frac{\sum_m 2s_m - 2l}{p_0}\right\} + \left[\frac{\sum_m 2s_m - 2l}{p_0}\right] + \lambda_{p_0}; \quad (3.5)$$

$$P_{p_0}^{XXZ}(\widetilde{\nu} \mid \mu) = \left[\frac{\sum_m 2s_m - 2l}{p_0}\right] + m_{p_0-1}(\nu).$$

Here $\mu = (2s_1, \ldots, 2s_m)$ and $\widetilde{\nu}$ is a pair $\widetilde{\nu} = (\nu, \lambda_{p_0})$, where ν is a partition such that $l(\nu) \le p_0 - 1$, $|\nu| + \lambda_{p_0} = l$. Relationship between λ from Definition 1 and ν is the following

$$m_j(\nu) = \lambda_j, \quad \text{i.e.} \quad \nu = (1^{\lambda_1} 2^{\lambda_2} \ldots (p_0 - 1)^{\lambda_{p_0-1}}).$$

Next, let us consider an integer $l \le \sum_m s_m$ and let $\nu \vdash l$ be a XXX–configuration. Let λ_{p_0} be an integer such that

$$2\sum s_m - 2l - p_0 < \lambda_{p_0} \le \sum_m s_m - l$$

and consider the pair $\widetilde{\nu} = (\nu, \lambda_{p_0})$. It is easy to check that

$$P_j^{XXZ}(\widetilde{\nu} \mid \mu) = \sum_m \min(j, 2s_m) - 2\sum_{k\le j} \nu'_k = P_j^{XXX}(\nu \mid \mu) \ge 0,$$

if $1 \le j < p_0 - 1$;

$$P_{p_0-1}^{XXZ}(\widetilde{\nu} \mid \mu) = \sum_m 2s_m - 2l + \lambda_{p_0} \ge 0;$$

$$P_{p_0}^{XXZ}(\widetilde{\nu} \mid \mu) = \lambda_{p_0-1} \ge 0.$$

Thus the pair $\widetilde{\nu} = (\nu, \lambda_{p_0})$ is a XXZ–configuration.

Furthermore it follows from our assumptions (namely, $\sum_m s_m < p_0$, $\lambda_{p_0} > 0$) that $\lambda_{p_0-1} = 0$ and both 1^-–strings and $(p_0 - 1)$–strings do not give a contribution to the space of XXZ–states. Thus we see that both XXX–configuration ν and XXZ–configuration $\widetilde{\nu} = (\nu, \lambda_{p_0})$ define the same number of states. Now, if $\widetilde{\nu} = (\nu, \mu)$ is a XXZ–configuration then ν is a XXX configuration as well. This is clear because (see (3.5))

$$P_j^{XXX}(\nu \mid \mu) \ge P_j^{XXZ}(\widetilde{\nu} \mid \mu), \quad 1 \le j \le p_0 - 1.$$

By the similar reasons if $(\widetilde{\nu}, \lambda_{p_0})$ is a XXZ–configuration, then for any integer k, $0 \le k \le \lambda_{p_0}$, the pair $(\widetilde{\nu}, \lambda_{p_0} - k)$ is also a XXZ–configuration. It follows from the above considerations that Π is a well–defined map. Furthermore, there exists one to one correspondence between the space of XXX–configurations and that of XXZ–configurations, namely,

$$\nu \leftrightarrow \widetilde{\nu} = (\nu, \lambda_{p_0}),$$

where $\lambda_{p_0} = [\sum_m s_m - |\nu|]$.

All others XXZ–configurations (ν, k) with $0 \le k < \sum_m s_m - |\nu| - 1$ give a contribution to the space of descendants for $\nu \leftrightarrow \widetilde{\nu}$.

■

References

1. Kirillov A.N., *Combinatorial identities and completeness of states for the generalized Heisenberg magnet,* Zap. Nauch. Sem. LOMI **131** (1984), 88-105.
2. Kirillov A.N., *On the Kostka–Green–Foulkes polynomials and Clebsch–Gordan numbers,* Journ. Geom. and Phys. **5** (1988), no. 3, 365-389.
3. Kirillov A.N., *Dilogarithm identities,* Progress of Theor. Phys. Suppl. **118** (1995), 61-142.
4. Kirillov A.N. and Liskova N.A., *Completeness of Bethe's states for generalized XXZ model, Int. Jour. Mod. Phys.* **7** (1992), 611-621.
5. Takahashi M. and Suzuki M., *One dimensional anisotropic Heisenberg model at finite temperature,* Progr. of Theor. Phys. **48** (1972), no. 6B, 2187-2209.
6. Kirillov A.N. and Reshetikhin N.Yu., *Exact solution of the integrable XXZ Heisenberg model with arbitrary spin,* J. Phys. A.: Math. Gen. **20** (1987), 1565-1597.

TRANSITION ON GROTHENDIECK POLYNOMIALS

A. LASCOUX
CNRS, Institut Gaspard Monge, Université de Marne-la-Vallée
77454 Marne-la-Vallée Cedex, France
E-mail: Alain.Lascoux@univ-mlv.fr

Grothendieck polynomials represent Schubert varieties in the Grothendieck ring of the flag manifold (for $Gl(n, \mathbb{C})$). We describe how general Grothendieck polynomials are related to those for Grasmann manifolds, which themselves are deformations of Schur functions.

1 Introduction

Given a vector bundle V of rank n, the Grothendieck ring of classes of vector bundles of the relative flag manifold $\mathcal{F}(V)$ is generated by the classes $a_1, \ldots, a_n$ of the so-called ***tautological line bundles*** on $\mathcal{F}(V)$.

The structure sheaf of a Schubert variety in $\mathcal{F}(V)$, having a finite resolution by vector bundles, can be expressed as a (Laurent) polynomial in the $a_i, 1/a_i$. Explicit representatives G_σ, $\sigma \in \mathfrak{S}_n$ were defined in [16,17] under the name "Grothendieck polynomials " (we recall their definition in Eq. 16). We shall also use ***codes*** of permutations rather than permutations, and in that case, note $G[J]$, $J \in \mathbb{N}^n$ (codes are defined in section 2).

I gave in [12,13] two different expressions of the $G[J]$, J partition, that I recall in the last section. As usual in the theory of symmetric functions, it is appropriate to let the number of indeterminates be infinite and to obtain the finite case by specializing $x_{n+1} = 0 = x_{n+2} = \cdots$. When using these conventions, we shall write partitions decreasingly and write g_λ for the symmetric function in an infinite number of variables obtained from the symmetric function $G[J]$ (in $x_1, \ldots, x_n$), with $J = [0 \leq j_1 \leq \cdots \leq j_n]$, $\lambda = [j_n, \ldots, j_1]$.

In general, introducing some equivalence (taking the "stable part") that can be defined in four different manners which will be described in section 5, and that we shall note $\overset{stab}{\sim}$, one can decompose any Grothendieck polynomial into a sum of Grasmannian ones

$$G_\sigma \overset{stab}{\sim} \sum c_\lambda g_\lambda \tag{1}$$

The similar decomposition for cohomology classes (represented by Schubert polynomials) embeds into the tradition of "Schubert calculus" (for example, intersection multiplicities of Schubert varieties, generalizing the so-called *Littlewood-Richardson rule*).

Buch [1] studies the algebra generated by the g_λ, and the decomposition (1) that he explicits for certain permutations. One motivation to write this text was to answer his question about positivity of the coefficients in (1) (once taking the correct normalization), and I thank him for having sent me his preprint.

Adapting what has been done for Schubert polynomials X_σ, $\sigma \in \mathfrak{S}_n$, in [15] and [18], we shall give in Eq. (32) a formula which can be translated into a tree which describes the multiplicities in (1). Moreover, this tree, as in the case of cohomology, gives a way to explicit the multiplicative structure of the Grothendieck ring of the Grassmannian, which is the problem considered by Buch[1].

To generate the tree, one needs only to describe the multiplication of Grothendieck polynomials by $1-1/a_i$, $i = 1, \ldots, n$, i.e. to describe the class of hyperplane sections of Schubert varieties. The similar rule for the cohomology ring is due to Monk (it was previously obtained by Chevalley) and leads to the explicit expression of Schubert polynomials as (positive) polynomials in the Chern classes of the tautological line bundles, as well as their stable part.

The cohomological computations can easily be extended to two sets of indeterminates, due to the extension of Monk's rule (cf.[10]), or translated into planar combinatorial constructions. The same is true for Grothendieck polynomials , but for simplicity, we shall restrict here to only one set of indeterminates.

Grothendieck polynomials are obtained from the action of the 0-Hecke algebra on the ring of polynomials. Let us emphasize that this action is not unique and depends upon several arbitrary parameters that we shall have to put into use (the action of the Iwahori-Hecke algebra of the symmetric group also depends upon several parameters, cf.[19]).

There exists quantum analogies of Grothendieck polynomials and we refer to the work of Kirillov[8] for their definition and properties.

The geometry of Grassmann varieties is well understood thanks to Schubert, Giambelli and their successors[9]. It is hoped that relating the Schubert subvarieties of a flag manifold to those of a Grassmannian will be of some help to understand those varieties, the singularities of which we still do not know how to describe [2].

In the present article, we do not consider Schubert varieties, but only the classes of their structure sheaves. Much of the combinatorics of the symmetric group that appear in my work with M.P. Schützenberger finds application to Schubert calculus, and I mentioned in this text the connections with [15–22].

2 0-Hecke algebra

The 0-Hecke algebra $\mathcal{H}$ of the symmetric group $\mathfrak{S}_n$ is the algebra generated by $T_1, \dots, T_{n-1}$ satisfying the braid relations

$$T_i T_{i+1} T_i = T_{i+1} T_i T_{i+1} \quad \& \quad T_i T_j = T_j T_i \ , \ |i-j| \neq 1$$

together with $T_i^2 = T_i$. Given arbitrary parameters $\alpha, \beta, \gamma, \delta \in \mathbb{C}$, such that $\alpha\delta - \beta\gamma = 1$, one gets an action of $\mathcal{H}$ on the ring $\mathbb{C}[a_1^{\pm}, \dots, a_n^{\pm}]$ by setting

$$T_i : f \mapsto f\, T_i := \frac{(\alpha a_i + \beta)(\gamma a_{i+1} + \delta) f - (\alpha a_{i+1} + \beta)(\gamma a_i + \delta) f^{s_i}}{a_i - a_{i+1}}$$

where s_i exchanges x_i, x_{i+1} (cf.[16],[19] for a more general statement; operators on polynomials will be denoted **on the right**).

Moreover, the elements $1 - T_i$ also satisfy the braid relations together with $(1-T_i)^2 = 1 - T_i$.

The operators

$$f \mapsto f \partial_i := \frac{f - f^{s_i}}{a_i - a_{i+1}}$$

are *Newton's divided differences.*

The operators

$$f \mapsto f \pi_i := f\, a_i\, \partial_i = \frac{a_i f - a_{i+1} f^{s_i}}{a_i - a_{i+1}}$$

and the operators

$$f \mapsto f \widehat{\pi}_i := f \partial_i\, a_{i+1} = \frac{a_{i+1} f - a_{i+1} f^{s_i}}{a_i - a_{i+1}}$$

are due to Demazure [3] and called *isobaric divided differences* in [16].

Let us also use the variables $x_i := 1 - 1/a_i$, and the corresponding operators that we shall distinguish by superscripts (for example, ∂_1^x sends x_1 onto 1, and $a_1 = 1/(1-x_1)$ onto $a_1 a_2$).

Since braid relations are satisfied, to any permutation σ are associated operators $T_\sigma, \pi_\sigma^a, \pi_\sigma^x, \partial_\sigma^a, \partial_\sigma^x, \widehat{\pi}_\sigma^a, \widehat{\pi}_\sigma^x$ which can be obtained by taking any reduced decomposition of σ and evaluating the associated product of elementary operators. In particular, the operators associated to the *maximal permutation* $\omega \in \mathfrak{S}_n$, $\omega := [n, \dots, 1]$ play a fundamental rôle.

We shall also need the isobaric divided differences corresponding to the shifted variables $a_1 - 1, a_2 - 1, \dots, a_n - 1$ that we shall denote $D_1, \dots, D_{n-1}$ to avoid any confusion with the preceding ones.

All these operators can be expressed in terms of Newton's divided differences only, but it greatly simplify computations, and comprehension, to introduce them instead of restricting to divided differences. Most of their properties directly arise from the following relations, which can be easily checked from their concrete action on polynomials or using the braid relations together with Leibniz formula: $f\,g\,\partial_i = f\,(g\partial_i) + (f\,\partial_i)\,g^{s_i}$.

For example, Eq. (2)–(7) are statements involving only two indices $\{i, i{+}1\}$. The operators commuting with multiplication by a function invariant under s_i, it is sufficient to check their action on the two polynomials 1 and x_i.

Lemma 1. *Divided differences satisfy the following relations :*

$$\pi_i^a = (1 - x_{i+1})\,\partial_i^x \;\; ; \;\; \widehat{\pi}_i^a = \partial_i^x\,(1 - x_i) \tag{2}$$

$$\partial_i^a = (1 - x_i)(1 - x_{i+1})\,\partial_i^x = \partial_i^x\,(1 - x_i)(1 - x_{i+1}) \tag{3}$$

$$\pi_i^x = (a_i - 1)a_{i+1}\,\partial_i^a \;\; ; \;\; \widehat{\pi}_i^x = \partial_i^a a_i(a_{i+1} - 1) \tag{4}$$

$$\partial_i^x = a_i a_{i+1}\,\partial_i^a = a_{i+1}\,\pi_i^a = \widehat{\pi}_i^a\,a_i = \partial_i^a a_i a_{i+1} \tag{5}$$

$$D_i = x_i(1 - x_{i+1})\,\partial_i^x = (a_i - 1)\,\partial_i^a \tag{6}$$

$$x_i\,\pi_i^a = \pi_i^a\,x_{i+1} + (1 - x_{i+1}) \;\; ; \;\; (x_i - 1)\,\pi_i^a = \widehat{\pi}_i^a\,(x_{i+1} - 1) \tag{7}$$

$$\pi_i^a\,D_{n-1}\cdots D_1 = D_{n-1}\cdots D_1\,\pi_{i+1}^a\;,\; i < n-1 \tag{8}$$

$$\widehat{\pi}_1\cdots\widehat{\pi}_{n-1}\,\pi_i = \pi_{i+1}\,\widehat{\pi}_1\cdots\widehat{\pi}_{n-1}\;,\; i < n-1 \tag{9}$$

$$D_\omega = (a_1 - 1)^{n-1}\cdots(a_n - 1)^0\,\partial_\omega^a = x_1^{n-1}\cdots x_n^0\,\pi_\omega^a \tag{10}$$

$$\pi_\omega^a = (1 - x_2)^1\cdots(1 - x_n)^{n-1}\,\partial_\omega^x \tag{11}$$

Only the last two relations require a comment. Indeed ∂_ω^a can be charaterized as the $\mathfrak{Sym}(a_1, \ldots, a_n)$-linear operator which annihilates all monomials $a^\lambda := a_1^{\lambda_1}\cdots a_n^{\lambda_n}$, $\lambda < \rho := [n{-}1, \ldots, 0]$ and sends a^ρ onto 1 (the preceding monomials are a basis of $\mathbb{C}[a_1^{\pm}, \ldots, a_n^{\pm}]$ as a free module over the ring $\mathfrak{Sym}$:= ring of symmetric functions in the a_i's or x_i's).

Using this fact, and the remark that all the operators that we have written are $\mathfrak{Sym}$-linear, one checks that

$$\partial_\omega^a = \sum_{\sigma\in\mathfrak{S}_n} (-1)^{\ell(\sigma)}\,\sigma\,\frac{1}{\prod_{i<j}(a_i - a_j)} = \frac{1}{\prod_{i<j}(a_i - a_j)}\sum_{\sigma\in\mathfrak{S}_n}\sigma \tag{12}$$

$$\pi^a_\omega = a^\rho \, \partial_\omega = \frac{1}{\prod_{i<j}(1 - a_j/a_i)} \sum_{\sigma\in\mathfrak{S}_n} \sigma \tag{13}$$

and, by change of variables $a_i \mapsto a_i - 1$,

$$D_\omega = \frac{1}{\prod_{i<j} 1 - \frac{aj-1}{a_i-1}} \sum_{\sigma\in\mathfrak{S}_n} \sigma = \prod_{i<j} \frac{a_i-1}{a_j-1} \sum_{\sigma\in\mathfrak{S}_n} \sigma = (a_1-1)^{n-1} \cdots (a_n-1)^0 \, \partial^a_\omega \tag{14}$$

More generally, one has, for $\alpha\delta - \beta\gamma = 1$,

$$T_\omega = \prod_{i<j} \frac{(\alpha a_i + \beta)(\gamma a_j + \delta)}{a_i - a_j} \sum_{\sigma\in\mathfrak{S}_n} \sigma = \prod_{i<j} (\alpha a_i + \beta)(\gamma a_j + \delta) \partial^a_\omega \tag{15}$$

In the above factorization, we consider T_ω as an operator on polynomials, not as an element of the Hecke algebra. Otherwise, we should have to work in the affine Hecke algebra.

The *Grothendieck polynomials* G_σ, $\sigma \in \mathfrak{S}_n$, are defined as follows:

$$\begin{cases} G_\omega = x_1^{n-1} \cdots x_n^0 = (1 - \frac{1}{a_1})^{n-1} \cdots (1 - \frac{1}{a_n})^0 \\ G_\sigma = \quad G_\omega \, \pi^a_{\omega\sigma} \end{cases} \tag{16}$$

Since Schubert polynomials are defined as the images under the ∂^x_σ of x^ρ, and since $\pi^a_i = (1 - x_{i+1}) \, \partial^x_i$, one sees that the term of smallest degree of G_σ is the Schubert polynomial (in the x_i's) of the same index.

Grothendieck polynomials , as well as Schubert polynomials, are a $\mathfrak{S}ym$-basis of $\mathbb{C}[a_1^\pm, \ldots, a_n^\pm] = \mathbb{C}[x_1^\pm, \ldots, x_n^\pm]$.

It is also convenient to index Grothendieck polynomials by codes of permutations, instead of permutations. The *code* of a permutation $\sigma = [\sigma^1, \ldots, \sigma^n] \in \mathfrak{S}_n$ is the vector $J = [j_1, \ldots, j_n]$ such that $j_i := \#\{k > i, \sigma^k < \sigma^i\}$. We shall write G_σ or $G[J]$ indifferently.

Grothendieck polynomials are stable with respect to the embeddings $\mathfrak{S}_n = \mathfrak{S}_n \times \mathfrak{S}_1 \hookrightarrow \mathfrak{S}_{n+1}$, and accordingly, one can freely add or suppress terminal 0's to codes without changing the polynomials : $G[J] = G[J0\ldots0]$.

Instead of starting from $G[\rho] = x^\rho$ for a given n, and letting n vary, one can as well rewrite the preceding definition as follows. Say that a vector $\lambda = [\lambda_1, \ldots, \lambda_k]$ is *dominant* if $\lambda_1 \geq \cdots \geq \lambda_k \geq 0$. Then

$$\begin{cases} G[\lambda] = x^\lambda & \lambda \text{ dominant} \\ G[\ldots, j_k, j_{k+1}, \ldots] \, \pi^a_k = G[\ldots, j_{k+1}-1, j_k, \ldots] & j_j > j_{k+1} \end{cases} \tag{17}$$

This convention allows to have all n's at the same time.

3 Symmetric Grothendieck Polynomials

On the Grothendieck ring of classes of vector bundles over a manifold act operators induced from the exterior powers and symmetric powers of vector bundles, that we shall denote Λ^i and S^i.

Explicitly, taking an extra variable z, the generating function of the S^i is such that

$$\sum_{i=0}^{\infty} z^i\, S^i(\mathbb{B} - \mathbb{B}') = \prod(1 - zb')/\prod(1 - zb) \tag{18}$$

if $\mathbb{B} = \sum b$, $\mathbb{B}' = \sum b'$ are finite sums of classes of line bundles.

More generally, given any partition $K = [k_1, \dots, k_n]$, $0 \le k_1 \le \cdots \le k_n$, one has a *Schur functor* S_K, the action of which on sums $\mathbb{B} = \sum_{i=1}^n b_i$ of line bundles is

$$S_K(\mathbb{B}) = \det \left|S^{k_j+j-i}(\mathbb{B})\right|_{1\le i,j\le n} = \det \left|b_i^{k_j+n-j}\right|_{1\le i,j\le n} \,/\, \det \left|b_i^{n-j}\right|_{1\le i,j\le n} \tag{19}$$

Still a little more generally again [13], Schur functors indexed by partitions in $\mathbb{N}^n$ can be considered as morphisms from the n-th tensor power of a λ-ring into itself :

$$S_K(\mathbb{B}^1, \dots, \mathbb{B}^n) := \det \left|S^{k_j+j-i}(\mathbb{B}^j)\right|_{1\le i,j\le n} \tag{20}$$

where $\mathbb{B}^1, \dots, \mathbb{B}^n$ are arbitrary elements of the λ-ring.

4 Stable part by restriction or symmetrization.

Let $k, n \in \mathbb{N}$, $k \ge n$, $I \in \mathbb{N}^k$. Then $G[0^{k-1}I]$ is symmetrical in $x_1, \dots, x_k$ and therefore the restriction

$$G[0^{k-1}I]\big|_{x_{k+1}=0=x_{k+2}=\cdots}$$

is a symmetric function $F[I]$ of k variables.

Similarly to the case of Schubert polynomials (cf.[15,25]) we shall adopt the common terminology and call $F[I]$ the *stable part* of $G[I]$, writing also F_σ if σ is any permutation of code $[I0\dots0]$.

One needs to check some stability with respect to k, since one has only required $k \ge n$. This will be clarified by the following property.

Theorem 2. *1) Let $I \in \mathbb{N}^n$. Then*

$$G[I]D_n \cdots D_1 = G[0I] \tag{21}$$

2) Let ω be the maximal permutation of $\mathfrak{S}_n$. Then

$$G[I]D_\omega = F[I] \in \mathfrak{S}ym(x_1, \dots, x_n) \tag{22}$$

Proof. $G[I]$ is the image of some $G[J]$, $J \in \mathbb{N}^n$ dominant, under some product of π_i^a, $i \leq n-1$. However, $G[J]$ can be written as a determinant (with $\mathbb{A}_k :=: a_1 + \cdots + a_k$) :

$$G[J] = G[J0] = (a_1 \cdots a_{n+1})^{-n} S_{n^{n+1}}(\mathbb{A}_{n+1} - 0, \mathbb{A}_n - j_n, \dots, \mathbb{A}_1 - j_1) .$$

Since $\forall i, j, k \in \mathbb{N}$, one has $S^k(\mathbb{A}_i - j) D_i = S^k(\mathbb{A}_{i+1} - j - 1)$ (D_i being the isobaric divided difference with respect to the pair $(a_i - 1, a_{i+1} - 1)$), each D_i in the product $D_n \cdots D_1$ operates in turn on a single column of the determinant, the others being symmetrical in $(a_i - 1, a_{i+1} - 1)$. Therefore the image of $G[J0]$ under $D_n \cdots D_1$ is

$$(a_1 \cdots a_{n+1})^{-n} S_{n^{n+1}}(\mathbb{A}_{n+1} - 0, \mathbb{A}_{n+1} - j_n - 1, \dots, \mathbb{A}_{n+1} - j_1 - 1)$$

i.e. is equal to the image of $G[j_1 + 1, \dots, j_n + 1, 0]$ under $\pi_n^a \cdots \pi_1^a$ which by definition is $G[0J]$.

Now, if $G[I] = G[J]\pi_i^a \pi_j^a \cdots$, then the commutations (8) show that

$$G[0I] = G[0J]\, \pi_{i+1}^a \pi_{j+1}^a \cdots = G[J]D_n \cdots D_1 \pi_{i+1}^a \pi_{j+1}^a \cdots = G[I]D_n \cdots D_1$$

and this proves Eq. (21).

To simplify indices, let us take $k = n = 4$ to attack Eq. (22) The morphism $G[I] \mapsto G[000I]$, $I \in \mathbb{N}^4$ is $D_\nu := (D_4 \cdots D_1)(D_5 \cdots D_1)(D_6 \cdots D_1)$. The specialization $x_5 = x_6 = x_7$ of $G[000I]$ is the same as the specialization of $G[000I]D_\eta$, with $\eta = s_6 s_5 s_4 s_6 s_5 s_6$. However, $\nu\eta = \omega_7$ (= the maximal element of $\mathfrak{S}_7$), and thus is equal to $\omega_4\,(\omega_4\omega_7)$. Now, given any symmetric function of $x_1, \dots, x_4$, applying to it any sequence of D_i and specializing $x_5 = 0 = x_6 = \cdots$ does not change the function. Therefore $G[I]\, D_{\omega_4}$, which indeed is symmetrical in $x_1, \dots, x_4$, coincides with $G[I]D_{\omega_7}\big|_{x_5=0=x_6=\cdots}$ □

Notice that if I is the code of a vexillary permutation, then $G[I0^{n-1}]$ has a determinantal expression in terms of the $S^k(\mathbb{A}_i - j)$, and the morphism $G[I0^{n-1}] \mapsto G[0^{n-1}I]$ consists in replacing in the determinantal expression of $G[I0^{n-1}]$ each term $S^k(\mathbb{A}_i - j)$ by $S^k(\mathbb{A}_{i+n-1} - j - n + 1)$. Specializing $x_{n+1}, x_{n+2}, \dots$ to 0 , that is, $a_{n+1}, a_{n+2}, \dots$ to 1, is obtained by putting $\mathbb{A}_{i+n} = \mathbb{A}_n + i$, and thus the stable part of $G[I]$ is a determinant in the $S^k(\mathbb{A}_n - j)$ (which is recognized to be the one expressing $G[J]$, with J= increasing reordering of I).

In other words, if $I \in \mathbb{N}^n$ is the code of a vexillary permutation, then $G[I]\, D_\omega = G[J]$. This leads to define, for a general $I \in \mathbb{N}^n$, a *G-key polynomial* $KG[I]$ by

$$\begin{cases} KG[I] = G[I] & I \text{ dominant} \\ KG[\ldots, j_k, j_{k+1}, \ldots]\, D_k = KG[\ldots, j_{k+1}, j_k, \ldots] & j_k > j_{k+1} \end{cases} \tag{23}$$

Using that $D_i D_\omega = D_\omega$, $i < n$, one gets that $KG[I]\, D_\omega = G[J]$, with $J=$ increasing reordering of I. Thus a decomposition of the type

$$G[I] = \sum c_{I,J} KG[J] \tag{24}$$

induces a decomposition

$$G[I]\, D_\omega = \sum c_{I,J} KG[J]\, D_\omega \tag{25}$$

of the stable part of $G[I]$ into Grassmannian Grothendieck polynomials.

In the case of Schubert polynomials, I gave with Schützenberger such a decomposition in terms of "Demazure characters", but it involves a combinatorics of tableaux which will be avoided here (cf.[21],[26]).

A Grothendieck polynomial indexed by a permutation $\sigma = \nu \times \eta$ belonging to a Young subgroup $\mathfrak{S}_m \times \mathfrak{S}_n \hookrightarrow \mathfrak{S}_{m+n}$ factorizes into two Grothendieck polynomials (proposition 6.7 of [14]). It is clear, on the definition by restriction, that the stable part of G_σ factorizes into the product of the stable parts of G_ν and G_η. In particular, if ν and η are vexillary, of respective codes J, K, which reorders into the partitions λ, μ, then

$$G_\sigma \overset{stab}{\sim} G[J]\, G[K] \overset{stab}{\sim} g_\lambda g_\mu \ ,$$

and therefore, as in the case of cohomology, any algorithm producing the stable part of a Grothendieck polynomial allows to describe products of Grassmannian Grothendieck polynomials. Instead of symmetrizing operators, or restrictions, we shall prefer, in the next section, using *transitions.*

Grothendieck polynomials satisfy a recursion with respect to $\mathbb{C}[x_1, \ldots, x_n] = \mathbb{C}[x_1] \otimes \mathbb{C}[x_2, \ldots, x_n]$ (cf.[17]) that Fomin and Kirillov[6] have written as a generating function in the 0-Hecke algebra. This allows them to define in the same fashion stable Grothendieck polynomials :

Let T_i be the generators of the 0-Hecke algebra (with $T_i^2 = -T_i$), and for any n, r: $n > r$, let

$$\mathcal{A}_{n,r}(x) := (1 + xT_{n-1}) \cdots (1 + xT_r)$$

Definition[6]**.** For any n, and any $\sigma \in \mathfrak{S}_n$, the *stable Grothendieck polynomials* $\widetilde{F}_\sigma$ are the coefficients of the generating function

$$\mathcal{A}_{n,1}(x_1)\cdots\mathcal{A}_{n,1}(x_n) = \sum\nolimits_{\sigma\in\mathfrak{S}_n} \widetilde{F}_\sigma T_\sigma \, . \tag{26}$$

On the other hand, the generating function[6] of Grothendieck polynomials G_σ, $\sigma \in \mathfrak{S}_{2n-1}$, is

$$\mathcal{G}_{2n-1} := \mathcal{A}_{2n-1,1}(x_1)\,\mathcal{A}_{2n-1,2}(x_2)\cdots\mathcal{A}_{2n-1,2n-1}(x_{2n-1}) \, . \tag{27}$$

Now, if $T_iT_j\cdots T_k$ is equal to T_σ, $\sigma \in \mathfrak{S}_n$, σ having code I, then $T_{i+n-1}T_{j+n-1}\cdots T_{k+n-1}$ is equal to $T_{\sigma'}$, σ' having code $[0^{n-1}I]$. Therefore, the coefficient of $T_{\sigma'}$ in the specialization $T_1 = 0 = \cdots = T_{n-1}$; $x_{n+1} = 0 = \cdots = x_{2n-1}$ of $\mathcal{G}_{2n-1}$ is the same as the coefficient of T_σ in $\mathcal{A}_{n,1}(x_1)\cdots\mathcal{A}_{n,1}(x_n)$, that is, one has the identity

$$\widetilde{F}_\sigma = G[0^{n-1}I]\,\big|_{x_{n+1}=0=\cdots=x_{2n-1}} = F_\sigma \, . \tag{28}$$

In fact, Grothendieck polynomials already appear at the level of the "twisted" group algebra of the symmetric group (i.e. the algebra generated by the s_i's, x_j's), as coefficients in the expansion of isobaric divided differences $\pi_\sigma\ \pi_\sigma$ in the basis of permutations (cf.[22]). Yang-Baxter equation in the 0-Hecke algebra also lead to similar expansion ([6], [11]).

5 Transition on Grothendieck polynomials

Let σ be a permutation J its code, and r be the length of J (i.e. $j_r > 0$, $j_i = 0\ \forall i > r$). Let ν be the permutation of code $[j_1,\ldots,j_{r-1},j_r-1]$. Recall that for what concerns Schubert polynomials in $x_1, x_2, \ldots$, one has the equation

$$X_\sigma = x_r X_\nu + \sum\nolimits_{i\in\mathcal{I}(\sigma)} X_{\tau_{ji}\nu} \, , \tag{29}$$

sum over all transpositions $\tau_{ji} : j := \nu_r, i \in \mathcal{I}(\sigma)$ (i.e. i such that such that $\ell(\tau_{ji}\nu) = \ell(\nu) + 1$).

These equations, called *transition* in [15], refine the Ehresmann-Bruhat order on the symmetric group and give the expression of Schubert polynomials as a positive sum of monomials; they also furnish the stable part of them as a sum of Schur functions, if one cancels the successive terms $x_r X_\nu$, when iterating Eq. (29).

One has similar transitions for Grothendieck polynomials . But first, one needs to explicit products of G_σ by x_r's. In [14], I described the $G_\sigma/(a_1\cdots a_r)$. It requires a small adaptation to get products by $1/a_r = 1 - x_r$ instead. On the other hand, I cannot relate products $G_\sigma(1-x_r)$ to the products $a^\lambda G_\sigma$, λ dominant, considered in [7] under the name *Pieri formula.* Geometrically these last products are classes of restriction of ample line bundles over Schubert subvarieties, and not intersections as in our case.

Let us keep the same hypotheses as in Eq. (29), ordering the values $i_1 < i_2 < \cdots < i_k$, $i \in \mathcal{I}(\sigma)$, appearing in the summation. Given any $\sigma \in \mathfrak{S}_n$, and any pair of integers $j, i \le n$, we shall write $\tau_{ji} \star G_\sigma$ for $G_{\tau_{ji}\sigma}$, considering that a transposition τ_{ji} acts on indices of Grothendieck polynomials present on its right.

Proposition 3. *Let σ be a permutation. With the hypothesis of Eq. (29), one has*

$$G_\sigma = G_\nu + (x_r - 1)\,(1-\tau_{ji_k})\star\cdots\star(1-\tau_{ji_1})\star G_\nu\ . \tag{30}$$

Proof. Suppose first that $\sigma_1 > \cdots \sigma_{r-1}$. If $j_{r-1} \neq 0$, then $G[j_1,\ldots,j_r] = x_1\cdots x_r G[j_1-1,\ldots,j_r-1]$ and one is reduced to study a polynomial of smaller degree. The fundamental case to study is $J = [j_1,\ldots,j_{r-2},0,j_r]$, $j_{r-2} > 0$.

By induction on r, we know that

$$\begin{aligned} G[j_1,\ldots,j_{r-2},j_r+1] = G[j_1,\ldots,j_{r-2},j_r] + \\ (x_{r-1}-1)(1-\tau_{ji_k})\star\cdots\star(1-\tau_{ji_2})\star G[j_1,\ldots,j_{r-2},j_r] \end{aligned} \tag{31}$$

the transpositions appearing in this formula being the same as in Eq. (30), but for τ_{j1} which is missing (the hypothesis $j_{r-2} > 0$ implies $i_1 = 1$).

Taking the image of (32) under π^u_{r-1} with the help of (7), one gets

$$\begin{aligned} &G[j_1,\ldots,j_{r-2},0,j_r] = G[j_1,\ldots,j_{r-2},0,j_r-1] \\ &+(1-\tau_{ji_k})\star\cdots\star(1-\tau_{ji_2})\star G[j_1,\ldots,j_{r-2},j_r,0]\widehat{\pi}^a_{r-1}(x_r-1) \\ &= G[j_1,\ldots,j_{r-2},0,j_r-1] \\ &+(1-\tau_{ji_k})\star\cdots\star(1-\tau_{ji_2})\star\big(G[j_1,\ldots,j_{r-2},0,j_r-1]-G[j_1,\ldots,j_{r-2},j_r,0]\big) \\ &= G[j_1,\ldots,j_{r-2},0,j_r-1] \\ &+(1-\tau_{ji_k})\cdots\star(1-\tau_{ji_2})\star(1-\tau_{j1})\star G[j_1,\ldots,j_{r-2},0,j_r-1]\,(x_r-1) \end{aligned}$$

which is the required expression for those special J.

A general $G[J']$, $J' \in \mathbb{N}^r$, will be obtained from a $G[J]$, $j_1 \ge \cdots \ge j_{r-1}$ by repeated use of the π^a_k, $i \le r-2$.

Suppose that Eq. (30) is true for a permutation σ. Let $p \le r-2$ be an integer such that $\nu_p > \nu_{p+1}$. Then, except when both ν_p and ν_{p+1} belong to $\mathcal{I}(\sigma)$, the image of (30) under π^a_p is a decomposition

$$G_{\sigma s_p} = G_{\nu s_p} + (x_r-1)\,(1-\tau_{ji_k})\star\cdots\star(1-\tau_{ji_1})\star G_{\nu s_p}$$

of the type required by the theorem. On the other hand, if $\nu_p = \beta, \nu_{p+1} = \alpha \in \mathcal{I}(\sigma)$, one writes the RHS of (30) as a summation

$$(x_r - 1) \sum \eta' \star (1 - \tau_{j\beta})(1 - \tau_{j\alpha}) \star \eta'' \star G_\nu \ ,$$

sum over certain pairs of permutations η', η''. Thus terms occur in four-tuples

$$(x_r - 1)\big(G_{\cdots\beta\alpha\cdots j\cdots} - G_{\cdots j\alpha\cdots\beta\cdots} - G_{\cdots\beta j\cdots\alpha\cdots} + G_{\cdots j\beta\cdots\alpha\cdots}\big) \ ,$$

writing only the indices at places $p, p+1, r$. Now, the image of such a term under π_p^a is

$$(x_r - 1)\big(G_{\cdots\alpha\beta\cdots j\cdots} - G_{\cdots\alpha j\cdots\beta\cdots}\big)$$

and therefore the decomposition of $G_{\sigma s_p}$ is obtained from the one of G_σ by replacing the double factor $(1 - \tau_{j\beta})(1 - \tau_{j\alpha})$ by the single one $(1 - \tau_{j\beta})$, and exchanging ν for νs_p. □

Specializing x_r to 0, one finally gets

Theorem 4. *1) Let σ be a permutation, $\sigma_1 = 1$. Using conventions (29), the stable part of G_σ is equal to that of*

$$\Big(1 - (1 - \tau_{ji_k}) \star \cdots \star (1 - \tau_{ji_1})\Big) \star G_\nu \ . \tag{32}$$

2) There exists positive integers m_J^σ such that

$$G_\sigma \overset{stab}{\sim} \sum\nolimits_J (-1)^{\ell(\sigma)-|J|} \, m_J^\sigma \, G[J] \ , \tag{33}$$

sum over all partitions J.

Proof. One gets 1) from a transition by specializing x_r to 0. Instead of having to embed $\mathfrak{S}_n$ into $\mathfrak{S}_1 \times \mathfrak{S}_n$, we have choosen the hypothesis $\sigma_1 = 1$.

From the theory of Schubert polynomials [18], one knows that all the permutations appearing when iterating sufficiently many times a Schubert-transition are vexillary. Now, in the case where σ is vexillary, the set $\mathcal{I}(\sigma)$ is of cardinality 1, and every long sequence of transitions starting from σ encounter a grassmannian permutation (with code obtained by reordering increasingly the code of σ).

There are more permutations appearing in Eq. (30) than Eq. (29). To adapt the argument of [18], one has to modify the weight function $(m-r)\ell(\sigma)$, where m is the first index such $\sigma_m > \sigma_{m+1}$. Buch proposes the function $\sum_{i \geq 1} 2^i j_{i+m}$, where $[j_1, j_2, ..]$ is the code of σ. This function vanishes for grasmannian permutations, and strictly decreases in a transition $\sigma \mapsto \eta$, η being any permutation appearing on the RHS of (30).

In summary, the stable part of G_σ belongs to the space generated by the stable parts of the G_γ, γ grassmannian. However, each product by a transposition appearing in a transition increases length (of permutations) by 1, and signs are as stated, since $\ell(\sigma) = |code(\sigma)|$ for any permutation σ (this positivity was conjectured by Buch [1], as we already mentioned in the introduction). □

As an example of decomposition given by theorem 4, let us take $\sigma = [1,5,2,4,3,9,6,7,8]$; this implies $J = [0,3,0,1,0,3]$, $r = 6$. One has

$$X_{152439678} = x_6 X_{152438679} + X_{15248367} + X_{15283467} + X_{18243567} ,$$

and therefore $\mathcal{I} = \{3,4,5\}$,

$$G_{152439678} = G_{152438679} + (x_6 - 1)(1 - \tau_{85}) \star (1 - \tau_{84}) \star (1 - \tau_{83}) \star G_{152438679} .$$

Expression (32) develops in this case into

$$\begin{aligned}(1 - \tau_{85}) &\star (1 - \tau_{84}) \star (1 - \tau_{83}) \star G_{152438679} \\ &= (1 - \tau_{85}) \star (1 - \tau_{84}) \big(G_{152438679} - G_{152483679}\big) \\ &= (1 - \tau_{85}) \star \big(G_{152438679} - G_{152483679} - G_{152834679} + G_{152843679}\big) \\ &= G_{152438679} - G_{152483679} - G_{152834679} + G_{152843679} - G_{182435679} \\ &+ G_{182453679} + G_{182835679} - G_{182543679} .\end{aligned}$$

Iterating transitions, and using codes now, one finally gets a sum

$$\begin{aligned}G[030103] \overset{stab}{\sim} &\ G[0601] + G[0502] + G[034] - G[053] - G[0602] + G[05011] \\ + &\ G[0412] + G[0331] - G[0431] - G[05021] - G[06011] \\ - &\ G[0512] - G[0341] + G[0531] + G[06021]\end{aligned}$$

of vexillary Grothendieck polynomials , which implies, by sorting indices,

$$\begin{aligned}G[030103] \overset{stab}{\sim} &\ g_{61} + g_{52} + g_{43} - g_{53} - g_{62} + g_{511} + g_{421} + g_{331} - 2g_{431} \\ - &\ 2g_{521} - g_{611} + g_{531} + g_{621} .\end{aligned}$$

One has a similar decomposition of the stable part of Schubert polynomials in terms of Schur functions (cf.[17],[4]). This decomposition is a direct by-product (just sorting increasingly all indices) of the decomposition of Schubert polynomials in terms of key polynomials (cf.[20],[26]), which involves finding all Young tableaux (in the alphabet $s_1, s_2, \dots$) which are reduced decompositions of a given permutation.

Fomin and Greene[5] obtain a decomposition of the stable part of G_σ in terms of Schur functions, by enumerating tableaux in the T_i's which are decompositions, in the 0-Hecke algebra, of a given T_σ. Reconstructing Eq. (33) from their result, however, is not possible since many cancellations occur :

Schur functions are not the appropriate symmetric functions for what concerns the Grothendieck ring.

We shall give elsewhere a decomposition of Grothendieck polynomials in terms of G-key polynomials. Let us mention that Grothendieck polynomials are given by a simple statistics on alternating sign matrices. This also will be written elsewhere.

6 Variations among determinantal expressions

There are many determinantal expressions of classes of Schubert subvarieties of Grassmannians, that is, of g_λ, λ partition, or of $G[J]$, J increasing.

Let us illustrate them on a concrete example, the case of a determinantal variety defined by the vanishing of minors of order 2 of a matrix of order 4. With the notations used in this text, the Grothendieck polynomial which is to be explicited is $G_{3412} = G[0022]$.

By definition, it is

$$(1-1/a_1)^5(1-1/a_2)^4(1-1/a_3)^1\,\pi^a_{4321} \tag{34}$$

$$= (1-1/a_1)^4(1-1/a_2)^4\,\pi^a_{4321} \tag{35}$$

$$= (a_1-1)^4(a_2-1)^4 a_3^4 a_4^4\,\pi^a_{4321}\,\frac{1}{(a_1a_2a_3a_4)^4} \tag{36}$$

$$= (a_1-1)^4(a_2-1)^4 a_3^4 a_4^4\, a_1^3 a_2^2 a_3^1\,\partial^a_{4321}\,\frac{1}{(a_1a_2a_3a_4)^4} \tag{37}$$

$$det\left|\, a_i^3(a_1-1)^4\,,\, a_i^2(a_1-1)^4\,,\, a_i^5\,,\, a_i^4 \,\right| \frac{(a_1a_2a_3a_4)^{-4}}{\Delta(\mathbb{A})} \tag{38}$$

with $\Delta(\mathbb{A}) := \prod_{i<j}(a_i - a_j)$.

The equality between Eq. (34) and Eq. (35) comes from the fact that $(1-1/a_1)\pi^a_1 = 1$, and that $\pi^a_{4321}\,\pi_1 = \pi^a_{4321}$.

Since ∂^a_{4321} is given by an alternating summation on the symmetric group, (37) is the expansion of (38). The quotient is a multi-Schur function

$$\begin{vmatrix} S^4(\mathbb{A}-4) & S^5(\mathbb{A}-4) & S^6(\mathbb{A}-4) & S^7(\mathbb{A}-4) \\ S^3(\mathbb{A}-4) & S^4(\mathbb{A}-4) & S^5(\mathbb{A}-4) & S^6(\mathbb{A}-4) \\ S^2(\mathbb{A}) & S^3(\mathbb{A}) & S^4(\mathbb{A}) & S^5(\mathbb{A}) \\ S^1(\mathbb{A}) & S^2(\mathbb{A}) & S^3(\mathbb{A}) & S^4(\mathbb{A}) \end{vmatrix} \frac{1}{(a_1a_2a_3a_4)^4} \tag{39}$$

and this the expression that is given in [12] (specialized to the case one of the vector bundles is trivial of rank 4).

In chapter 2 of my thesis [13], I used determinants with entries more general than symmetric powers. In the case considered, it gives:

$$G[0022] = \begin{vmatrix} 1 - \frac{1}{a_1a_2a_3a_4} & a_1 + a_2 + a_3 + a_4 - 4 \\ \frac{1}{a_1a_2a_3a_4}(4 - \frac{1}{a_1} - \frac{1}{a_2} - \frac{1}{a_3} - \frac{1}{a_4}) & 1 - \frac{1}{a_1a_2a_3a_4} \end{vmatrix} . \quad (40)$$

One can see, more easily than using Grothendieck polynomials , that this last determinant is equal to the class of the complex resolving submaximal minors (*Eagon-Northcott complex*).

Using now variables x_i's, and relation (11), one has:

$$G[0022] = x_1^5 x_2^4 (1 - x_2) x_3 (1 - x_3)^2 (1 - x_4)^3 \, \partial^x_{4321} \quad (41)$$

$$= \left| x_i^5 \; x_i^4 (1 - x_i) \; x_i (1 - x_i)^2 \; (1 - x_i)^3 \right| \frac{1}{\Delta(X)} \quad (42)$$

$$= \pm \begin{vmatrix} S^2(X) & S^3(X) & S^4(X) & S^5(X) \\ S^2(X-1) & S^3(X-1) & S^4(X-1) & S^5(X-1) \\ S^0(X-2) & S^1(X-2) & S^2(X-2) & S^3(X-2) \\ S^0(X-3) & S^1(X-3) & S^2(X-3) & S^3(X-3) \end{vmatrix} . \quad (43)$$

The equivalence of (38) and (42), or (39) and (43) is in fact a mere change of variables $x_i \mapsto 1 - \frac{1}{a_i}$, using

$$\Delta(X) = \prod((1 - \frac{1}{a_i}) - (1 - \frac{1}{a_j})) = (-1)^{\binom{n}{2}} \Delta(\mathbb{A})(a_1 \cdots a_n)^{-n+1}.$$

The expression (43) is given by Lenart [24](th. 2.4).

Proposition 3.11 of [14] states that, more generally, Grothendieck polynomials are still given by determinants in the case of *vexillary permutations.*

Note, however, that when the partition J has repeated parts, then one can slightly modify the determinant expressing $G[J]$ without changing its value. This is already seen at the level of the equality of (34) and (35).

References

1. A. Buch, *A Littlewood-Richardson rule for the K-theory of Grassmannians*, math.AG/0004137 (2000).
2. M. Demazure, *Désingularisation des variétés de Schubert généralisées*, Ann. Sci. École Norm. Sup. (4) **7**, 53–88 (1974).
3. M. Demazure, *Une nouvelle formule des caractères*, Bull. Sc. M. **98**, 163–172 (1974).
4. P.H. Edelman, C. Greene, *Balanced tableaux*, Adv. in Math. **63**, 42–99 (1987).
5. S. Fomin and C. Greene, *Noncommutative Schur functions and their applications*, Discrete Math. **193**, 179–200 (1998).
6. S. Fomin and A. Kirillov, *Grothendieck polynomials and the Yang-Baxter equation*, 6th Conf. on Formal Power Series and Alg. Comb., DIMACS, Rutgers, p. 183–190 (1994).
7. W. Fulton, A. Lascoux, *A Pieri formula in the Grothendieck ring of a flag bundle*, Duke Math. **76**, 711–729 (1994).
8. A. Kirillov, *Quantum Grothendieck polynomials*, CRM Proc. Lecture Notes **22**, p. 215-226, Amer. Math. Soc. (1999).
9. S. Kleiman, D. Laksov , *Schubert calculus*, Am. Math. Monthly **79**, 1061–1082 (1972).
10. A. Konhert, S. Veigneau, *Using Schubert basis to compute with multivariate polynomials*, Adv. Appl. Math. **19**, 45–60 (1997).
11. A. Lascoux, B. Leclerc and J.Y. Thibon, *Flag Varieties and the Yang-Baxter Equation* , Letters in Math. Phys. **40**, 75–90 (1997).
12. A. Lascoux, *Puissances extérieures, déterminants et cycles de Schubert*, Bull S.M.F. **102**, 161–179 (1974).
13. A. Lascoux, *Polynômes symétriques, Foncteurs de Schur et Grassmanniennes*, Thèse, Université Paris 7 (1977).
14. A. Lascoux, *Anneau de Grothendieck de la variété de drapeaux*, in "The Grothendieck Festchrift", vol III, 1–34, Birkhäuser(1990).
15. A. Lascoux & M.P. Schützenberger, *Polynômes de Schubert*, Comptes Rendus **294**, 447 (1982).
16. A. Lascoux & M.P. Schützenberger, *Symmetry and Flag manifolds*, in Invariant Theory, Springer L.N. **996**, 118–144 (1983).
17. A. Lascoux & M.P. Schützenberger, *Structure de Hopf de l'anneau de cohomologie et de l'anneau de Grothendieck d'une variété de drapeaux*, Comptes Rendus **295**, 629 (1982).
18. A. Lascoux & M.P. Schützenberger, *Schubert polynomials and the Littlewood-Richardson rule*, Letters in Math.Phys. **10**, 111–124 (1985).

19. A. Lascoux & M.P. Schützenberger, *Symmetrization operators on polynomial rings*, Funct.Anal. **21**, 77–78 (1987).
20. A. Lascoux & M.P. Schützenberger, *Tableaux and non commutative Schubert Polynomials*, Funk.Anal **23**, 63–64 (1989).
21. A. Lascoux & M.P. Schützenberger, *Keys and Standard Bases*, "Invariant Theory and Tableaux", IMA Vol. in Maths and Appl. **19**,125–144, Springer (1990).
22. A. Lascoux & M.P. Schützenberger, *Algèbre des différences divisées*, Discrete Maths **99**, 165–179 (1992).
23. C. Lenart, *Noncommutative Schubert calculus and Grothendieck polynomials*, Adv. in Math. **143**, 159–183 (1999).
24. C. Lenart, *Combinatorial aspects of the K-theory of Grassmannians* Annals of Comb **4**, 67–82 (2000).
25. I. G. Macdonald, *Notes on Schubert polynomials*, Pub. du LACIM, Université du Québec à Montréal, (1991).
26. V. Reiner, M. Shimozono, *Key polynomials and a flagged Littlewood-Richardson rule*, J. Comb. Th. A **70**, 107–143 (1995).

TABLEAU REPRESENTATION FOR MACDONALD'S NINTH VARIATION OF SCHUR FUNCTIONS

JUN NAKAGAWA
Department of Mathematics, Kobe University,
Rokko, Kobe 657-8501, Japan
E-mail: nakagawa@math.kobe-u.ac.jp

MASATOSHI NOUMI
Department of Mathematics, Kobe University,
Rokko, Kobe 657-8501, Japan
E-mail: noumi@math.sci.kobe-u.ac.jp

MIKI SHIRAKAWA
Department of Mathematics, Kobe University,
Rokko, Kobe 657-8501, Japan
E-mail: sirakawa@math.kobe-u.ac.jp

YASUHIKO YAMADA
Department of Mathematics, Kobe University,
Rokko, Kobe 657-8501, Japan
E-mail: yamaday@math.kobe-u.ac.jp

The ninth variation of Schur functions associated with a general $N \times N$ matrix is studied. In particular, two types of tableau representations for them are constructed by using decomposition of matrices and properties of minor determinants.

Introduction

The purpose of this paper is to give several constructions for Macdonald's ninth variation for Schur functions [4] such as the Weyl formula, the Jacobi-Trudi formula and the tableau representation. As for the terminology of partitions and symmetric functions, we basically follow the textbook of Macdonald [5] (see also Chapter I, §3, Examples 21 and §5, Example 30, as for the ninth variation).

Let $h_k^{(r)}$ be a set of variables indexed by $k = 1, 2, \ldots$ and $r \in \mathbb{Z}$. For convenience, define $h_0^{(r)} = 1$ and $h_k^{(r)} = 0$ $(k < 0)$ for all $r \in \mathbb{Z}$. For a given partition $\lambda = (\lambda_1, \lambda_2, \ldots)$, I.G. Macdonald [4] defined the *ninth variation of Schur function* S_λ by the Jacobi-Trudi formula

$$S_\lambda = \det(h^{(1-j)}_{\lambda_i - i + j})_{1 \le i,j \le l(\lambda)}, \tag{0.1}$$

where $l(\lambda)$ stands for the number of parts of λ. It is convenient to extend this definition to

$$S_\lambda^{(r)} = \det(h_{\lambda_i-i+j}^{(r+1-j)})_{1\le i,j\le l(\lambda)} \qquad (r \in \mathbb{Z}), \tag{0.2}$$

so that $S_\lambda = S_\lambda^{(0)}$. By specializing the variables $h_k^{(r)}$, $S_\lambda^{(r)}$ provide with many variations of Schur functions [4].

In this paper, we define the ninth variation of Schur functions $S_\lambda^{(r)}$ associated with a general matrix X in terms of the Weyl formula. We then prove the Jacobi-Trudi formula and the Giambelli formula by using appropriate decompositions of matrices. By means of elementary properties of minor determinants, we also construct tableau representations for $S_\lambda^{(r)}$. We remark that such tableau representations as we discuss in Section 3 appeared for the first time in the work of A.N. Kirillov and N.Y. Reshetikhin on spin chain models (see also the paper by V.V. Bazhanov and N.Y. Reshetikhin [2]).

Throughout this paper, we use the following notation of minor determinants for an $N \times N$ matrix $X = (x_{ij})_{1\le i,j\le N}$:

$$\xi_{j_1,\ldots,j_r}^{i_1,\ldots,i_r}(X) = \det(x_{i_a,j_b})_{1\le a,b\le r}. \tag{0.3}$$

For two subsets $I, J \subset \{1,\ldots,N\}$ with $|I| = |J| = r$, we define $\xi_J^I(X)$ to be

$$\xi_J^I(X) = \xi_{j_1,\ldots,j_r}^{i_1,\ldots,i_r}(X), \tag{0.4}$$

where $i_1 < i_2 < \ldots < i_r$ and $j_1 < j_2 < \ldots < j_r$ are the sequences obtained by arranging the elements of I and J in the increasing order, respectively. When $I = \{1,\ldots,r\}$, we simply write $\xi_J(X) = \xi_{j_1,\ldots,j_r}(X)$ for $\xi_J^{\{1,\ldots,r\}}(X) = \xi_{j_1,\ldots,j_r}^{1,\ldots,r}(X)$. Recall that, as for the product of matrices, the minor determinants are determined by the expansion

$$\xi_J^I(XY) = \sum_K \xi_K^I(X)\,\xi_J^K(Y). \tag{0.5}$$

This formula plays the key role in constructing tableau representations for $S_\lambda^{(r)}$.

1 Weyl formula and Jacobi-Trudi formula

In this section, we define the ninth variation of Schur functions for a general $N \times N$ matrix X, and prove two types of Jacobi-Trudi formulas and the Giambelli formula for them. (We remark that a similar argument can be carried out for a general matrix X with the infinite indexing set $\mathbb{N}$ or $\mathbb{Z}$, with appropriate modifications.)

In the following, we fix an $N \times N$ matrix $X = (x_{ij})_{1\le i,j\le N}$, and assume that X is *generic* in the sense that the principal minor determinants $\xi^{1,\dots,i}_{1,\dots,i}(X)$ $(i = 1,\dots,N)$ do not vanish; otherwise, simply regard x_{ij} as indeterminates. Taking two nonnegative integers r, s with $r+s=N$, let $\lambda = (\lambda_1, \lambda_2, \dots)$ be a partition contained in the $r \times s$ rectangle (s^r). This condition is equivalent to saying that $l(\lambda) \le r$ and $l(\lambda') = \lambda_1 \le s$, where λ' is the conjugate partition of λ. To such a λ we associate a subset J of $\{1,\dots,N\}$ with $|J| = r$ by setting

$$J = \{j_1, \dots, j_r\}, \quad j_a = \lambda_{r+1-a} + a \quad (a = 1, \dots, r). \tag{1.6}$$

This subset J, or the increasing sequence of indices $(j_1,\dots,j_r)$, is called the *Maya diagram* for λ. Note that this correspondence gives rise to a bijection between the partitions λ contained in the $r \times s$ rectangle (s^r) and the subsets of $\{1,\dots,N\}$ of cardinality r.

We define the *ninth variation of Schur function* $S^{(r)}_\lambda$ *for* X in terms of the *Weyl formula*

$$S^{(r)}_\lambda = \frac{\xi^{1,\dots,r}_{j_1,\dots,j_r}(X)}{\xi^{1,\dots,r}_{1,\dots,r}(X)}, \tag{1.7}$$

as the ratio of minor determinants of X, specified by the Maya diagram $J = \{j_1 < \dots < j_r\}$ for λ. As for the partitions λ of one row and of one column, we use the notation

$$\begin{aligned} h^{(r)}_k &= S^{(r)}_{(k)} = \frac{\xi^{1,\dots,r}_{1,\dots,r-1,r+k}(X)}{\xi^{1,\dots,r}_{1,\dots,r}(X)} \qquad (k = 0,1,\dots,s), \\ e^{(r)}_k &= S^{(r)}_{(1^k)} = \frac{\xi^{1,\dots,r}_{1,\dots,\widehat{r-k+1},\dots,r+1}(X)}{\xi^{1,\dots,r}_{1,\dots,r}(X)} \qquad (k = 0,1,\dots,r), \end{aligned} \tag{1.8}$$

respectively. For convenience we define $h^{(r)}_k = e^{(r)}_k = 0$ for $k < 0$.

Theorem 1.1 (Jacobi-Trudi formula) *For each partition* λ *contained in the rectangle* (s^r)*, the ninth variation of Schur function* $S^{(r)}_\lambda$ *for* X*, defined by Eq. (1.7), has the Jacobi-Trudi formulas*

$$S^{(r)}_\lambda = \det\left(h^{(r+1-j)}_{\lambda_i - i + j}\right)_{1\le i,j\le r}, \quad S^{(r)}_\lambda = \det\left(e^{(r-1+j)}_{\lambda'_i - i + j}\right)_{1\le i,j\le s}. \tag{1.9}$$

Proof. Consider the Gauss decomposition of X:

$$X = X_-\, X_0\, X_+, \tag{1.10}$$

where X_-, X_0 and X_+ are lower unitriangular, diagonal and upper unitriangular, respectively. Note that these three matrices are determined uniquely as matrices with entries in the field of rational functions in the variables x_{ij} $(1 \leq i, j \leq N)$. Furthermore, the entries of these matrices are given explicitly as follows:

$$(X_-)_{ij} = \frac{\xi^{1,\ldots,j-1,i}_{1,\ldots,j-1,j}(X)}{\xi^{1,\ldots,j}_{1,\ldots,j}(X)} \qquad (i \geq j),$$

$$(X_0)_{ii} = \frac{\xi^{1,\ldots,i}_{1,\ldots,i}(X)}{\xi^{1,\ldots,i-1}_{1,\ldots,i-1}(X)} \qquad (i = 1, 2, \ldots, N), \tag{1.11}$$

$$(X_+)_{ij} = \frac{\xi^{1,\ldots,i-1,i}_{1,\ldots,i-1,j}(X)}{\xi^{1,\ldots,i}_{1,\ldots,i}(X)} \qquad (i \leq j).$$

Also the inverse matrix X_+^{-1} of the upper triangular component is given by

$$(X_+^{-1})_{ij} = (-1)^{j-i} \frac{\xi^{1,\ldots,j-1}_{1,\ldots,\widehat{i},\ldots,j}(X)}{\xi^{1,\ldots,j-1}_{1,\ldots,j-1}(X)} \qquad (i \leq j). \tag{1.12}$$

It is worthwhile to notice that

$$(X_+)_{ij} = h^{(i)}_{j-i}, \quad (X_+^{-1})_{ij} = (-1)^{j-i}\, e^{(j-1)}_{j-i} \qquad (i \leq j). \tag{1.13}$$

We take a partition $\lambda \subset (s^r)$ and the increasing sequence of indices $(j_1, \ldots, j_r)$ defined by Eq. (1.6). Then the minor determinant $\xi^{1,\ldots,r}_{j_1,\ldots,j_r}(X)$ is computed via the Gauss decomposition as

$$\xi^{1,\ldots,r}_{j_1,\ldots,j_r}(X) = \xi^{1,\ldots,r}_{1,\ldots,r}(X_0)\, \xi^{1,\ldots,r}_{j_1,\ldots,j_r}(X_+). \tag{1.14}$$

Since

$$\xi^{1,\ldots,r}_{1,\ldots,r}(X_0) = \prod_{i=1}^{r} \frac{\xi^{1,\ldots,i}_{1,\ldots,i}(X)}{\xi^{1,\ldots,i-1}_{1,\ldots,i-1}(X)} = \xi^{1,\ldots,r}_{1,\ldots,r}(X), \tag{1.15}$$

we have

$$S^{(r)}_\lambda = \frac{\xi^{1,\ldots,r}_{j_1,\ldots,j_r}(X)}{\xi^{1,\ldots,r}_{1,\ldots,r}(X)} = \xi^{1,\ldots,r}_{j_1,\ldots,j_r}(X_+). \tag{1.16}$$

By Eq. (1.13), the right-hand side is rewritten as

$$\begin{aligned} \xi^{1,\ldots,r}_{j_1,\ldots,j_r}(X_+) &= \det((X_+)_{a,j_b})_{1\leq a,b\leq r} \\ &= \det(h^{(a)}_{j_b-a})_{1\leq a,b\leq r} \\ &= \det(h^{(r+1-a)}_{j_{r+1-b}-(r+1-a)})_{1\leq a,b\leq r} \\ &= \det(h^{(r+1-a)}_{\lambda_b-b+a})_{1\leq a,b\leq r}. \end{aligned} \tag{1.17}$$

Hence we have the first Jacobi-Trudi formula

$$S_\lambda^{(r)} = \xi^{1,\dots,r}_{j_1,\dots,j_r}(X_+) = \det(h^{(r+1-a)}_{\lambda_b-b+a})_{1\le a,b\le r}. \tag{1.18}$$

We now consider the increasing sequence $(k_1, k_2, \dots, k_s)$ of indices, complementary to $(j_1, \dots, j_r)$:

$$k_a = r + a - \lambda'_a \quad (a = 1, \dots, s). \tag{1.19}$$

Note that

$$\{j_1, \dots, j_r\} \cup \{k_1, \dots, k_s\} = \{1, \dots, r, r+1, \dots, r+s = N\}. \tag{1.20}$$

The minor determinant $\xi^{1,\dots,r}_{j_1,\dots,j_r}(X_+)$ is then expressed as

$$\xi^{1,\dots,r}_{j_1,\dots,j_r}(X_+) = (-1)^{|\lambda|}\,\xi^{k_1,\dots,k_s}_{r+1,\dots,r+s}(X_+^{-1}) \tag{1.21}$$

in terms of the inverse matrix, where $|\lambda| = \sum_{a=1}^{r} j_a - a = \sum_{a=1}^{s} r + a - k_a$. Again by Eq. (1.13), we compute

$$\begin{aligned} \xi^{k_1,\dots,k_s}_{r+1,\dots,r+s}(X_+^{-1}) &= \det((X_+^{-1})_{k_a,r+b})_{1\le a,b\le s} \\ &= \det((-1)^{r+b-k_a}\, e^{(r+b-1)}_{r+b-k_a})_{1\le a,b\le s} \\ &= (-1)^{|\lambda|} \det(e^{(r+b-1)}_{\lambda'_a-a+b})_{1\le a,b\le s}. \end{aligned} \tag{1.22}$$

Hence we have

$$S_\lambda^{(r)} = (-1)^{|\lambda|}\,\xi^{k_1,\dots,k_s}_{r+1,\dots,r+s}(X_+^{-1}) = \det(e^{(r+b-1)}_{\lambda'_a-a+b})_{1\le a,b\le s}, \tag{1.23}$$

which gives the second Jacobi-Trudi formula of Eq. (1.9). Q.E.D.

Let μ be a partition contained in λ, and let I be the corresponding Maya diagram with $|I| = r$. The ninth variation of skew Schur function $S^{(r)}_{\lambda/\mu}$ is then defined by

$$S^{(r)}_{\lambda/\mu} = \xi^I_J(X_+) = (-1)^{|\lambda/\mu|}\,\xi^{J^c}_{I^c}(X_+^{-1}), \tag{1.24}$$

where I^c, J^c are the complements of I, J in $\{1, \dots, N\}$, respectively. This definition can be written as

$$S^{(r)}_{\lambda/\mu} = \det\left(h^{(\mu_j+r-j+1)}_{\lambda_i-\mu_j-i+j}\right)_{1\le i,j\le r} = \det\left(e^{(r-1-\mu'_j+j)}_{\lambda'_i-\mu'_j-i+j}\right)_{1\le i,j\le s}. \tag{1.25}$$

By a similar method, one can also prove the Giambelli formula for $S^{(r)}_\lambda$. We introduce the following notation for the hooks $(p\,1^q)$:

$$h^{(r)}_{p,q} = S^{(r)}_{(p\,1^q)} = \frac{\xi^{1,\dots,r}_{1,\dots,\widehat{r-q},\dots,r,r+p}(X)}{\xi^{1,\dots,r}_{1,\dots,r}(X)} \quad (1 \le p \le s,\ 0 \le q \le r-1). \tag{1.26}$$

For a given partition $\lambda \subset (s^r)$, we define the two sequences of indices

$$p_1 > p_2 > \cdots > p_t > 0, \quad q_1 > q_2 > \cdots > q_t \geq 0 \tag{1.27}$$

by the condition

$$\begin{aligned} \{j_1, \ldots, j_r\} \setminus \{1, \ldots, r\} &= \{r + p_t, \ldots, r + p_1\} \\ \{1, \ldots, r\} \setminus \{j_1, \ldots, j_r\} &= \{r - q_1, \ldots, r - q_t\}. \end{aligned} \tag{1.28}$$

Note that the partition λ is recovered as the union on t hooks $(p_a\, 1^{q_a})$ $(a = 1, \ldots, t)$ located on the main diagonal. Equivalently,

$$\lambda = (p_1 - 1, \ldots, p_t - 1 | q_1, \ldots, q_t) \tag{1.29}$$

in the Frobenius notation [5].

Theorem 1.2 (Giambelli formula) *With the two sequences $(p_1, \ldots, p_t)$ and $(q_1, \ldots, q_t)$ of indices defined as above, $S_\lambda^{(r)}$ is expressed as*

$$S_\lambda^{(r)} = \det\left(h_{p_i, q_j}^{(r)}\right)_{1 \leq i,j \leq t}. \tag{1.30}$$

Proof. We consider the $r \times N$ matrix $Y = (x_{i,j})_{1 \leq i \leq r, 1 \leq j \leq N}$ and decompose Y into the form $Y = T\,Z$, where $T = (x_{i,j})_{1 \leq i,j \leq r}$ and $Z = T^{-1}Y$. The matrix Z is determined as follows: $Z_{ij} = \delta_{ij}$ for $1 \leq i, j \leq r$ and

$$Z_{ij} = (-1)^{r-i} \frac{\xi^{1,\ldots,r}_{1,\ldots,\widehat{i},\ldots,r,j}(X)}{\xi^{1,\ldots,r}_{1,\ldots,r}(X)} = (-1)^{r-i} h^{(r)}_{j-r,r-i} \tag{1.31}$$

for $1 < i < r$ and $r + 1 \leq j \leq N$. Via this decomposition, we have

$$\xi^{1,\ldots,r}_{j_1,\ldots,j_r}(X) = \xi^{1,\ldots,r}_{1,\ldots,r}(T)\, \xi^{1,\ldots,r}_{j_1,\ldots,j_r}(Z). \tag{1.32}$$

Since $\xi^{1,\ldots,r}_{1,\ldots,r}(T) = \xi^{1,\ldots,r}_{1,\ldots,r}(X)$, we have

$$\begin{aligned} S_\lambda^{(r)} &= \xi^{1,\ldots,r}_{j_1,\ldots,j_r}(Z) \\ &= (-1)^{|q| + t(t-1)/2} \xi^{r-q_1,\ldots,r-q_t}_{r+p_t,\ldots,r+p_1}(Z) \\ &= (-1)^{|q|} \det(Z_{r-q_i, r+p_j})_{1 \leq i,j \leq t} \\ &= \det(h^{(r)}_{p_j,q_i})_{1 \leq i,j \leq t}, \end{aligned} \tag{1.33}$$

as desired. Q.E.D.

Remark. Let us take the Vandermonde matrix $X = (x_i^{j-1})_{1 \leq i,j \leq N}$. In this special case, our ninth variation of Schur function $S_\lambda^{(r)}$ for X is given by

$$S_\lambda^{(r)} = s_\lambda(x_1, \ldots, x_r), \tag{1.34}$$

for $0 \le r \le N$ and $\lambda \subset (s^r)$. In particular, we have

$$\begin{aligned} h_k^{(r)} &= h_k(x_1, \ldots, x_r) \quad (0 \le k \le s), \\ e_k^{(r)} &= e_k(x_1, \ldots, x_r) \quad (0 \le k \le r), \end{aligned} \tag{1.35}$$

where h_k and e_k denotes the complete homogeneous symmetric functions and the elementary symmetric functions in the usual sense, respectively. Then Theorem 1.1 implies

$$s_\lambda(x_1, \ldots, x_r) = \det\left(h_{\lambda_i - i + j}(x_1, \ldots, x_{r-j+1})\right)_{1 \le i,j \le l(\lambda)}. \tag{1.36}$$

Fix a positive integer n and set $l = l(\lambda)$ and $m = l(\lambda') = \lambda_1$. We take r and s so that $r \ge n + l - 1$ and $s \ge m + l - 1$. Then by setting $x_{n+1} = \cdots = x_N = 0$ $(N = r + s)$ in Eq. (1.36), we obtain the usual Jacobi-Trudi formula

$$s_\lambda(x_1, \ldots, x_n) = \det\left(h_{\lambda_i - i + j}(x_1, \ldots, x_n)\right)_{1 \le i,j \le l} \tag{1.37}$$

for Schur functions in n variables. Similarly, from

$$s_\lambda(x_1, \ldots, x_r) = \det\left(e_{\lambda'_i - i + j}(x_1, \ldots, x_{r+j-1})\right)_{1 \le i,j \le l(\lambda')}, \tag{1.38}$$

we obtain

$$s_\lambda(x_1, \ldots, x_r) = \det\left(e_{\lambda'_i - i + j}(x_1, \ldots, x_{r+j-1})\right)_{1 \le i,j \le m}. \tag{1.39}$$

(Take $r \ge \max\{l + m - 1, n\}$, $s \ge m$ and set $x_{n+1} = \cdots = x_N = 0$.) We also remark that, from Eq. (1.25), we have

$$S_{\lambda/\mu}^{(r)} = \det\left(h_{\lambda_i - \mu_j - i + j}(x_1, \ldots, x_{\mu_j + r - j + 1})\right)_{1 \le i,j \le l(\lambda)}, \tag{1.40}$$

hence,

$$\begin{aligned} S_{\lambda/\mu}^{(r)}\big|_{x_{n+1} = \cdots = x_N = 0} &= \det\left(h_{\lambda_i - \mu_j - i + j}(x_1, \ldots, x_n)\right)_{1 \le i,j \le l(\lambda)} \\ &= s_{\lambda/\mu}(x_1, \ldots, x_n) \end{aligned} \tag{1.41}$$

for any $\mu \subset \lambda$, provided that $r \ge n + l - 1$ and $s \ge m + l - 1$.

2 Tableau representation

As we have seen in the previous section, the ninth variation of Schur function $S_\lambda^{(r)}$ for a general matrix X is represented by a minor determinant $\xi_J(X_+)$ of the upper triangular component X_+ in the Gauss decomposition of X. Suppose furthermore that X_+ is decomposed into the product

$$X_+ = U_1 U_2 \cdots U_m \tag{2.42}$$

of several upper unitriangular matrices U_k of particular form. Let I, J be two subsets of $\{1,\dots,N\}$ of cardinality r. Then for each I, J with $|I|=|J|=r$, the minor determinant $\xi^I_J(X_+)$ is expressed as a sum

$$\xi^I_J(X_+) = \sum_{K_1,\dots,K_{m-1}} \xi^I_{K_1}(U_1)\xi^{K_1}_{K_2}(U_2)\cdots\xi^{K_{m-1}}_J(U_m) \tag{2.43}$$

over all sequences $(K_1,\dots,K_{m-1})$ of indexing sets of cardinality r. In particular, we have

$$S^{(r)}_\lambda = \sum_{K_1,\dots,K_{m-1}} u^1_{IK_1}u^2_{K_1K_2}\cdots u^m_{K_{m-1}J}, \tag{2.44}$$

where $I=\{1,\dots,r\}$ and $u^k_{AB}=\xi^A_B(U_k)$. If the matrices U_k are chosen so that the minor determinants u^k_{AB} can be described nicely, the formula Eq. (2.44) will possibly give rise to a combinatorial expression for $S^{(r)}_\lambda$. In this way, one can provide with various combinatorial formulas for $S^{(r)}_\lambda$, depending on the choice of the decomposition Eq. (2.42). Note also that, as is often the case, the sequences of indexing sets, over which the summation is taken, can be interpreted in terms of some other types of combinatorial objects such as non-intersecting paths (in the sense of I. Gessel and G. Viennot [3]) and Young tableaux. What we will discuss in this section concerns only the upper triangular matrix X_+, and does not depend on the components X_-, X_0 of the Gauss decomposition Eq. (1.10) of X.

We now consider two types of upper unitriangular matrices for the building blocks of the decomposition Eq. (2.42) of X_+:

$$\begin{aligned}(1): U &= (1+u_1E_{12})(1+u_2E_{23})\cdots(1+u_{N-1}E_{N-1,N}),\\ (2): V &= (1+v_{N-1}E_{N-1,N})\cdots(1+v_2E_{23})(1+v_1E_{12}),\end{aligned} \tag{2.45}$$

where $E_{ij}=(\delta_{ia}\delta_{jb})_{1\le a,b\le N}$ denote the matrix units. Note that $V=U^{-1}$ if $v_i=-u_i$ for all i. In this sense, these two types of matrices are *dual* to each other. Note also that the matrix V is in fact 2-diagonal:

$$V = 1+\sum_{i=1}^{N-1} v_i\,E_{i,i+1}. \tag{2.46}$$

Let I, J be two subsets of $\{1,\dots,N\}$ with $|I|=|J|=r$ and μ, λ the corresponding partitions. Consider first an elementary matrix $1+uE_{i,i+1}$ $(1\le i\le N-1)$. Then there are only two cases where the minor determinant $\xi^I_J(1+uE_{i,i+1})$ is nonzero: (1) $I=J$; or (2) $i\in I$, $i+1\notin I$ and J is obtained from I by replacing $i+1$ for i. In these two cases, $\xi^I_J(1+uE_{i,i+1})$ is equal to 1 or u, respectively. From this fact, one can easily prove

Lemma 2.1 *Let I, J and μ, λ as above.*
(1) *The minor determinant $\xi^I_J(U)$ of U in Eq. (2.45) vanish unless $\mu \subset \lambda$ and the skew diagram λ/μ is a horizontal strip. If λ/μ is a horizontal strip, one has*

$$\xi^I_J(U) = \prod_{(a,b)\in\lambda/\mu} u_{r-a+b}. \tag{2.47}$$

(2) *The minor determinant $\xi^I_J(V)$ of V in Eq. (2.45) vanish unless $\mu \subset \lambda$ and the skew diagram λ/μ is a vertical strip. If λ/μ is a vertical strip, one has*

$$\xi^I_J(V) = \prod_{(a,b)\in\lambda/\mu} v_{r-a+b}. \tag{2.48}$$

The conditions on the skew diagram λ/μ in Lemma 2.1 can be interpreted in terms of the Maya diagrams $I = \{i_1 < \cdots < i_r\}$ and $J = \{j_1 < \cdots < j_r\}$. Firstly, the condition that $\mu \subset \lambda$ and that λ/μ is a horizontal strip holds if and only if

$$i_1 \le j_1 < i_2 \le j_2 < \cdots < i_r \le j_r. \tag{2.49}$$

This condition can also be represented graphically as follows.

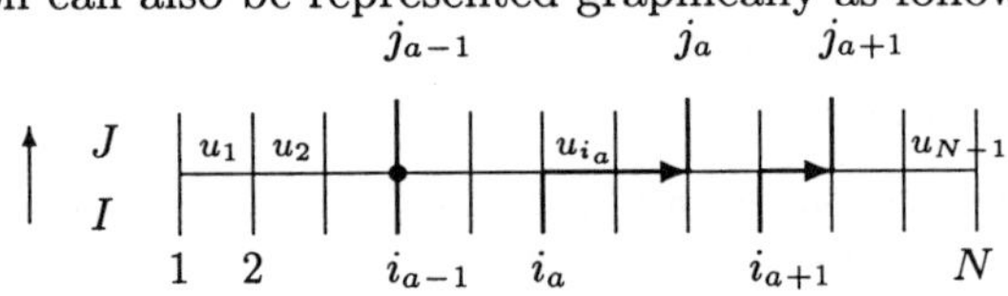

(2.50)

The minor determinant $\xi^I_J(U)$ also has an alternative expression

$$\xi^I_J(U) = \prod_{a=1}^{r} u_{i_a} u_{i_a+1} \cdots u_{j_a-1}, \tag{2.51}$$

as the product of the u-variables attached to the horizontal line segments that the paths pass through in the figure above. Similarly, the condition for the skew diagram λ/μ to be a vertical strip is equivalent to

$$j_a = i_a \quad \text{or} \quad i_a + 1 \quad \text{for} \quad a = 1, \ldots, r. \tag{2.52}$$

By using the graphical expression

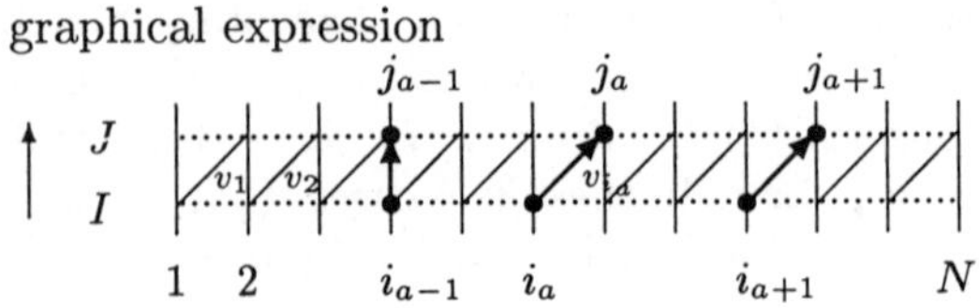

(2.53)

the minor determinant $\xi_J^I(V)$ is determined as the product of the v-variables attached to the slanted line segments lying on the paths:

$$\xi_J^I(V) = \prod_{1 \le a \le r : j_a = i_a + 1} v_{i_a}. \tag{2.54}$$

We now suppose that X_+ is expressed as the product

$$X_+ = U_1 U_2 \cdots U_m, \tag{2.55}$$

where

$$U_k = (1 + u_k^{(1)} E_{12})(1 + u_k^{(2)} E_{23}) \cdots (1 + u_k^{(N-1)} E_{N-1,N}) \tag{2.56}$$

for $k = 1, \ldots, m$. It is convenient to arrange the variables $u_k^{(i)}$ graphically as in Fig. 1. For two sets I, J of indices with $|I| = |J| = r$, we have the expression of the minor determinant

$$\xi_J^I(X_+) = \sum_{K_1, \ldots, K_{m-1}} \xi_{K_1}^I(U_1) \xi_{K_2}^{K_1}(U_2) \cdots \xi_J^{K_{m-1}}(U_m). \tag{2.57}$$

Let μ and λ be the partitions corresponding to the Maya diagrams I and J, respectively. Also, let $\lambda^{(k)}$ be the partition corresponding to K_k for $k = 0, 1, \ldots, m$, setting $K_0 = I$ and $K_m = J$. Note that a sequence $(K_0, K_1, \ldots, K_m)$ with $|K_k| = r$ contributes nontrivially to the sum Eq. (2.57) only if each pair (K_{k-1}, K_k) satisfies the condition of Lemma 2.1, (1) for each $k = 1, \ldots, m$. Such sequences of subsets $(K_0, K_1, \ldots, K_m)$ are in one-to-one correspondence with the sequences of partitions

$$\mu = \lambda^{(0)} \subset \lambda^{(1)} \subset \cdots \subset \lambda^{(m)} = \lambda, \tag{2.58}$$

such that $\lambda^{(k)}/\lambda^{(k-1)}$ $(k = 1, \ldots, m)$ are horizontal strips. Representing such sequences of partitions in terms of the semi-standard tableaux (column strict tableaux[5]) of shape λ/μ with labels $\{1, \ldots, m\}$, we obtain the tableau representation for the ninth variation of skew Schur function $S_{\lambda/\mu}^{(r)}$:

$$S_{\lambda/\mu}^{(r)} = \sum_{T \in \mathrm{SSTab}_m(\lambda/\mu)} \prod_{(a,b) \in \lambda/\mu} u_{T(a,b)}^{(r-a+b)}. \tag{2.59}$$

We recall that the semi-standard tableau for the sequence Eq. (2.58) is defined as the mapping $T : \lambda/\mu \to \{1, \ldots, m\}$ such that

$$T(a,b) = k \qquad \Longleftrightarrow \qquad (a,b) \in \lambda^{(k)}/\lambda^{(k-1)}, \tag{2.60}$$

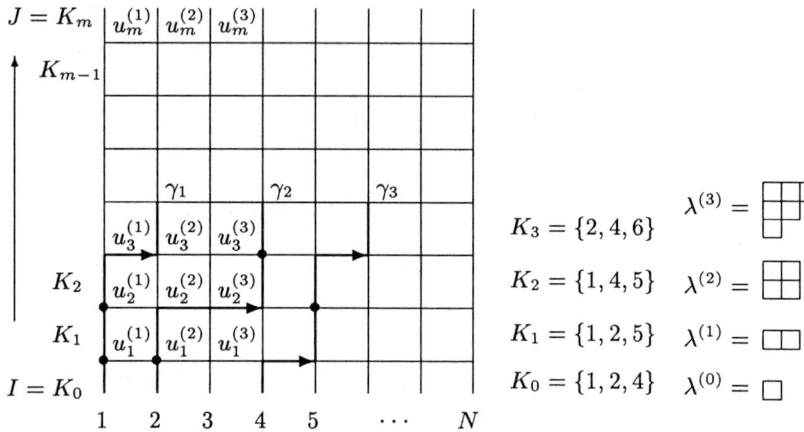

Figure 1. Non-intersecting paths (1)

for each $(a,b) \in \lambda/\mu$. In terms of the graphical expression Eq. (2.50), it is rewritten as the summation

$$S^{(r)}_{\lambda/\mu} = \sum_{(\gamma_1,\ldots,\gamma_r)} u_{\gamma_1} \cdots u_{\gamma_r} \tag{2.61}$$

over all r-tuple of non-intersecting paths $(\gamma_1,\ldots,\gamma_r)$ connecting $(i_1,\ldots,i_r)$ to $(j_1,\ldots,j_r)$, as in Fig. 1. Here, for a path γ, u_γ denotes the product of all u-variables $u_k^{(i)}$ attached to the horizontal line segments lying on γ.

We also formulate the other type of tableau representation associated with the matrices of the form V of Eq. (2.45). Suppose that X_+ is decomposed into

$$X_+ = V_1\, V_2 \,\cdots\, V_m, \tag{2.62}$$

where

$$V_k = (1 + v_k^{(N-1)} E_{N-1,N}) \cdots (1 + v_k^{(2)} E_{23})(1 + v_k^{(1)} E_{12}), \tag{2.63}$$

From this decomposition, we obtain the tableau representation

$$S^{(r)}_{\lambda/\mu} = \sum_{T \in \mathrm{SSTab}_m(\lambda'/\mu')} \prod_{(a,b)\in\lambda'/\mu'} v^{(r+a-b)}_{T(a,b)}. \tag{2.64}$$

It is also represented as a sum

$$S^{(r)}_{\lambda/\mu} = \sum_{(\gamma_1,\ldots,\gamma_r)} v_{\gamma_1} \cdots v_{\gamma_r} \tag{2.65}$$

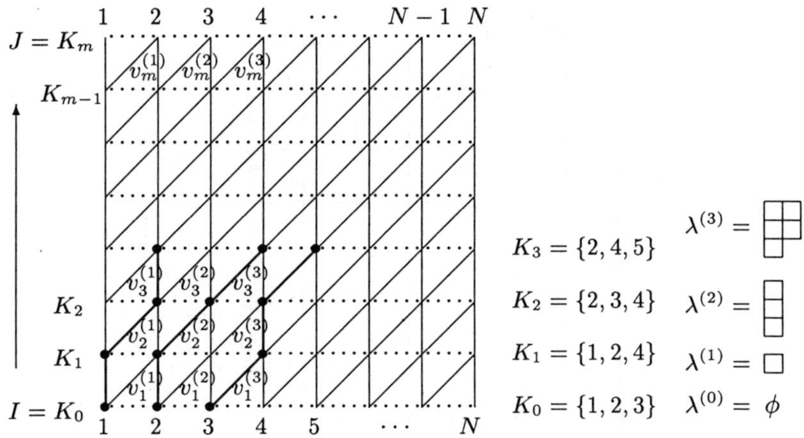

Figure 2. Non-intersecting paths (2)

over all r-tuple of non-intersecting paths $(\gamma_1, \ldots, \gamma_r)$ connecting $(i_1, \ldots, i_r)$ to $(j_1, \ldots, j_r)$, as in Fig. 2. Here, v_γ denotes the product of all v-variables $v_k^{(i)}$ attached to the slanted line segments lying on γ.

3 Specifying the tableau variables

In this section, we discuss a particular choice of decompositions Eq. (2.55) and Eq. (2.62) such that the variables $u_k^{(i)}$ and $v_k^{(i)}$ for the tableau representations can be specified in terms of the matrix X or X_+ itself.

For each pair (a, b) of nonnegative integers with $a + b \leq N$, define

$$Q_a^b = \xi_{b+1,\ldots,b+a}^{1,\ldots,a}(X) = \det\left(x_{i,b+j}\right)_{1\leq i,j\leq a}, \tag{3.66}$$

so that

$$Q_0^b = 1, \quad \frac{Q_a^b}{Q_a^0} = S_{(b^a)}^{(a)}. \tag{3.67}$$

We also define the variables

$$x_a^{(b)} = \frac{Q_a^{b+1}\, Q_{a-1}^b}{Q_{a-1}^{b+1}\, Q_a^b} \tag{3.68}$$

for $a \geq 1$ $b \geq 0$ with $a + b \leq N - 1$. We set $x_a^{(b)} = 0$ for $b < 0$. (In these definitions, we have tacitly assumed that the denominators do not vanish.)

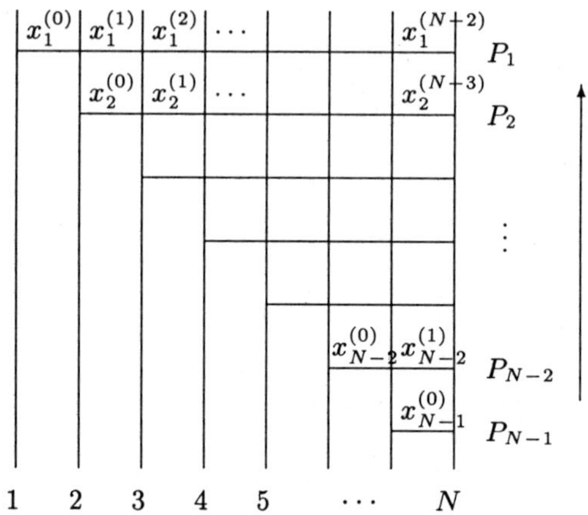

Figure 3. Arrangement of x-variables (1)

Proposition 3.1 *The upper triangular matrix X_+ is decomposed into the product*

$$X_+ = P_{N-1} \cdots P_1, \tag{3.69}$$

where

$$P_k = (1 + x_k^{(0)} E_{k,k+1})(1 + x_k^{(1)} E_{k+1,k+2}) \cdots (1 + x_k^{(N-k-1)} E_{N-1,N}). \tag{3.70}$$

This type of decomposition of matrices is investigated extensively in the work of A. Berenstein, S. Fomin and A. Zelevinsky [1]. Here we indicate an idea for a proof of the proposition above. The fact that a generic upper unitriangular matrix has a decomposition of this form can be shown by induction on N, for example. Supposing that X_+ has a decomposition in the form above, consider the tableau representation Eq. (2.59) for the diagram Fig. 3. Then one sees immediately that, for $I = \{1, \ldots, a\}$ and $J = \{b+1, \ldots, b+a\}$, there is only one r-tuple of non-intersecting paths connecting I to J:

$$\frac{Q_a^b}{Q_a^0} = S_{(b^a)}^{(a)} = \prod_{i=1}^{a} x_i^{(0)} x_i^{(1)} \cdots x_i^{(b-1)}, \tag{3.71}$$

so that

$$\frac{Q_a^b}{Q_{a-1}^b} = x_a^{(0)} x_a^{(1)} \cdots x_a^{(b-1)}. \tag{3.72}$$

Hence we have the expression Eq. (3.68) for the x-variables.

Theorem 3.2 *With the x-variables defined as above, the ninth variation of Schur function* $S_\lambda^{(r)}$ *and of skew Schur function* $S_{\lambda/\mu}^{(r)}$ *have the tableau representations*

$$S_\lambda^{(r)} = \sum_{T\in \mathrm{SSTab}_r(\lambda)} \prod_{(a,b)\in\lambda} x_{r+1-T(a,b)}^{(T(a,b)-a+b-1)}, \tag{3.73}$$

and

$$S_{\lambda/\mu}^{(r)} = \sum_{T\in \mathrm{SSTab}_{r+\mu_1}(\lambda/\mu)} \prod_{(a,b)\in\lambda/\mu} x_{r+\mu_1+1-T(a,b)}^{(T(a,b)-\mu_1-a+b-1)} \tag{3.74}$$

respectively, where $\mu \subset \lambda \subset (s^r)$.

Proof. This theorem follows from the general tableau representation Eq. (2.59) for the arrangement of x-variables indicated in Fig. 3. Formula Eq. (2.59) for these x-variables reads as

$$S_{\lambda/\mu}^{(r)} = \sum_{T\in \mathrm{SSTab}_{N-1}(\lambda/\mu)} \prod_{(a,b)\in\lambda/\mu} x_{N-T(a,b)}^{(T(a,b)-N+r-a+b)}, \tag{3.75}$$

since $u_k^{(i)} = x_{N-k}^{(i+k-N)}$. Note that, in this expression, the sum is taken over all tableaux $T \in \mathrm{SSTab}_{N-1}$ such that

$$T(a,b) \geq N-r+a-b \quad \text{for any} \quad (a,b)\in\lambda/\mu. \tag{3.76}$$

This condition implies

$$T(a,b) \geq N-r-\mu_1 \quad \text{for any} \quad (a,b)\in\lambda/\mu. \tag{3.77}$$

By renaming the labels $\{N-r-\mu_1, \ldots, N-1\}$ as $\{1,\ldots,r+\mu_1\}$, we obtain Eq. (3.74), hence, Eq. (3.73) as the special case when $\mu_1 = 0$. Q.E.D.

We remark that, in Eq. (3.73), one has $T(a,b)-a+b-1 \geq 0$ for any $T \in \mathrm{SSTab}_r(\lambda)$ and $(a,b)\in\lambda$, since $T(a,b)\geq a$. It means that the tableau representation Eq. (3.73) for $S_\lambda^{(r)}$ has as many monomial terms as the semi-standard tableaux in $\mathrm{SSTab}_r(\lambda)$. In Eq. (3.74) for the case of skew-tableaux, not all $T\in \mathrm{SSTab}_{r+\mu_1}(\lambda/\mu)$ contribute in general; the summation is taken over all T such that $T(a,b)\geq \mu_1+a-b+1$ for all $(a,b)\in\lambda/\mu$.

Remark. In the special case where $X = (x_i^{j-1})_{1\leq i,j\leq N}$, one has

$$Q_a^b = (x_1\cdots x_a)^b \prod_{1\leq i<j\leq a}(x_j-x_i), \quad x_a^{(b)} = x_a \tag{3.78}$$

for all a, b. Hence, Eq. (3.73) recovers the well-known tableau representation for the Schur function $s_\lambda(x_r, x_{r-1},\ldots,x_1)$. Also, from Eq. (3.74) together

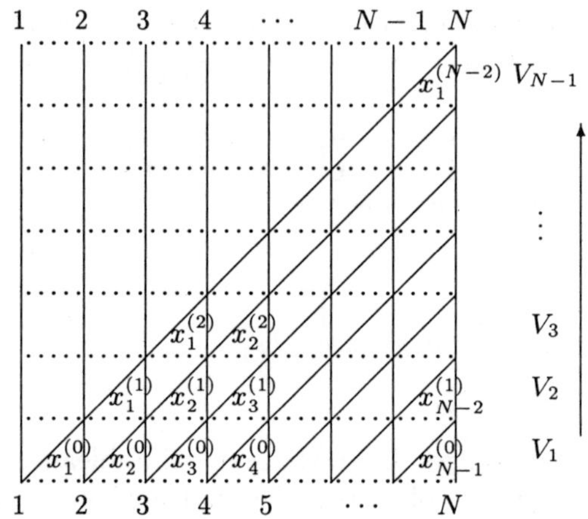

Figure 4. Arrangement of x-variables (2)

with Eq. (1.41), we obtain the tableau representation for the skew Schur function

$$s_{\lambda/\mu}(x_1, \ldots, x_n) = \sum_{T \in \mathrm{SSTab}_n(\lambda/\mu)} \prod_{(a,b)\in\lambda/\mu} x_{n+1-T(a,b)} \tag{3.79}$$

by shifting the labels appropriately.

By rearranging the order of elementary matrices in the decomposition Eq. (3.69), we obtain the following decomposition of X_+:

$$X_+ = V_1 V_2 \cdots V_{N-1}, \tag{3.80}$$

where

$$V_k = (1 + x_{N-k}^{(k-1)} E_{N-1,N}) \cdots (1 + x_2^{(k-1)} E_{k+1,k+2})(1 + x_1^{(k-1)} E_{k,k+1}) \tag{3.81}$$

for $k = 1, \ldots, N-1$, with the same x-variables Eq. (3.68). From the tableau representation of Eq. (2.64), we now obtain

$$S_{\lambda/\mu}^{(r)} = \sum_{T \in \mathrm{SSTab}_{N-1}(\lambda'/\mu')} \prod_{(a,b)\in\lambda'/\mu'} x_{r+a-b+1-T(a,b)}^{(T(a,b)-1)}. \tag{3.82}$$

This formula can be regarded as the dual version of Eq. (3.73), Eq. (3.74). Note that the summation is in fact taken over all $T \in \mathrm{SSTab}_{N-1}(\lambda'/\mu')$ such that

$$T(a,b) \leq r + a - b \quad \text{for all} \quad (a,b) \in \lambda'/\mu'. \tag{3.83}$$

References

1. A. Berenstein, S. Fomin and A. Zelevinsky, Parametrizations of canonical bases and totally positive matrices, Adv. in Math. **122**(1996), 49–149.
2. V.V. Bazhanov and N.Y. Reshetikhin, Scattering amplitudes in offcritical models and RSOS integrable models, in *Common trends in mathematics and quantum field theories, Kyoto, 1990*, Progr. Theoret. Phys. Suppl. **102**(1990), 301–318.
3. I. Gessel and G. Viennot, Binomial determinants, paths, and hook length formulae. Adv. in Math. **58**(1985), 300–321.
4. I.G. Macdonald, Schur functions: Theme and variations, Publ. I.R.M.A. Strasbourg, Acte 28e, Séminaire Lotharingien, 5–39, 1992.
5. I.G. Macdonald, *Symmetric Functions and Hall Polynomials*, Second Edition, Oxford University Press, 1995.

T-ANALOGUE OF THE Q-CHARACTERS OF FINITE DIMENSIONAL REPRESENTATIONS OF QUANTUM AFFINE ALGEBRAS

HIRAKU NAKAJIMA
Department of Mathematics, Kyoto University
Kyoto 606-8502, Japan
Email: nakajima@kusm.kyoto-u.ac.jp

Frenkel-Reshetikhin introduced q-characters of finite dimensional representations of quantum affine algebras [6]. We give a combinatorial algorithm to compute them for all simple modules. Our tool is t-analogue of the q-characters, which is similar to Kazhdan-Lusztig polynomials, and our algorithm has a resemblance with their definition. We need the theory of quiver varieties for the definition of t-analogues and the proof. But it appear only in the last section. The rest of the paper is devoted to an explanation of the algorithm, which one can read without the knowledge about quiver varieties. A proof is given only in part. A full proof will appear elsewhere.

1 The quantum loop algebra

Let $\mathfrak{g}$ be a simple Lie algebra of type ADE over $\mathbb{C}$, $\mathbf{L}\mathfrak{g} = \mathfrak{g} \otimes \mathbb{C}[z, z^{-1}]$ be its loop algebra, and $\mathbf{U}_q(\mathbf{L}\mathfrak{g})$ be its quantum universal enveloping algebra, or the quantum loop algebra for short. It is a subquotient of the quantum affine algebra $\mathbf{U}_q(\widehat{\mathfrak{g}})$, i.e., without central extension and degree operator. Let I be the set of simple roots, P be the weight lattice, and P^* be its dual lattice (all for $\mathfrak{g}$). The algebra has the so-called Drinfeld's new realization: It is a $\mathbb{C}(q)$-algebra with generators q^h, $e_{k,r}$, $f_{k,r}$, $h_{k,n}$ ($h \in P^*$, $k \in I$, $r \in \mathbb{Z}$, $n \in \mathbb{Z} \setminus \{0\}$) with certain relations (see e.g., [1, 12.2]).

The algebra $\mathbf{U}_q(\mathbf{L}\mathfrak{g})$ is a Hopf algebra, where the coproduct is defined using the Drinfeld-Jimbo realization of $\mathbf{U}_q(\mathbf{L}\mathfrak{g})$. So a tensor product $M \otimes_{\mathbb{C}(q)} M'$ of $\mathbf{U}_q(\mathbf{L}\mathfrak{g})$-modules M, M' has a structure of a $\mathbf{U}_q(\mathbf{L}\mathfrak{g})$-module.

Let $\mathbf{U}_\varepsilon(\mathbf{L}\mathfrak{g})$ be its specialization at $q = \varepsilon \in \mathbb{C}^*$. For precise definition of the specialization, we first introduce an integral form $\mathbf{U}_q^{\mathbb{Z}}(\mathbf{L}\mathfrak{g})$ of $\mathbf{U}_q(\mathbf{L}\mathfrak{g})$ and set $\mathbf{U}_\varepsilon(\mathbf{L}\mathfrak{g}) = \mathbf{U}_q^{\mathbb{Z}}(\mathbf{L}\mathfrak{g}) \otimes_{\mathbb{Z}[q,q^{-1}]} \mathbb{C}$, where $\mathbb{Z}[q, q^{-1}] \to \mathbb{C}$ is given by $q^{\pm 1} \mapsto \varepsilon^{\pm 1}$. See [3] for detail. But we assume ε is *not* a root of unity in this paper. So we just replace q by ε in the definition of $\mathbf{U}_q(\mathbf{L}\mathfrak{g})$.

The quantum loop algebra $\mathbf{U}_q(\mathbf{L}\mathfrak{g})$ contains the quantum enveloping algebra $\mathbf{U}_q(\mathfrak{g})$ for the finite dimensional Lie algebra $\mathfrak{g}$ as a subalgebra. The specialization $\mathbf{U}_\varepsilon(\mathbf{L}\mathfrak{g})$ contains the specialization $\mathbf{U}_\varepsilon(\mathfrak{g})$ of $\mathbf{U}_q(\mathfrak{g})$.

1.1 *Finite dimensional representations of* $\mathbf{U}_\varepsilon(\mathbf{L}\mathfrak{g})$

The algebra $\mathbf{U}_\varepsilon(\mathbf{L}\mathfrak{g})$ contains a commutative subalgebra generated by q^h, $h_{k,n}$ ($h \in P^*$, $k \in I$, $n \in \mathbb{Z} \setminus \{0\}$). Let us introduce generating functions $\psi_k^\pm(z)$ ($k \in I$) by

$$\psi_k^\pm(z) \stackrel{\text{def.}}{=} q^{\pm h_k} \exp\left(\pm(q - q^{-1}) \sum_{m=1}^{\infty} h_{k,\pm m} z^{\mp m}\right).$$

A $\mathbf{U}_\varepsilon(\mathbf{L}\mathfrak{g})$-module M is called of *type* 1 if M has a weight space decomposition as a $\mathbf{U}_\varepsilon(\mathfrak{g})$-module:

$$M = \bigoplus_{\lambda \in P} M(\lambda), \qquad M(\lambda) = \left\{ m \in M \,\middle|\, q^h * m = \varepsilon^{\langle h, \lambda \rangle} m \right\}.$$

We will only consider type 1 modules in this paper.

A type 1 module M is an *l-highest weight module* ('l' stands for the loop) if there exists a vector $m_0 \in M$ such that

$$e_{k,r} * m_0 = 0, \qquad \mathbf{U}_\varepsilon(\mathbf{L}\mathfrak{g})^- * m_0 = M,$$
$$\psi_k^\pm(z) * m_0 = \Psi_k^\pm(z) m_0 \quad \text{for } k \in I$$

for some $\Psi_k^\pm(z) \in \mathbb{C}[[z^\mp]]$. The pair of the I-tuple $(\Psi^+(z), \Psi^-(z)) = (\Psi_k^+(z), \Psi_k^-(z))_{k \in I} \in (\mathbb{C}[[z^\mp]]^I)^2$ is called the *l-highest weight* of M, and m_0 is called the *l-highest weight vector.*

Theorem 1.1.1 (Chari-Pressley [2]). (1) *Every finite-dimensional simple* $\mathbf{U}_\varepsilon(\mathbf{L}\mathfrak{g})$*-module of type* 1 *is an l-highest weight module, and its l-highest weight is given by*

$$\Psi_k^\pm(z) = \varepsilon^{\deg P_k} \left(\frac{P_k(\varepsilon^{-1}/z)}{P_k(\varepsilon/z)} \right)^\pm \tag{1.1.2}$$

for some polynomials $P_k(u) \in \mathbb{C}[u]$ *with* $P_k(0) = 1$. *Here* $(\ )^\pm \in \mathbb{C}[[z^\mp]]$ *denotes the expansion at* $z = \infty$ *and* 0 *respectively.*

(2) *Conversely, for given* $P_k(u)$ *as above, there exists a finite-dimensional simple l-highest weight* $\mathbf{U}_\varepsilon(\mathbf{L}\mathfrak{g})$*-module* M *of type* 1 *such that the l-highest weight is given by the above formula.*

Assigning to M *the* I*-tuple* $P = (P_k)_{k \in I} \in \mathbb{C}[u]^I$ ($P_k(0) = 1$) *defines a bijection between the set of all* P*'s and the set of isomorphism classes of finite-dimensional simple* $\mathbf{U}_\varepsilon(\mathbf{L}\mathfrak{g})$*-modules of type* 1.

We denote by L_P the simple $\mathbf{U}_\varepsilon(\mathbf{L}\mathfrak{g})$-module associated to P. We call P the *Drinfeld polynomial.* For the abuse of terminology, we also say 'P is the l-highest weight of L_P'.

Since $\mathbb{C}\langle q^h,\ h_{k,n}\rangle$ is a commutative subalgebra of $\mathbf{U}_\varepsilon(\mathbf{L}\mathfrak{g})$, any $\mathbf{U}_\varepsilon(\mathbf{L}\mathfrak{g})$-module M decomposes into a direct sum $M = \bigoplus M(\Psi^+, \Psi^-)$ of generalized eigenspaces, where

$$M(\Psi^+, \Psi^-) \stackrel{\text{def.}}{=} \left\{ m \in M \,\middle|\, (\psi_k^\pm(z) - \Psi_k^\pm(z)\,\mathrm{Id})^N * m = 0 \text{ for } k \in I \text{ and sufficiently large } N \right\}$$

for $\Psi_k^\pm(z) \in \mathbb{C}[[z^\mp]]$. The pair of the I-tuple $(\Psi^+, \Psi^-) = (\Psi_k^+, \Psi_k^-)_{k\in I}$ is called an *l-weight*, and $M(\Psi^+, \Psi^-)$ is called an *l-weight space* of M if $M(\Psi^+, \Psi^-) \neq 0$.

Theorem 1.1.3 (Frenkel-Reshetikhin [6]**).** *Any l-weight of any finite dimensional* $\mathbf{U}_\varepsilon(\mathbf{L}\mathfrak{g})$*-module* M *of type* 1 *has the following form*

$$\Psi_k^\pm(z) = \varepsilon^{\deg Q_k - \deg R_k} \left(\frac{Q_k(\varepsilon^{-1}/z) R_k(\varepsilon/z)}{Q_k(\varepsilon/z) R_k(\varepsilon^{-1}/z)} \right)^\pm \tag{1.1.4}$$

for some polynomials

$$Q_k(u) = \prod_{i=1}^{s_k} (1 - a_{ki} u), \quad R_k(u) = \prod_{j=1}^{r_k} (1 - b_{kj} u).$$

Again for the abuse of terminology, we also say 'Q/R is an l-weight of M'. We denote the l-weight space $M(\Psi^+, \Psi^-)$ by $M(Q/R)$.

Frenkel-Reshetikhin [6] defined the q-character of M by

$$\chi_q(M) \stackrel{\text{def.}}{=} \sum_{Q/R} \dim M(Q/R) \prod_{k\in I} \prod_{i=1}^{s_k} \prod_{j=1}^{r_k} Y_{k,a_{ki}} Y_{k,b_{kj}}^{-1}.$$

Theorem 1.1.5 (Frenkel-Reshetikhin [6]**).** (1) *χ_q defines an injective ring homomorphism from the Grothendieck ring* $\operatorname{Rep} \mathbf{U}_\varepsilon(\mathbf{L}\mathfrak{g})$ *of finite dimensional* $\mathbf{U}_\varepsilon(\mathbf{L}\mathfrak{g})$*-modules of type* 1 *to* $\mathbb{Z}[Y_{k,a}^\pm]_{k\in I, a\in\mathbb{C}^*}$ *(a ring of Laurent polynomials in infinitely many variables).*

(2) *If we compose a map $Y_{k,a}^\pm \mapsto y_k^\pm$ (forgetting 'spectral parameters'), it gives the usual character of the restriction of M to a* $\mathbf{U}_\varepsilon(\mathfrak{g})$*-module.*

Definition 1.1.6. A monomial $\prod_{k\in I} \prod_{i=1}^{s_k} \prod_{j=1}^{r_k} Y_{k,a_{ki}} Y_{k,b_{kj}}^{-1}$ appearing in the q-character χ_q is called *l-dominant* if $r_k = 0$ for all k, i.e., a product of positive powers of $Y_{k,c}$'s or 1.

If L_P is the simple $\mathbf{U}_\varepsilon(\mathbf{L}\mathfrak{g})$-module with l-highest weight P, its q-character contains an l-dominant monomial corresponding to the l-highest weight. We denote it by m_P. Its coefficient in $\chi_q(L_P)$ is 1.

Since $\{L_P\}_P$ forms a basis of $\operatorname{Rep}\mathbf{U}_\varepsilon(\mathbf{L}\mathfrak{g})$, we have the following useful condition for the simplicity of a finite dimensional $\mathbf{U}_\varepsilon(\mathbf{L}\mathfrak{g})$-module M of type 1:

$$\text{If } \chi_q(M) \text{ contains only one } l\text{-dominant term, then } M \text{ is simple.} \qquad (1.1.7)$$

1.2 Example

We give examples of q-characters.

If $\mathfrak{g} = A_n$, we have an evaluation homomorphism $\mathrm{ev}_a\colon \mathbf{U}_\varepsilon(\mathbf{L}\mathfrak{g}) \to \mathbf{U}_\varepsilon(\mathfrak{g})$ corresponding to $\mathbf{L}\mathfrak{g} \to \mathfrak{g}$; $z \mapsto a$ (Jimbo). Hence pullbacks of simple $\mathbf{U}_\varepsilon(\mathfrak{g})$-modules are simple $\mathbf{U}_\varepsilon(\mathbf{L}\mathfrak{g})$-modules.

Example 1.2.1. Let $\mathfrak{g} = A_1 = \mathfrak{sl}_2$ and V be the 2-dimensional simple $\mathbf{U}_\varepsilon(\mathbf{L}\mathfrak{g})$-module. Then the q-character of $M_a = \mathrm{ev}_a(M)$ is given by[a]

$$\chi_q(M_a) = Y_{1,a} + Y_{1,a\varepsilon^2}^{-1}.$$

Since χ_q is a ring homomorphism, we have

$$\begin{aligned}\chi_q(M_a \otimes M_b) &= \left(Y_{1,a} + Y_{1,a\varepsilon^2}^{-1}\right)\left(Y_{1,b} + Y_{1,b\varepsilon^2}^{-1}\right)\\ &= Y_{1,a}Y_{1,b} + Y_{1,a\varepsilon^2}^{-1}Y_{1,b} + Y_{1,a}Y_{1,b\varepsilon^2}^{-1} + Y_{1,a\varepsilon^2}^{-1}Y_{1,b\varepsilon^2}^{-1}.\end{aligned}$$

If $b \neq a\varepsilon^2, a\varepsilon^{-2}$, then $M_a \otimes M_b$ is simple by the criterion (1.1.7).

If $b = a\varepsilon^2$ or $a\varepsilon^{-2}$, then the second or third term becomes 1. In fact, it is known that $M_a \otimes M_{a\varepsilon^2}$ decomposes (in $\operatorname{Rep}\mathbf{U}_\varepsilon(\mathbf{L}\mathfrak{g})$) to a sum $M'_a \oplus M''$, where M' is the 3-dimensional simple $\mathbf{U}_\varepsilon(\mathbf{L}\mathfrak{g})$-module, and M'' is the trivial module. Thus we have

$$\begin{aligned}\chi_q(M_a \otimes M_{a\varepsilon^2}) &= \chi_q(M'_a) + \chi_q(M'')\\ &= Y_{1,a}Y_{1,a\varepsilon^2} + Y_{1,a}Y_{1,a\varepsilon^4}^{-1} + Y_{1,a\varepsilon^2}^{-1}Y_{1,a\varepsilon^4}^{-1} + 1.\end{aligned}$$

See also Examples 4.1.6, 6.1.3, 7.2.1.

2 Standard modules

2.1

In [15] we defined a family of finite dimensional $\mathbf{U}_\varepsilon(\mathbf{L}\mathfrak{g})$-modules of type 1 and called them *standard modules.* They are parametrized by the I-tuples $P = (P_k)_{k\in I} \in \mathbb{C}[u]^I$ exactly as simple modules. We denote by M_P associated to P. The definition will be recalled in §8, but we give here their algebraic identification due to Varagnolo-Vasserot [16].

[a]This can be checked directly. But it also follows from Theorem 5.2.1 below.

Definition 2.1.1. We say L_P an *l-fundamental representation* if

$$P_k(u) = \begin{cases} 1 - su & \text{if } k = k_0, \\ 1 & \text{otherwise}, \end{cases}$$

for some $s \in \mathbb{C}^*$ and $k_0 \in I$. We denote L_P by $L(\Lambda_{k_0})_s$. (Λ_k is the k-th fundamental weight of $\mathfrak{g}$.)

For $s \in \mathbb{C}^*$ and a finite sequence $(k_\alpha)_\alpha = (k_1, k_2, \dots)$ in I and a sequence $(n_\alpha)_\alpha = (n_1 \geq n_2 \geq \dots)$ of integers, we set

$$M(s; (k_\alpha)_\alpha, (n_\alpha)_\alpha) \stackrel{\text{def.}}{=} L(\Lambda_{k_1})_{\varepsilon^{n_1} s} \otimes L(\Lambda_{k_2})_{\varepsilon^{n_2} s} \otimes \cdots .$$

Note that $\mathbf{U}_\varepsilon(\mathbf{L}\mathfrak{g})$ is *not* cocommutative Hopf algebra, so the tensor product depends on the ordering of factors.

Theorem 2.1.2 (Varagnolo-Vasserot [16]**).** (1) *A standard module M is isomorphic to a module of the form (finite tensor product)*

$$\begin{aligned} &\bigotimes_i M(s^i; (k^i_{\alpha_i})_{\alpha_i}, (n^i_{\alpha_i})_{\alpha_i}) \\ &= M(s^1; (k^1_{\alpha_1})_{\alpha_1}, (n^1_{\alpha_1})_{\alpha_1}) \otimes M(s^2; (k^2_{\alpha_2})_{\alpha_2}, (n^2_{\alpha_2})_{\alpha_2}) \otimes \cdots \end{aligned}$$

such that $s^i/s^j \notin \varepsilon^{\mathbb{Z}}$ for $i \neq j$ and $n^i_1 \geq n^i_2 \geq \dots$ for each i.

(2) *The above tensor product is independent of the ordering of the factors $M(s^i; (k^i_{\alpha_i})_{\alpha_i}, (n^i_{\alpha_i})_{\alpha_i})$.*

(3) *The I-tuple of polynomials P corresponding to M is the product of Drinfeld polynomials of l-fundamental representations appearing as factors of M.*

Note that if P is given, we can define a module M of the above form by decomposing P into a product of Drinfeld polynomials of l-fundamental representations. Thus we may denote the above module by M_P.

The following properties of M_P were shown in [15]:

(1) $\{M_P\}$ is a basis of $\operatorname{Rep} \mathbf{U}_\varepsilon(\mathbf{L}\mathfrak{g})$.

(2) M_P is an l-highest weight module with l-highest weight P (i.e., given by (1.1.2)).

(3) L_P is the unique simple quotient of M_P.

(4) M_P depends 'continuously' on P in a certain sense. For example, $\dim M_P$ is independent of P.

(5) for a generic P, $M_P \cong L_P$.

Conjecturally M_P is isomorphic to the specialization of the module $V^{\max}(\lambda)$, introduced by Kashiwara [8], and further studied by Chari-Pressley [4].

3 t-analogues of q-characters

A main tool in this paper is a t-analogue of the q-character:

$$\chi_{q,t}\colon \operatorname{Rep}\mathbf{U}_\varepsilon(\mathbf{L}\mathfrak{g}) \to \mathbb{Z}[t,t^{-1}][Y_{k,a}^{\pm}]_{k\in I, a\in\mathbb{C}^*}.$$

This is a homomorphism of additive groups, not of rings, and has the property $\chi_{q,t=1} = \chi_q$. We define $\chi_{q,t}$ for all standard modules M_P. Since $\{M_P\}_P$ is a basis of $\operatorname{Rep}\mathbf{U}_\varepsilon(\mathbf{L}\mathfrak{g})$, we can extend it linearly to any finite dimensional $\mathbf{U}_\varepsilon(\mathbf{L}\mathfrak{g})$-modules.

For the definition we need geometric constructions of standard modules, so we will postpone it to §8.3. We give an alternative definition, which is conjecturally the same as the geometric definition.

3.1 A conjectural definition

Let $M = M_P$ be a standard module, Q/R be an l-weight of M, $M(Q/R)$ be the corresponding l-weight space. Define a filtration on $M(Q/R)$ by

$$0 = M^{-1}(Q/R) \subset M^0(Q/R) \subset M^1(Q/R) \subset \cdots$$
$$M^n(Q/R) \overset{\text{def.}}{=} \bigcap_k \operatorname{Ker}(\psi_k^{\pm}(z) - \Psi_k^{\pm}(z)\,\mathrm{id})^{n+1}.$$

Conjecture 3.1.1. *The t-analogue $\chi_{q,t}(M_P)$, defined geometrically in §8.3, is equal to*

$$\chi_{q,t}(M_P) = \sum_{Q/R}\sum_n t^{2n-d(Q/R,P)} \dim\left(M^n(Q/R)/M^{n-1}(Q/R)\right)\, m_{Q/R},$$

where $d(Q/R,P)$ is an integer (determined explicitly from Q/R, P by (5.1.2) *below), and $m_{Q/R}$ is a monomial in $Y_{k,a}^{\pm}$ corresponding to the l-weight space $M(Q/R)$.*

This definition makes sense for any finite dimensional modules, but is *not* well-defined on the Grothendieck group $\operatorname{Rep}\mathbf{U}_\varepsilon(\mathbf{L}\mathfrak{g})$. Thus the above does not hold for simple modules.

3.2

A main result of this paper is a combinatorial algorithm for computing $\chi_{q,t}(M_P)$ and $[M_P : L_Q]$. It is divided into three steps:

Step 1 Compute $\chi_{q,t}$ for all l-fundamental representations.

Step 2 Compute $\chi_{q,t}(M_P)$ for all standard modules M_P.

Step 3 Express the multiplicity $[M_P : L_Q]$ in terms of $\chi_{q,t}(M_R)$ for various R.

Step 1 is a modification of Frenkel-Mukhin's algorithm [5] for computing χ_q of l-fundamental representations. Step 2 is nothing but a study of $\chi_{q,t}$ of tensor products of l-fundamental representations. Although $\chi_{q,t}$ is not a ring homomorphism, $\chi_{q,t}$ of tensor products is given by a simply modified multiplication. For the proof we use an idea in [13]. Step 3 was essentially done in [15].

4 Step 3

We start with Step 3. The algorithm is similar to the definition of Kazhdan-Lusztig polynomials [9]. It is also similar to the algorithm for computing the transition matrix between the canonical basis and the PBW basis of type ADE [10].

4.1

Let

$$A_{k,a} \overset{\text{def.}}{=} Y_{k,a\varepsilon} Y_{k,a\varepsilon^{-1}} \prod_{l:l\neq k} Y_{l,a}^{c_{kl}},$$

where c_{kl} is the (k,l)-entry of the Cartan matrix.

Definition 4.1.1. (1) Let m, m' be monomials in $Y_{k,a}^{\pm}$ ($k \in I$, $a \in \mathbb{C}^*$). We define an ordering $\leq$ among monomials by

$$m \leq m' \iff \frac{m'}{m} \text{ is a monimial in } A_{k,a}^{-1} \ (k \in I,\ a \in \mathbb{C}^*).$$

Here a monomial in $A_{k,a}^{-1}$ means a product of nonnegative powers of $A_{k,a}^{-1}$. It does not contain any factors $A_{k,a}$.

(2) If $\Psi^{\pm}, \Psi'^{\pm}$ are l-weights of finite dimensional $\mathbf{U}_\varepsilon(\mathbf{L}\mathfrak{g})$-modules, or $Q/R, Q'/R'$ are related to l-weights by (1.1.4), we write $\Psi^{\pm} \leq \Psi'^{\pm}$, $Q/R \leq Q'/R'$ if the corresponding monomials m, m' satisfy $m \leq m'$.

Recall that $\chi_q(L_P)$ contains an l-dominant monomial m_P corresponding to the highest weight vector. It is known that any monomial m appearing $\chi_q(L_P)$, $\chi_q(M_P)$ satisfies $m \leq m_P$ ([5, 4.1], [15, 13.5.2]).

Let

$$c_{QP}(t) \stackrel{\text{def.}}{=} \text{the coefficient of } m_Q \text{ in } \chi_{q,t}(M_P).$$

Then $(c_{QP}(t))_{P,Q}$ is upper-triangular and $c_{PP}(t) = 1$ by the above mentioned result.

Let $(c^{QP}(t))$ be the inverse matrix $(c_{QP}(t))^{-1}$. Let

$$u_{RP}(t) \stackrel{\text{def.}}{=} \sum_Q c^{RQ}(t^{-1}) c_{QP}(t).$$

Let $\overline{}$ be the involution on $\mathbb{Z}[t, t^{-1}]$ given by $t^{\pm 1} \mapsto t^{\mp 1}$.

Lemma 4.1.2 (Lusztig [10, 7.10]**).** *There exists a unique solution $Z_{QP}(t) \in \mathbb{Z}[t^{-1}]$ $(Q \leq P)$ of*

$$Z_{RP}(t) = \sum_{Q: R \leq Q \leq P} \overline{Z_{RQ}(t)} u_{QP}(t), \tag{4.1.3}$$

$$Z_{PP}(t) = 1, \quad Z_{QP}(t) \in t^{-1}\mathbb{Z}[t^{-1}] \text{ for } Q < P. \tag{4.1.4}$$

This lemma is proved by induction, and holds in a general setting. Lusztig has been using this (or its variant) in many places.

Theorem 4.1.5. *The multiplicity $[M_P : L_Q]$ of a simple module L_Q in a standard module M_P is equal to $Z_{QP}(1)$.*

The proof will be given in §8.4.

Example 4.1.6. Let $\mathfrak{g} = A_1$, $M_a = \mathrm{ev}_a^*(M)$ where M is the 2-dimensional simple $\mathbf{U}_\varepsilon(\mathfrak{g})$-module as before. By steps 1,2 explained below[b] we have

$$\chi_{q,t}(M_{a\varepsilon^2} \otimes M_a) = Y_{1,a}Y_{1,a\varepsilon^2} + Y_{1,a}Y_{1,a\varepsilon^4}^{-1} + Y_{1,a\varepsilon^2}^{-1}Y_{1,a\varepsilon^4}^{-1} + t^{-1}1. \tag{4.1.7}$$

Let $P(u) = (1 - au)(1 - a\varepsilon^2 u)$ (i.e., $M_P = M_{a\varepsilon^2} \otimes M_a$), $Q(u) = 1$ (i.e. $M_Q =$ trivial module). Then the above algorithm gives us $Z_{QP}(t) = t^{-1}$.

5 Step 1

5.1 Some definitions

Let M_P be a standard module. Let m_P be the monomial corresponding to the l-highest weight vector. Let $M_P(Q/R)$ be an l-weight space

[b] or direct calculation for the definition (8.3.1)

as before. We denote by $m_{Q/R}$ the corresponding monomial. We define $w_{k,a}(P), v_{k,a}(Q/R,P) \in \mathbb{Z}_{\geq 0}, u_{k,a}(Q/R) \in \mathbb{Z}$ by

$$m_P = \prod_{k\in I, a\in\mathbb{C}^*} Y_{k,a}^{w_{k,a}(P)},$$

$$m_{Q/R} = m_P \prod_{k\in I, a\in\mathbb{C}^*} A_{k,a}^{-v_{k,a}(Q/R,P)} = \prod_{k\in I, a\in\mathbb{C}^*} Y_{k,a}^{u_{k,a}(Q/R)}.$$

Suppose two standard modules M_{P^1}, M_{P^2} and l-weight spaces $M_{P^1}(Q^1/R^1) \subset M_{P^1}$, $M_{P^2}(Q^2/R^2) \subset M_{P^2}$ are given. We define

$$\begin{aligned} &d(Q^1/R^1, P^1; Q^2/R^2, P^2) \\ &\stackrel{\text{def.}}{=} \sum_{k,a} \left(v_{k,a}(Q^1/R^1, P^1) u_{k,a\varepsilon^{-1}}(Q^2/R^2) + w_{k,a\varepsilon}(P^1) v_{k,a}(Q^2/R^2, P^2) \right). \end{aligned} \tag{5.1.1}$$

We also define

$$d(Q/R, P) \stackrel{\text{def.}}{=} d(Q/R, P; Q/R, P). \tag{5.1.2}$$

We denote $d(Q^1/R^1, P^1; Q^2/R^2, P^2)$ also by $d(m_{Q^1/R^1}, m_{P^1}; m_{Q^2/R^2}, m_{P^2})$.

We need the following modification of $\chi_{q,t}$. Write $\chi_{q,t}(M_P) = \sum_m a_m(t)\, m$, where m is a monomial and $a_m(t)$ is its coefficient. Let

$$\widetilde{\chi_{q,t}}(M_P) \stackrel{\text{def.}}{=} \sum_m t^{d(m,m_P)} a_m(t)\, m, \tag{5.1.3}$$

where $d(m, m_P)$ is defined in (5.1.2).[c]

5.2

Frenkel-Mukhin [5, 5.1,5.2] proved that the image of the q-character χ_q is contained in

$$\bigcap_{k\in I} \left(\mathbb{Z}[Y_{l,a}^{\pm}]_{l\neq k, a\in\mathbb{C}^*} \otimes \mathbb{Z}[Y_{k,b}(1 + A_{k,b\varepsilon}^{-1})]_{b\in\mathbb{C}^*} \right).$$

We have the t-analogue of this result, replacing $(1 + A_{k,b\varepsilon}^{-1})^n$ by

$$\left(1 + A_{k,b\varepsilon}^{-1}\right)_t^n \stackrel{\text{def.}}{=} \sum_{r=0}^{n} t^{r(n-r)} \begin{bmatrix} n \\ r \end{bmatrix}_t A_{k,b\varepsilon}^{-r},$$

where $\left[{n \atop r} \right]_t$ is the t-binomial coefficient. More precisely, we have

[c]In fact, $d(m, m_P)$ is determined from $a_m(t)$ so that $t^{d(m,m_P)} a_m(t)$ is a polynomial in t with nonzero constant term.

Theorem 5.2.1. (1) *For each* $k \in I$, $\widetilde{\chi_{q,t}}(M_P)$ *is expressed as a linear combination of*

$$\prod_i Y_{k,b_i}^{n_i}\left(1+A_{k,b_i\varepsilon}^{-1}\right)_t^{n_i} = Y_{k,b_1}^{n_1}\left(1+A_{k,b_1\varepsilon}^{-1}\right)_t^{n_1} Y_{k,b_2}^{n_2}\left(1+A_{k,b_2\varepsilon}^{-1}\right)_t^{n_2}\cdots$$

with coefficients in $\mathbb{Z}[t][Y_{l,a}^{\pm}]_{l\neq k, a\in\mathbb{C}^*}$, *where* $b_i \in \mathbb{C}^*$, $n_i \in \mathbb{Z}_{>0}$ *with* $b_i \neq b_j$ *for* $i \neq j$.

(2) *If* L_P *is an* l*-fundamental representation (and hence* $M_P = L_P$*), then* $\chi_{q,t}(M_P)$ *contains no* l*-dominant monomials other than* m_P *and the condition above uniquely determines* $\chi_{q,t}(M_P)$.

Remark 5.2.2. The statements (1),(2) for $t = 1$ was proved by Frenkel-Mukhin [5, Theorem 5.9]. The proof of (2) is the same for $t = 1$ and the general case, as illustrated in the following examples. In this sense, (2) should also be creditted to them.

5.3 Graph

We give few examples of $\chi_{q,t}$ of l-fundamental representations determined by the above theorem.

We attach to each standard module M_P, an oriented colored graph Γ_P. (It is a slight modification of the graph in [6, 5.3].) The vertices are monomials in $\chi_{q,t}(M_P)$. We draw an colored edge $\xrightarrow{k,a}$ from m_1 to m_2 if $m_2 = m_1 A_{k,a}^{-1}$. We also write the multiplicity of the monomials in $\chi_{q,t}(M_P)$.

Example 5.3.1. Let $\mathfrak{g} = A_3 = \mathfrak{sl}_4$ and $M_P = L(\Lambda_2)_1$. Then the corresponding graph Γ_P is

$$\begin{array}{ccccccc}
Y_{2,1} & \xrightarrow{2,\varepsilon} & Y_{1,\varepsilon}Y_{2,\varepsilon^2}^{-1}Y_{3,\varepsilon} & \xrightarrow{1,\varepsilon^2} & Y_{1,\varepsilon^3}^{-1}Y_{3,\varepsilon} & & \\
 & & \downarrow{\scriptstyle 3,\varepsilon^2} & & \downarrow{\scriptstyle 3,\varepsilon^2} & & \\
 & & Y_{1,\varepsilon}Y_{3,\varepsilon^3}^{-1} & \xrightarrow{1,\varepsilon^2} & Y_{1,\varepsilon^3}^{-1}Y_{2,\varepsilon^2}Y_{3,\varepsilon^3}^{-1} & \xrightarrow{2,\varepsilon^3} & Y_{2,\varepsilon^4}^{-1}.
\end{array}$$

Let us explain how we determine this graph inductively. We start with the l-highest weight $Y_{2,1}$. We know that its coefficient is 1. Applying Theorem 5.2.1(1) with $k = 2$, we get $Y_{1,\varepsilon}Y_{2,\varepsilon^2}^{-1}Y_{3,\varepsilon}$ with coefficient 1. Then we apply Theorem 5.2.1(1) with $k = 1$ to get $Y_{1,\varepsilon^3}^{-1}Y_{3,\varepsilon}$. And so on. All multiplicities are 1 in this case.

For $\mathfrak{g} = A_n$, it is known that the coefficients of $\chi_{q,t}(L(\Lambda_k)_a)$ are all 1.[d] Thus $\chi_{q,t}(L(\Lambda_k)_a) = \chi_{q,t=1}(L(\Lambda_k)_a)$.

[d] More generally, if the coefficients of α_k in the highest root is 1, then the same holds. This

Example 5.3.2. Let $\mathfrak{g} = D_4$ and $M_P = L(\Lambda_2)_1$. The graph Γ_P is Figure 1. It is known that the restriction of M_P to a $\mathbf{U}_\varepsilon(\mathfrak{g})$-module is a direct sum of the adjoint representation and the trivial representation. This fact is reflected in $\chi_{q,t}(M_P)$ where $Y_{2,\varepsilon^2}Y_{2,\varepsilon^4}^{-1}$ has the coefficient $[2]_t$ and all others has 1. Note that the number of monomials is 28, which is the dimension of the adjoint representation. See also Example 7.2.3 below.

Let us give a more complicated example.

Example 5.3.3. Let $\mathfrak{g} = A_2$ and $M_P = L(\Lambda_2)_\varepsilon^{\otimes 2} \otimes L(\Lambda_1)_1$. Although this is *not* an l-fundamental representation, $\chi_{q,t}(M_P)$ has no l-dominant terms other than m_P, so the condition Theorem 5.2.1(1) gives us $\chi_{q,t}$. The graph is Figure 2.

Remark 5.3.4. As we can see in above examples, the crystal graphs are subgraphs of Γ_P. The set of vertices is the same, but the set of arrows is smaller. We would like to discuss this further elsewhere.

6 Step 2

6.1

Let $M_P = M(s^1; (k^1_{\alpha_1}), (n^1_{\alpha_1}))\otimes M(s^2; (k^2_{\alpha_2}), (n^2_{\alpha_2}))\otimes\cdots$ be a standard module with $s^i/s^j \notin \varepsilon^{\mathbb{Z}}$ as in Theorem 2.1.2.

Proposition 6.1.1. *We have*

$$\chi_{q,t}(M_P) = \chi_{q,t}(M(s^1; (k^1_{\alpha_1})_{\alpha_1}, (n^1_{\alpha_1})_{\alpha_1}))\chi_{q,t}(M(s^2; (k^2_{\alpha_2})_{\alpha_2}, (n^2_{\alpha_2})_{\alpha_2}))\cdots$$

if $s^i/s^j \notin \varepsilon^{\mathbb{Z}}$ for $i \neq j$.

Thus it is enough to study

$$\chi_{q,t}(M(s; (k_\alpha)_\alpha, (n_\alpha)_\alpha)) = \chi_{q,t}(L(\Lambda_{k_1})_{\varepsilon^{n_1}s} \otimes L(\Lambda_{k_2})_{\varepsilon^{n_2}s} \otimes \cdots).$$

Let

$$\chi_{q,t}(L(\Lambda_{k_\alpha})_{\varepsilon^{n_\alpha}s})) = \sum_{r_\alpha} a_{m_\alpha,r_\alpha}(t)m_{\alpha,r_\alpha},$$

where m_{α,r_α} is a monomial in $Y^{\pm}_{k,a}$ and $a_{m_\alpha,r_\alpha}(t) \in \mathbb{Z}[t,t^{-1}]$ is its coefficient.

If $t = 1$, $\chi_{q,1}$ is a ring homomorphism, hence we have

$$\chi_{q,1}(M(a; (k_\alpha)_\alpha, (n_\alpha)_\alpha)) = \sum_{r_1,r_2,\dots} \prod_\alpha a_{m_r,r_\alpha}(1)m_{\alpha,r_\alpha}.$$

---

result easily follows from the theory of quiver varieties. **Exercise**: Check this using the above algorithm.

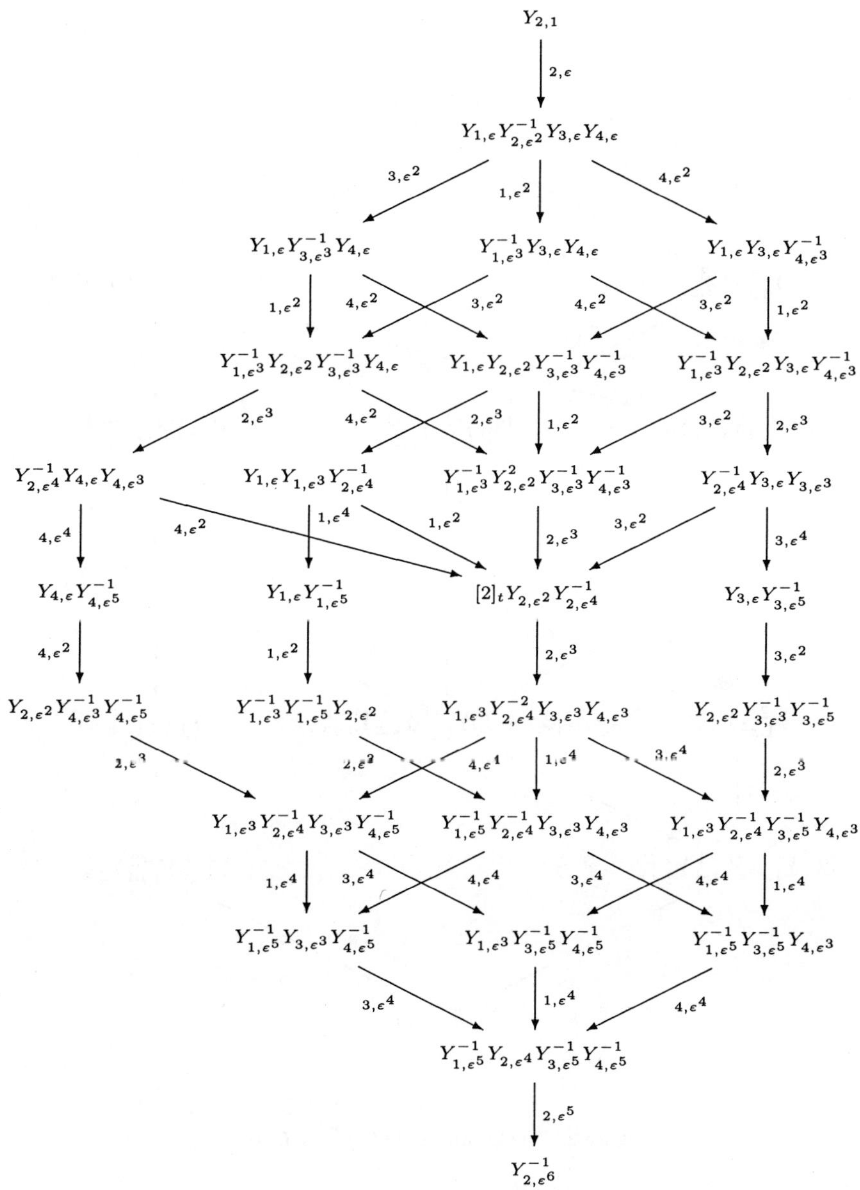

Figure 1. The graph for $L(\Lambda_2)_1$

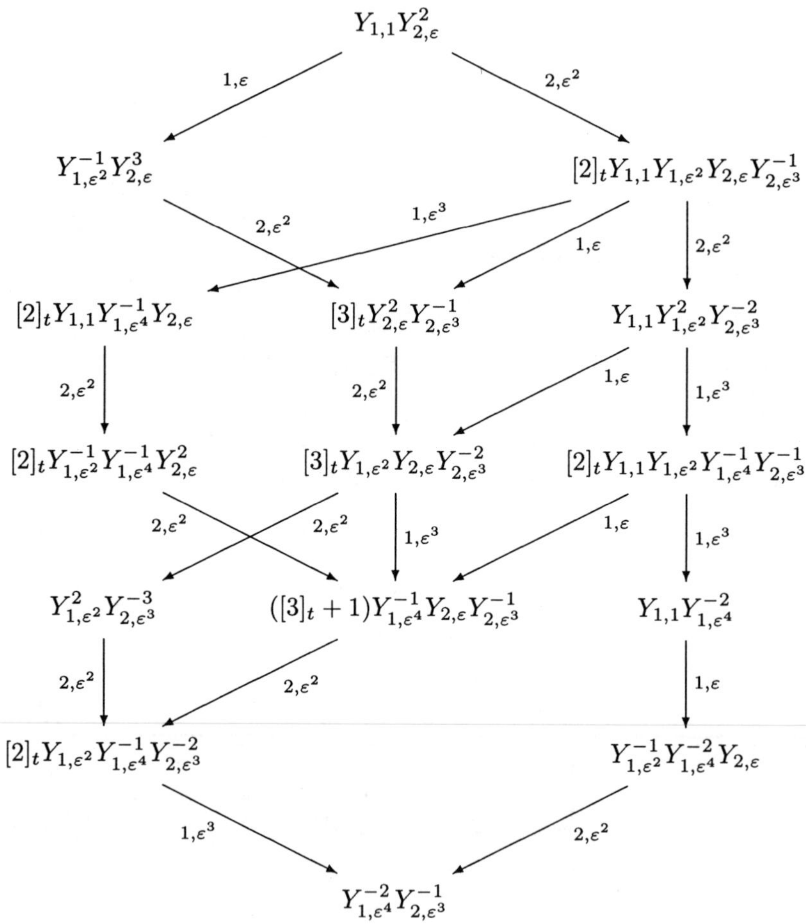

Figure 2. The graph for $L(\Lambda_2)_\varepsilon^{\otimes 2} \otimes L(\Lambda_1)_1$

Theorem 6.1.2. *Let P^α be the Drinfeld polynomial of $L(\Lambda_{k_\alpha})_{\varepsilon^{n_\alpha}s}$. Then we have*

$$\chi_{q,t}(M(a;(k_\alpha),(n_\alpha))) \\ = \sum_{r_1,r_2,\dots} t^{\sum_{\alpha,\beta} \pm d(m_{\alpha,r_\alpha},m_{P^\alpha};m_{\beta,r_\beta},m_{P^\beta})} \prod_\alpha a_{m_r,r_\alpha}(t) m_{\alpha,r_\alpha},$$

where the sign for $d(m_{\alpha,r_\alpha},m_{P^\alpha};m_{\beta,r_\beta},m_{P^\beta})$ is $-$ if $\alpha \le \beta$ and $+$ otherwise.
Example 6.1.3. For $\mathfrak{g} = A_1$, we have

$$d(Y_{1,a\varepsilon^2}^{-1}, Y_{1,a}; Y_{1,a}, Y_{1,a}) = 1, \qquad d(Y_{1,a}, Y_{1,a}; Y_{1,a}^{-1}, Y_{1,a\varepsilon^{-2}}) = 1$$

and all others are 0. Then we get (4.1.7).

If $P = (1 - au)^n$, we get

$$\chi_{q,t}(M_P) = \sum_{r=0}^{n} \begin{bmatrix} n \\ r \end{bmatrix}_t Y_{1,a}^{n-r} Y_{1,a\varepsilon^2}^{-r}$$

from $\chi_{q,t}(L(\Lambda_1)_a) = Y_{1,a} + Y_{1,a\varepsilon^2}^{-1}$. This also follows directly from the definition (8.3.1) below. The t-binomial coefficients appear as Poincaré polynomials of Grassmann manifolds.

7 Restriction to $\mathbf{U}_\varepsilon(\mathfrak{g})$

Finite dimensional simple $\mathbf{U}_\varepsilon(\mathfrak{g})$-modules are classified by highest weights. Let $\operatorname{Res} M_P$ be the restriction of a standard module M_P to a $\mathbf{U}_\varepsilon(\mathfrak{g})$-module. It decomposes into a sum of various simple modules. Once $\chi_q(M_P)$ is computed, the character of $\operatorname{Res} M_P$ is given by replacing $Y_{k,a}^\pm$ by $y_k^\pm$ (Theorem 1.1.5(2)). Combining with the knowledge of characters of simple finite dimensional $\mathbf{U}_\varepsilon(\mathfrak{g})$-modules, we can determine the multiplicity of simple modules in $\operatorname{Res} M_P$.

Characters of simple finite dimensional $\mathbf{U}_\varepsilon(\mathfrak{g})$-modules are the same as that of simple $\mathfrak{g}$-modules, hence are known. However, we express them in terms of $\chi_{q,t}$ in this section.

7.1

For a dominant weight $\mathbf{w} = \sum w_k \Lambda_k$ we denote by $L_{\mathbf{w}}$ the simple highest weight $\mathbf{U}_\varepsilon(\mathfrak{g})$-module with the highest weight $\mathbf{w}$.

We consider a standard module M_P with $\deg P_k = w_k$. By the 'continuity' of M_P on P, $\operatorname{Res} M_P$ depends only on $w_k = \deg P_k$, and not on P itself. Let us denote the multiplicity of $L_{\mathbf{w}'}$ in $\operatorname{Res} M_P$ by $Z_{\mathbf{w}',\mathbf{w}}$, i.e.,

$$\operatorname{Res} M_P = \bigoplus_{\mathbf{w}'} L_{\mathbf{w}'}^{\oplus Z_{\mathbf{w}',\mathbf{w}}}.$$

We will give a formula expressing $Z_{\mathbf{w}',\mathbf{w}}$ in terms of $\chi_{q,t}(M_P)$. Although we can give algorithm for arbitrary P in principle, the following choice will make the formula simple.

Choose and fix orientations of edges in the Dynkin diagram. We define integer $m(k)$ for each vertex k so that $m(k) - m(l) = 1$ if we have an oriented edge from k to l, i.e., $k \to l$. Then we define P by

$$P_k(u) = (1 - u\varepsilon^{m(k)})^{w_k}.$$

Let $\widetilde{\chi_{q,t}}(M_P)$ as in (5.1.3). Let $\widetilde{\chi_t}(M_P) \in \mathbb{Z}[t] \otimes \mathbb{Z}[t_k^{\pm}]_{k\in I}$ be a t-analogue of the ordinary character which is obtained from $\widetilde{\chi_{q,t}}(M_P)$ by sending $Y_{k,a}^{\pm}$ to $y_k^{\pm}$.

For another dominant weight $\mathbf{w}' = \sum w'_k \Lambda_k$, let

$$c_{\mathbf{w}',\mathbf{w}}(t) \overset{\text{def.}}{=} \text{the coefficient of } \prod y_k^{w'_k} \text{ in } \widetilde{\chi_t}(M_P).$$

The matrix $(c_{\mathbf{w}',\mathbf{w}}(t))_{\mathbf{w}',\mathbf{w}}$ is upper-triangular with respect to the usual order on weights, and diagonal entries are all 1.

Theorem 7.1.1. *$c_{\mathbf{w}',\mathbf{w}}(0)$ is the weight multiplicity of $\mathbf{w}'$ in the highest weight module $L_{\mathbf{w}}$ with the highest weight $\mathbf{w}$.*

This is just a simple rephrasing of a main result in [12,14]. The proof will be given in §8.5.

Note that $c_{\mathbf{w}',\mathbf{w}}(1)$ gives the weight multiplicity of $\mathbf{w}'$ in $\operatorname{Res} M_P$ since $\widetilde{\chi_{t=1}}$ is the ordinary character. Thus we have

$$c_{\mathbf{w}'',\mathbf{w}}(1) = \sum_{\mathbf{w}'} c_{\mathbf{w}'',\mathbf{w}'}(0)\, Z_{\mathbf{w}',\mathbf{w}}.$$

This equation determines the multiplicity $Z_{\mathbf{w}',\mathbf{w}}$ only from the knowledge of $\chi_{q,t}$.

According to a conjecture of Lusztig [11] together with a formula (8.5.1) below, $c_{\mathbf{w}',\mathbf{w}}(t)$ should be written by ferminonic form of Hatayama el al. [7]. More precisely, we should have $\sum_{\mathbf{w}'} c^{\mathbf{w}'',\mathbf{w}'}(0)\, c_{\mathbf{w}',\mathbf{w}}(t) = M(\mathbf{w},\mathbf{w}'',t^2)$, where $(c^{\mathbf{w}'',\mathbf{w}'}(0))$ is the inverse matrix of $(c_{\mathbf{w}'',\mathbf{w}'}(0))$. See [11] for the definition of $M(\mathbf{w},\mathbf{w}',q)$. Although this formula can be checked in many examples, the complexity of the combinatorics prevent us from proving it in full generality. Conjecturally $M(\mathbf{w},\mathbf{w}',q=1)$ gives us the multiplicities of the restriction of M_P (Kirillov-Reshetikhin[e]). Thus the conjecture is compatible with our

[e] In fact, they consider more general modules, not necessarily standard modules.

result in this section.

7.2

Example 7.2.1. Let $\mathfrak{g} = A_1$ and $\mathbf{w} = 2\Lambda_1$. We take $P = (1-u)^2$ by the above choice. By Example 6.1.3, we have

$$\widetilde{\chi}_t(M_P) = y_1^2 + (1+t^2) + y_1^{-2}.$$

Thus $Z_{0,\mathbf{w}} = 1$. Since $\operatorname{Res}(M_P) = L_{\Lambda_1} \otimes L_{\Lambda_1} = L_{2\Lambda_1} \oplus L_0$, this is the correct answer !

Example 7.2.2. Let $\mathfrak{g} = A_3$ and $\mathbf{w} = \Lambda_2$. By Example 5.3.1 all the coefficients of $\chi_{q,t}(L(\Lambda_2)_1)$ are 1. Hence $\operatorname{Res} M_P = \operatorname{Res} L(\Lambda_2)_1$ is simple as a $\mathbf{U}_\varepsilon(\mathfrak{g})$-module.

Example 7.2.3. Let $\mathfrak{g} = D_4$, $\mathbf{w} = \Lambda_2$, $\mathbf{w}' = 0$. By Example 5.3.2 we have $c_{\mathbf{w}',\mathbf{w}}(t) = 4 + t^2$. Thus $Z_{\mathbf{w}',\mathbf{w}} = 1$, i.e. $\operatorname{Res}(L(\Lambda_2)_1) = L_{\Lambda_2} \oplus L_0$.

8 Quiver varieties

In this section, we give the definition of $\chi_{q,t}$ and prove Theorems 4.1.5, 7.1.1. As we mentioned, those proofs are essentially given in [12,14] and [15] respectively. The only things we do here are translation of results into the language of $\chi_{q,t}$. We believe that this section gives good introductions to [12,14,15].

8.1

Let $\mathbf{w} = \sum w_k \Lambda_k$ ($w_k \in \mathbb{Z}_{\geq 0}$) be a dominant weight of the finite dimensional Lie algebra $\mathfrak{g}$. In [12,14,15], we have attached to each $\mathbf{w}$, a map $\pi\colon \mathfrak{M}(\mathbf{w}) \to \mathfrak{M}_0(\infty, \mathbf{w})$ with the following properties:

(1) $\mathfrak{M}(\mathbf{w})$ is a finite disjoint union of nonsingular quasi-projective varieties of various dimensions.

(2) $\mathfrak{M}_0(\infty, \mathbf{w})$ is an affine algebraic variety.

(3) π is a projective morphism.

(4) There exist actions of $G_{\mathbf{w}} \times \mathbb{C}^*$ on $\mathfrak{M}(\mathbf{w})$ and $\mathfrak{M}_0(\infty, \mathbf{w})$ such that π is equivariant.

(5) $\mathfrak{M}_0(\infty, \mathbf{w})$ is a cone, and the vertex (denoted by 0) is the unique fixed point of the $\mathbb{C}^*$-action (restriction of $G_{\mathbf{w}} \times \mathbb{C}^*$-action to the second factor).

Here $G_{\mathbf{w}} = \prod_{k \in I} \mathrm{GL}(w_k, \mathbb{C})$.

We consider the fiber product

$$Z(\mathbf{w}) \stackrel{\text{def.}}{=} \mathfrak{M}(\mathbf{w}) \times_{\mathfrak{M}_0(\infty,\mathbf{w})} \mathfrak{M}(\mathbf{w}).$$

The convolution product makes the (Borel-Moore) homology group $H_*(Z(\mathbf{w}), \mathbb{C})$ into an associative (noncommutative) algebra. One of main results in [14] is a construction of a surjective algebra homomorphism

$$\mathbf{U}(\mathfrak{g}) \to H_{\mathrm{top}}(Z(\mathbf{w}), \mathbb{C}),$$

where $\mathbf{U}(\mathfrak{g})$ is the universal enveloping algebra of $\mathfrak{g}$ (**NB**: not a 'quantum' version). Here $H_{\mathrm{top}}(\ )$ means the degree $= \dim_{\mathbb{R}} Z(\mathbf{w})$ part of the homology group. More precisely, we take degree $=$ dimension part on each connected components of $Z(\mathbf{w})$, and then make the direct sum. Note that the the dimension differs on various components.

Let $\mathfrak{L}(\mathbf{w}) = \pi^{-1}(0)$. It is known that $\mathfrak{M}(\mathbf{w})$ has a holomorphic symplectic form such that $\mathfrak{L}(\mathbf{w})$ is a lagrangian subvariety. The convolution makes $H_{\mathrm{top}}(\mathfrak{L}(\mathbf{w}), \mathbb{C})$ (the top degree part of the Borel-Moore homology group, in the same sense as above) into an $H_{\mathrm{top}}(Z(\mathbf{w}), \mathbb{C})$-module. It is a $\mathbf{U}(\mathfrak{g})$-module by the above homomorphism. By [14, 10.2] it is the simple finite dimensional $\mathbf{U}(\mathfrak{g})$-module $L_{\mathbf{w}}$ with highest weight $\mathbf{w}$. And connected components $\mathfrak{M}(\mathbf{v}, \mathbf{w})$ of $\mathfrak{M}(\mathbf{w})$ are parametrized by vectors $\mathbf{v} = \sum v_k \alpha_k$ (α_k is the kth simple root of $\mathfrak{g}$) so that

$$H_{\mathrm{top}}(\mathfrak{L}(\mathbf{w}), \mathbb{C}) = \bigoplus_{\mathbf{v}} H_{\mathrm{top}}(\mathfrak{M}(\mathbf{v}, \mathbf{w}) \cap \mathfrak{L}(\mathbf{w}), \mathbb{C})$$

is the weight space decomposition of the simple highest weight module $L_{\mathbf{w}}$, where $H_{\mathrm{top}}(\mathfrak{M}(\mathbf{v}, \mathbf{w}) \cap \mathfrak{L}(\mathbf{w}), \mathbb{C})$ has weight $\mathbf{w} - \mathbf{v}$. In particular, $\mathbf{v} = 0$ corresponds to the highest weight vector. In fact, $\mathfrak{M}(0, \mathbf{w})$ is consisting of a single point.

The space $\mathfrak{M}_0(\infty, \mathbf{w})$ has a stratification

$$\mathfrak{M}_0(\infty, \mathbf{w}) = \bigcup \mathfrak{M}_0^{\mathrm{reg}}(\mathbf{v}, \mathbf{w}),$$

where $\mathbf{v}$ runs over the set of vectors such that $\mathbf{w} - \mathbf{v}$ is a weight of $L_{\mathbf{w}}$ which is *dominant* [12, §3].

8.2

Let us give the $\mathbf{U}_q(\mathbf{L}\mathfrak{g})$-version of the construction of the previous subsection.

We use the following notation: Let $R(G)$ denote the representation ring of a linear algebraic group G. If G acts a quasi-projective variety X, $K^G(X)$ denotes the Grothendieck group of G-equivariant coherent sheaves on X.

The representation ring $R(G_{\mathbf{w}} \times \mathbb{C}^*)$ of $G_{\mathbf{w}} \times \mathbb{C}^*$ is isomorphic to the tensor product $R(G_{\mathbf{w}}) \otimes_{\mathbb{Z}} R(\mathbb{C}^*)$. Moreover, $R(\mathbb{C}^*)$ is isomorphic to $\mathbb{Z}[q, q^{-1}]$, where q is the canonical 1-dimensional representation of $\mathbb{C}^*$.

The convolution makes the Grothendieck group $K^{G_{\mathbf{w}} \times \mathbb{C}^*}(Z(\mathbf{w}))$ into a $R(G_{\mathbf{w}} \times \mathbb{C}^*) = R(G_{\mathbf{w}})[q, q^{-1}]$-algebra. One of main results in [15] is a construction of an algebra homomorphism

$$\mathbf{U}_q^{\mathbb{Z}}(\mathbf{L}\mathfrak{g}) \otimes_{\mathbb{Z}} R(G_{\mathbf{w}}) \to K^{G_{\mathbf{w}} \times \mathbb{C}^*}(Z(\mathbf{w}))/\operatorname{torsion}.$$

By the equivariance of π, $\mathfrak{L}(\mathbf{w}) = \pi^{-1}(0)$ is invariant under $G_{\mathbf{w}} \times \mathbb{C}^*$. The convolution makes $K^{G_{\mathbf{w}} \times \mathbb{C}^*}(\mathfrak{L}(\mathbf{w}))$ into a $K^{G_{\mathbf{w}} \times \mathbb{C}^*}(Z(\mathbf{w}))$-module. Moreover, it is free of finite rank over $R(G_{\mathbf{w}} \times \mathbb{C}^*)$ [15, §7]. It is a $\mathbf{U}_q^{\mathbb{Z}}(\mathbf{L}\mathfrak{g}) \otimes_{\mathbb{Z}} R(G_{\mathbf{w}})$-module by the above homomorphism. By [15, §13], it contains a vector m_0 such that

$$\begin{gathered} e_{k,r} * m_0 = 0, \qquad \left(\mathbf{U}_q^{\mathbb{Z}}(\mathbf{L}\mathfrak{g})^- \otimes_{\mathbb{Z}} R(G_{\mathbf{w}})\right) * m_0 = K^{G_{\mathbf{w}} \times \mathbb{C}^*}(\mathfrak{L}(\mathbf{w})), \\ \psi_k^{\pm}(z) * m_0 = q^{w_k} \left(\frac{\bigwedge_{-1/qz} q^{-1} W_k}{\bigwedge_{-q/z} q^{-1} W_k} \right)^{\pm} * m_0 \quad \text{for } k \in I. \end{gathered} \tag{8.2.1}$$

The right hand side of the third equation needs an explanation: First W_k is the vector representation of $\mathrm{GL}(w_k, \mathbb{C})$, considered as a $G_{\mathbf{w}} \times \mathbb{C}^*$-module. Then $\bigwedge_u V = \sum u^i \bigwedge^i V$. Since $\bigwedge_{-q/z} q^{-1} W_k$ is $1 - (1/z) W_k + \dots$ (1 is the trivial module), we can define $\left(\bigwedge_{-q/z} q^{-1} W_k\right)^{-1}$ as a formal power series in $1/z$. This gives us the case $(\ )^+$ of the above formula. In the case $(\ )^-$, we expand as $\bigwedge_{-q/z} q^{-1} W_k = (-1/z)^{w_k} \left(\bigwedge^{w_k} W_k - z \bigwedge^{w_k - 1} W_k + \cdots\right)$. Then $\bigwedge^{w_k} W_k$ is an invertible element, we can also define $\left(\bigwedge_{-q/z} q^{-1} W_k\right)^{-1}$. The vector m_0 is the canonical generator of $K^{G_{\mathbf{w}} \times \mathbb{C}^*}(\mathfrak{M}(0, \mathbf{w}))$. (Recall $\mathfrak{M}(0, \mathbf{w})$ is a point.)

The module $K^{G_{\mathbf{w}} \times \mathbb{C}^*}(\mathfrak{L}(\mathbf{w}))$ should be considered as a 'universal' standard module since standard modules are obtained from it by specializations as we explain now.

Let $a = (s, \varepsilon) \in G_{\mathbf{w}} \times \mathbb{C}^*$ be a semisimple element. It defines a homomorphism $\chi_a \colon R(G_{\mathbf{w}} \times \mathbb{C}^*) \to \mathbb{C}$ by sending a representation to the value of the character at a. Then

$$K^{G_{\mathbf{w}} \times \mathbb{C}^*}(\mathfrak{L}(\mathbf{w})) \otimes_{R(G_{\mathbf{w}} \times \mathbb{C}^*)} \mathbb{C} \tag{8.2.2}$$

is a module over $\mathbf{U}_\varepsilon(\mathbf{L}\mathfrak{g}) = \mathbf{U}_q^{\mathbb{Z}}(\mathbf{L}\mathfrak{g}) \otimes_{\mathbb{Z}[q, q^{-1}]} \mathbb{C}$. By (8.2.1) it is a finite-dimensional l-highest weight module. This is the *standard module* M_P, where $P_k(u) = \chi_a(\bigwedge_{-u} q^{-1} W_k)$. Note that the set of conjugacy classes of $a = (s, \varepsilon)$

bijectively corresponds to the set of I-tuple of polynomials P with $\deg P_k = w_k$.

8.3

Let A be the Zariski closure of $a^{\mathbb{Z}}$ in $G_{\mathbf{w}} \times \mathbb{C}^*$. It is an abelian reductive group. We have $K^{G_{\mathbf{w}}\times\mathbb{C}^*}(\mathfrak{L}(\mathbf{w})) \otimes_{R(G_{\mathbf{w}}\times\mathbb{C}^*)} R(A) \cong K^A(\mathfrak{L}(\mathbf{w}))$ [15, §7]. Since χ_a factors through $R(A)$, the standard module M_P is isomorphic to $K^A(\mathfrak{L}(\mathbf{w})) \otimes_{R(A)} \mathbb{C}$. By Thomason's localization theorem, it is isomorphic to $K(\mathfrak{L}(\mathbf{w})^A) \otimes_{\mathbb{Z}} \mathbb{C}$, where $\mathfrak{L}(\mathbf{w})^A$ is the fixed point set. Furthermore, it is isomorphic to $H_*(\mathfrak{L}(\mathbf{w})^A, \mathbb{C})$ via the Chern character homomorphism [15, §7].

Let $\pi^A \colon \mathfrak{M}(\mathbf{w})^A \to \mathfrak{M}_0(\infty, \mathbf{w})^A$ be the restriction of the map $\pi\colon \mathfrak{M}(\mathbf{w}) \to \mathfrak{M}_0(\infty, \mathbf{w})$ to the fixed point set. Let $\mathfrak{M}(\mathbf{w})^A = \bigsqcup_\rho \mathfrak{M}(\rho)$ be the decomposition into connected components. Each $\mathfrak{M}(\rho)$ is a nonsingular quasi-projective variety. Then we have the direct sum decomposition

$$M_P \cong H_*(\mathfrak{L}(\mathbf{w})^A, \mathbb{C}) \cong \bigoplus_\rho H_*(\mathfrak{M}(\rho) \cap \mathfrak{L}(\mathbf{w}), \mathbb{C}).$$

In [15, §13, §14] we have shown that this is the l-weight space decomposition of M_P. In particular, the index ρ can be considered as an l-weight of M_P. Thus we have arrived at a geometric interpretation of χ_q:

$$\chi_q(M_P) = \sum_\rho \dim H_*(\mathfrak{M}(\rho) \cap \mathfrak{L}(\mathbf{w}), \mathbb{C})\; m_\rho,$$

where m_ρ is the monomial corresponding to the l-weight ρ.

Now we define the t-analogue $\chi_{q,t}$ by

$$\chi_{q,t}(M_P) \overset{\text{def.}}{=} \sum_\rho \sum_k \dim H_k(\mathfrak{M}(\rho) \cap \mathfrak{L}(\mathbf{w}), \mathbb{C})\; t^{k-\dim_{\mathbb{C}} \mathfrak{M}(\rho)}\, m_\rho. \qquad (8.3.1)$$

By [15, §14] we have a stratification

$$\mathfrak{M}_0(\infty, \mathbf{w})^A = \bigcup_\rho \mathfrak{M}_0^{\mathrm{reg}}(\rho),$$

consisting of nonsingular locally closed subvarieties. Here the index set $\{\rho\}$ is the subset of the above index set consisting of *l-dominant* l-weights.

8.4 Proof of Theorem 4.1.5

The l-highest weight P is fixed throughout the proof. Thus the dominant weight vector $\mathbf{w}$ and the element $a = (s, \varepsilon) \in G_{\mathbf{w}}$ are fixed.

We change the notation now. If ρ corresponds to an l-weight space $M_P(Q/R)$, we denote above $\mathfrak{M}(\rho)$ by $\mathfrak{M}(Q/R, P)$. We also denote by $\mathfrak{M}_0^{\text{reg}}(Q, P)$ for above $\mathfrak{M}_0^{\text{reg}}(\rho)$ if ρ corresponds to an l-dominant l-weight Q. Thus we have

$$\mathfrak{M}(\mathbf{w})^A = \bigsqcup_{Q/R} \mathfrak{M}(Q/R, P), \quad \mathfrak{M}_0(\infty, \mathbf{w})^A = \bigcup_{Q} \mathfrak{M}_0^{\text{reg}}(Q, P).$$

In this notation $H_*(\mathfrak{M}(P,P) \cap \mathfrak{L}(\mathbf{w}), \mathbb{C})$ is the l-highest weight space. Since $\mathfrak{M}(0, \mathbf{w})$ is a single point as we explained, we have $\mathfrak{M}(P,P) = \mathfrak{M}(0, \mathbf{w})$. We also have $\mathfrak{M}_0^{\text{reg}}(P, P) = \{0\}$.

Lemma 8.4.1. (1) $\dim_{\mathbb{C}} \mathfrak{M}(Q/R, P) = d(Q/R, P)$. $\dim_{\mathbb{C}} \mathfrak{M}_0^{\text{reg}}(Q, P) = d(Q, P)$.

(2) *If* $\mathfrak{M}_0^{\text{reg}}(Q, P) \subset \overline{\mathfrak{M}_0^{\text{reg}}(R, P)}$, *then* $R \leq Q$.

(3) *Choose* $x \in \mathfrak{M}_0^{\text{reg}}(Q, P)$. *Then* $(\pi^A)^{-1}(x) \cap \mathfrak{M}(S/T, P)$ *is isomorphic to* $\mathfrak{M}(S/T, Q) \cap \mathfrak{L}(\mathbf{w})$.

Proof. (1) The first equation is the dimension formula [15, 4.1.6]. The second equation follows from $\dim_{\mathbb{C}} \mathfrak{M}_0^{\text{reg}}(Q, P) = \dim_{\mathbb{C}} \mathfrak{M}(Q, P)$, which is clear from the definition [15, §4].

(2),(3) The results are known or trivial for $Q = P$. Now use the transversal slice at $x \in \mathfrak{M}_0^{\text{reg}}(Q, P)$ [15, §3] to reduce a general case to this case. □

Let $D^b(\mathfrak{M}_0(\infty, \mathbf{w})^A)$ be the bounded derived category of complexes of sheaves such that cohomology sheaves are constant along each stratum $\mathfrak{M}_0^{\text{reg}}(Q, P)$. Let $IC(\mathfrak{M}_0^{\text{reg}}(Q, P))$ be the intersection homology complex associated with the constant local system $\mathbb{C}_{\mathfrak{M}_0^{\text{reg}}(Q,P)}$ on $\mathfrak{M}_0^{\text{reg}}(Q, P)$. By using the transversal slice [15, §3], one can check that it is an object in $D^b(\mathfrak{M}_0(\infty, \mathbf{w})^A)$. Let $\mathbb{C}_{\mathfrak{M}(Q/R,P)}$ be the constant local system on $\mathfrak{M}(Q/R, P)$. Then $\pi^A_*(\mathbb{C}_{\mathfrak{M}(Q/R,P)})$ is an object of $D^b(\mathfrak{M}_0(\infty, \mathbf{w})^A)$ again by the transversal slice argument. Using the decomposition theorem of Beilinson-Bernstein-Deligne, we have shown that there exists an isomorphism in $D^b(\mathfrak{M}_0(\infty, \mathbf{w})^A)$:

$$\pi^A_*(\mathbb{C}_{\mathfrak{M}(R,P)}[\dim_{\mathbb{C}} \mathfrak{M}(R, P)]) \cong \bigoplus_{Q,k} L_{Q,k}(R, P) \otimes IC(\mathfrak{M}_0^{\text{reg}}(Q, P))[k] \quad (8.4.2)$$

for some vector space $L_{Q,k}(R, P)$ [15, 14.3.2]. Since $\pi^A(\mathfrak{M}(R, P)) \subset \overline{\mathfrak{M}_0^{\text{reg}}(R, P)}$ by definition [15, §4], the summation runs over $Q \geq R$ by Lemma 8.4.1. Let

$$L_{RQ}(t) \stackrel{\text{def.}}{=} \sum_k \dim L_{Q,k}(R, P)\, t^{-k}.$$

Applying the Verdier duality to the both hand side of (8.4.2) and using the self-duality of $\pi_*^A(\mathbb{C}_{\mathfrak{M}(R,P)}[\dim_{\mathbb{C}} \mathfrak{M}(R,P)])$ and $IC(\mathfrak{M}_0^{\mathrm{reg}}(Q,P))$, we find $L_{RQ}(t) = \overline{L_{RQ}(t)}$.

Choose a point x_Q from $\mathfrak{M}_0^{\mathrm{reg}}(Q,P)$ for each stratum. Let $i_{x_Q}\colon \{x_Q\} \to \mathfrak{M}_0(\infty,\mathbf{w})^A$ denote the inclusion. Consider

$$H^k(i_{x_Q}^! \pi_*^A \mathbb{C}_{\mathfrak{M}(R,P)}[\dim \mathfrak{M}(R,P)]) = H_{\dim_{\mathbb{C}} \mathfrak{M}(R,P)-k}((\pi^A)^{-1}(x_Q) \cap \mathfrak{M}(R,P), \mathbb{C}$$

By Lemma 8.4.1(3) this is isomorphic to $H_{\dim_{\mathbb{C}} \mathfrak{M}(R,P)-k}(\mathfrak{M}(R,Q) \cap \mathfrak{L}(\mathbf{w}), \mathbb{C})$. Therefore we have

$$\begin{aligned}&\sum_k \dim H^k(i_{x_Q}^! \pi_*^A \mathbb{C}_{\mathfrak{M}(R,P)}[\dim \mathfrak{M}(R,P)])\, t^{\dim_{\mathbb{C}} \mathfrak{M}(Q,P)-k} \\ &= \sum_d \dim H_d(\mathfrak{M}(R,Q) \cap \mathfrak{L}(\mathbf{w}), \mathbb{C})\, t^{d+\dim_{\mathbb{C}} \mathfrak{M}(Q,P)-\dim_{\mathbb{C}} \mathfrak{M}(R,P)} = c_{RQ}(t),\end{aligned} \tag{8.4.3}$$

where we used $\dim_{\mathbb{C}} \mathfrak{M}(R,P) - \dim_{\mathbb{C}} \mathfrak{M}(Q,P) = \dim_{\mathbb{C}} \mathfrak{M}(R,Q)$ in the last equality.

By [15, 14.3.10], we have

$$[M_Q : L_R] = \dim H^*(i_{x_Q}^! IC(\mathfrak{M}_0^{\mathrm{reg}}(R,P))).$$

(In fact, we defined the standard module M_Q as $H_*((\pi^A)^{-1}(x_Q), \mathbb{C})$ in [15, §13], which apriori depends on P. Thus the definition coincides only when $Q = P$. However, by using the transversal slice, we can show that the right hand side is the same for both definitions. cf. Lemma 8.4.1.)

Let

$$Z_{RQ}(t) \stackrel{\text{def.}}{=} \sum_k \dim H^k(i_{x_Q}^! IC(\mathfrak{M}_0^{\mathrm{reg}}(R,P)))\, t^{\dim \mathfrak{M}_0^{\mathrm{reg}}(Q,P)-k}.$$

We have $[M_Q : L_R] = Z_{RQ}(1)$. By the defining property of the intersection homology, $Z_{RQ}(t)$ satisfies (4.1.4).

Substituting (8.4.2) into (8.4.3), we get

$$c_{SQ}(t) = \sum_R L_{SR}(t) Z_{RQ}(t).$$

Now $L_{SR}(t) = \overline{L_{SR}(t)}$ implies (4.1.3). This completes the proof of Theorem 4.1.5.

8.5 Proof of Theorem 7.1.1

By the result explained in §8.1, the weight multiplicity of $\mathbf{w}'$ in $L_{\mathbf{w}}$ is equal to

$$\dim H_{\mathrm{top}}(\mathfrak{M}(\mathbf{w}-\mathbf{w}',\mathbf{w})\cap\mathfrak{L}(\mathbf{w}),\mathbb{C}).$$

The assertion follows from more general formula

$$\widetilde{\chi}_t(M_P)=\sum_{\mathbf{w}'}\sum_{d}H_d(\mathfrak{M}(\mathbf{w}-\mathbf{w}',\mathbf{w})\cap\mathfrak{L}(\mathbf{w}),\mathbb{C})\,t^{\dim_{\mathbb{C}}\mathfrak{M}(\mathbf{w}-\mathbf{w}',\mathbf{w})-d}\prod_k y_k^{w'_k}. \tag{8.5.1}$$

Note that $\mathfrak{M}(\mathbf{w}-\mathbf{w}',\mathbf{w})\cap\mathfrak{L}(\mathbf{w})$ is a lagrangian subvariety in $\mathfrak{M}(\mathbf{w}-\mathbf{w}',\mathbf{w})$, so we have top $=\dim_{\mathbb{C}}\mathfrak{M}(\mathbf{w}-\mathbf{w}',\mathbf{w})$.

In order to prove (8.5.1), we use [12, 5.7], where the Betti numbers are given in terms of those of fixed point components. It looks almost the same as above. However, there is one significant difference. The $\mathbb{C}^*$-action used there is different from our $\mathbb{C}^*$-action used here, defined in [15, §2]. This is the reason why we choose P and corresponding $a=(s,\varepsilon)$ as explained in §7. Then $A=\overline{a^{\mathbb{Z}}}$ is isomorphic to $\mathbb{C}^*$ and the action is the same as the $\mathbb{C}^*$-action considered in [12, §5].

We decompose $\mathfrak{M}(\mathbf{w})^A=\bigsqcup\mathfrak{M}(\rho)$ into connected components as before. By [12, 5.7] we have[f]

$$\dim H_d(\mathfrak{M}(\mathbf{w}-\mathbf{w}',\mathbf{w})\cap\mathfrak{L}(\mathbf{w}),\mathbb{C})=\sum_{\rho}\dim H_{\dim_{\mathbb{C}}\mathfrak{M}(\mathbf{w}-\mathbf{w}',\mathbf{w})-d}(\mathfrak{M}(\rho),\mathbb{C}),$$

where the summation runs over the set of ρ such that the corresponding monomial m_ρ is sent to $\prod_k y_k^{w'_k}$ after $Y_{k,a}\to y_k$. The $\mathbb{C}^*$-action makes $\mathfrak{M}_0(\infty,\mathbf{w})^{\mathbb{C}^*}=\{0\}$, so $\mathfrak{M}(\rho)=\mathfrak{M}(\rho)\cap\mathfrak{L}(\mathbf{w})$. Hence the above expression coincides with the definition of the coefficients $\widetilde{\chi}_t$.

Acknowledgements

We would like to thank E. Frenkel and E. Mukhin for explanations of their algorithm computing q-characters of fundamental representations.

The author are partially supported by by the Grant-in-aid for Scientific Research (No.11740011), the Ministry of Education, Japan.

---

[f] In fact, this formula even holds for general P if we replace $\dim_{\mathbb{C}}\mathfrak{M}(\mathbf{w}-\mathbf{w}',\mathbf{w})-d$ by a suitable degree. However, this degree shift is given by a complicated expression in ρ. So our choice of P is most economical.

References

1. V. Chari and A. Pressley, *A guide to quantum groups*, Cambridge University Press, Cambridge, 1994.
2. ———, *Quantum affine algebras and their representations*, Representations of groups (Banff, AB, 1994), Amer. Math. Soc., Providence, RI, 1995, pp. 59–78.
3. ———, *Quantum affine algebras at roots of unity*, Represent. Theory **1** (1997), 280–328 (electronic).
4. ———, *Integrable and Weyl modules for quantum affine* sl_2, preprint, arXiv:math.QA/0007123.
5. E. Frenkel and E. Mukhin, *Combinatorics of q-characters of finite-dimensional representations of quantum affine algebras*, preprint, arXiv:math.QA/9911112.
6. E. Frenkel and N. Reshetikhin, *The q-characters of representations of quantum affine algebras and deformations of* $\mathcal{W}$*-algebras*, Recent developments in quantum affine algebras and related topics (Raleigh, NC, 1998), Contemp. Math., 248, Amer. Math. Soc., Providence, RI, 1999, pp. 163–205.
7. G. Hatayama, A. Kuniba, M. Okado, T. Takagi and Y. Yamada, *Remarks on fermionic formula*, Recent developments in quantum affine algebras and related topics (Raleigh, NC, 1998), Contemp. Math., 248, Amer. Math. Soc., Providence, RI, 1999, pp. 243–291.
8. M. Kashiwara, *Crystal bases of modified quantized enveloping algebra*, Duke Math. J. **73** (1994), no. 2, 383–413.
9. D. Kazhdan and G. Lusztig, *Representations of Coxeter groups and Hecke algebras*, Invent. Math. **53** (1979), no. 2, 165–184.
10. G. Lusztig, *Canonical bases arising from quantized enveloping algebras*, J. Amer. Math. Soc. **3** (1990), no. 2, 447–498.
11. ———, *Ferminonic form and Betti numbers*, preprint, arXiv:math.QA/0005010.
12. H. Nakajima, *Instantons on ALE spaces, quiver varieties, and Kac-Moody algebras*, Duke Math. J. **76** (1994), no. 2, 365–416.
13. ———, *Homology of moduli spaces instantons on ALE spaces I*, J. Differential Geom. **40** (1994), 105–127.
14. ———, *Quiver varieties and Kac-Moody algebras*, Duke Math. J. **91** (1998), no. 3, 515–560.
15. ———, *Quiver varieties and finite dimensional representations of quantum affine algebras*, preprint, arXiv:math.QA/9912158, to appear in J. Amer. Math. Soc.

16. M. Varagnolo and E. Vasserot, *Standard modules of quantum affine algebras*, preprint, arXiv:math.QA/0006084.

GENERALIZED HÖLDER'S THEOREM FOR MULTIPLE GAMMA FUNCTION

MICHITOMO NISHIZAWA

Department of Mathematics, School of Science and Engineering
Waseda University
3-4-1, Ohkubo Shinjuku-ku, Tokyo 169-8555, Japan
E-mail: mnishi@mn.waseda.ac.jp

We improve Barnes' proof of nonexistence of algebraic differential equations for his multiple gamma function. It is also proved that multiple q-gamma function satisfies no algebraic differential equation over a certain field. Furthermore, we consider an example of differential relations between several Barnes' multiple gamma functions in a special case.

1 Introduction

In 1904, Barnes[2] introduced his multiple gamma function which satisfies functional relation

$$\frac{\Gamma_r(z+\omega_j|\underline{\omega})}{\Gamma_r(z|\underline{\omega})} = \frac{\rho_{r-1}(\underline{\omega}(j))}{\Gamma_{r-1}(z|\underline{\omega}(j))} \tag{1}$$

where $\underline{\omega} = (\omega_1, \omega_2, \cdots \omega_r)$ is a set of complex parameters, $\underline{\omega}(j) := (\omega_1, \cdots, \omega_{j-1}, \omega_{j+1}, \cdots, \omega_r)$ and $\rho_r(\underline{\omega}(j))$ depends only on $\underline{\omega}(j)$. In the case $r = 1$, $\Gamma_1(z|(\omega_1))$ is equal to Euler's gamma function $\Gamma(z/\omega_1)$ up to an elementary factor.

After his discovery, many mathematicians have studied this function: Hardy[4,5] studied this function from his viewpoint of theory of elliptic functions, and Shintani[20,21] applied it to the study on the Kronecker limit formula for zeta functions attached to certain algebraic fields. Recently, this function plays an important role in studies on quantum integral systems[8,10,11,12,19] and on q-analysis with $|q| = 1$[14,16,22].

In this article, we consider differential relations between these functions. It is well known that Euler's gamma function satisfies no algebraic differential equation. It is called *Hölder's theorem.* As a generalization of it, Barnes[3] gave a proof that his multiple gamma function satisfies no algebraic differential equation. He introduced an weight for algebraic differential equations and showed that any algebraic differential equation with nonzero weight can be transformed to some equation with weight 1 by using the functional relation for the multiple gamma function. His result, which corresponds to Lemma 5 in this article, is very important. However it should be considered as the first step of induction. In order to prove the generalized Hölder's theorem, we have to construct the induction completely. Furthermore, on defining this weight, he did not make clear the function field which he considered. There would remain some ambiguous parts in his proof.

In this article, we reconstruct a proof of the generalized Hölder's theorem by using a weight for multivariable polynomials over $\mathbf{C}(z)$. This method can be applied to its generalization for a q-analogue of multiple gamma function introduced by the author[13]. We can see that this q-analogue satisfies no algebraic differential equation over $\mathbf{C}(z, q^z)$. Although there is no differential relation for a single multiple gamma function, there exists differential relation between several multiple gamma functions in a special case. In the case when all ω_j $(j = 1, \cdots, r)$ are equal, we can derive an differential relation between multiple gamma function by using an infinite sum representation of the logarithmic derivative of the multiple gamma function. As a corollary of this relation, we can prove generalized Hölder's theorem more easily in this case. We remark that this case is closely related Vignéras multiple gamma function[25]. She introduced meromorphic functions $G_r(z)$ which satisfy the following relation:

$$\begin{aligned} &G_r(z+1) = G_{r-1}(z)G_r(z), \qquad G_r(1) = 1 \\ &\frac{d^2}{dz^2}\log G_r(z+1) \geq 0 \quad \text{for} \quad z \geq 0, \qquad G_0(z) = z. \end{aligned} \tag{2}$$

This function have been applied to represent the determinant of the Laplacians[24,25,26]. $G_r(z)$ is related with $\Gamma_r(z)$ by the relation

$$G_r(z) = (\text{elementary factor}) \times \Gamma(z|(1, \cdots, 1)).$$

This article is organized as follows: In Section 2, we give a survey on Barnes' multiple gamma function. In Section 3, we give a proof of generalized Hölder's theorem and remark its generalization to q-analogue. In Section 4, we derive a differential relation between $\Gamma_r(z|(\omega_1, \cdots, \omega_r))$ and $\Gamma_{r+1}(z|(\omega_1, \cdots, \omega_{r+1}))$ in the case when $\omega_1 = \cdots = \omega_r = \omega_{r+1}$.

2 Barnes' Multiple Gamma Function

We summarize some facts about Barnes' multiple gamma function. We follow the notation in Barnes[2]. Let $\underline{\omega} := (\omega_1, \omega_2, \cdots, \omega_r)$ be a set of complex numbers lying the same side of some straight line across the origin on the complex plane.

Definition 1 *We define $\zeta(s,z|\underline{\omega})$ and $\Gamma_r(z|\underline{\omega})$ by the following formula:*

$$\zeta_r(s,z|\underline{\omega}) := \sum_{\underline{m}\in(\mathbf{Z}_{\geq 0})^r} (z+\underline{m}\cdot\underline{\omega})^{-s},$$

$$\Gamma_r(z|\underline{\omega}) := \rho_r(\underline{\omega}) \exp\left[\frac{\partial}{\partial s}\zeta_r(s,z|\underline{\omega})\right]_{s=0}$$

where $\underline{m} := (m_1, m_2, \cdots, m_r)$, $\underline{m}\cdot\underline{\omega} := m_1\omega_1 + \cdots + m_r\omega_r$, and

$$\log\rho_r(\underline{\omega}) := -\lim_{z\to 0}\left\{\left[\frac{\partial}{\partial s}\zeta_r(s,z|\underline{\omega})\right]_{s=0} + \log z\right\}. \quad ▮$$

We recall the basic facts concerning Barnes' multiple gamma function.

Proposition 2 (Barnes) (i) $\Gamma_r(z|\underline{\omega})$ *satisfies a functional relation* (1) *where* $\underline{\omega}(i) := (\omega_1, \omega_2, \cdots \omega_{i-1}, \omega_{i+1}, \cdots, \omega_r)$.
(ii) $\Gamma_r(z|\underline{\omega})$ *have an infinite product representation*

$$\Gamma_r(z|\underline{\omega})^{-1} = \exp\left(\sum_{k=1}^{r}\frac{\gamma_{r,k}}{k!}z^k\right) z \prod_{\underline{m}}{}' \left[\left(1+\frac{z}{\underline{m}\cdot\underline{\omega}}\right)\exp\left(\sum_{l=1}^{r}\frac{1}{l}\frac{(-z)^l}{(\underline{m}\cdot\underline{\omega})^l}\right)\right]$$

where $\gamma_{r,k}$ $(k=1,\cdots,r)$ are scalars and $\prod'$ means that $\underline{m}$ runs over all nonnegative integer except $\underline{m}=(0,0,\cdots,0)$.
(iii) $\Gamma_r(z|\underline{\omega})$ *is symmetric with respect to $\underline{\omega} := (\omega_1, \cdots, \omega_r)$.* ▮

We define $\psi_r(z|\underline{\omega})$ as the logarithmic derivative of $\Gamma_r(z|\underline{\omega})$. From Proposition 2 (ii), it follows that

$$\begin{aligned}\psi_r(z|\underline{\omega}) &:= \frac{d}{dz}\log\Gamma_r(z|\underline{\omega}) \\ &= -\sum_{k=1}^{r}\frac{\gamma_{r,k}}{(k-1)!}z^{k-1} - \frac{1}{z} - \sum_{\underline{m}}{}'\left[\frac{1}{z+\underline{m}\cdot\underline{\omega}} + \sum_{l=1}^{r}\frac{(-z)^{l-1}}{(\underline{m}\cdot\underline{\omega})^l}\right],\end{aligned}$$

where $\sum'$ means that $\underline{m}$ runs over all elements of $(\mathbf{Z}_{\geq 0})^r - \{(0,0,\cdots,0)\}$. From Proposition 2 (i), $\psi_r(z|\omega)$ satisfies the relation

$$\psi_r(z|\underline{\omega}) - \psi_r(z+\omega_j|\underline{\omega}) = \psi_{r-1}(z|\underline{\omega}(j)). \tag{3}$$

We observe that $\psi_r(z|\underline{\omega})$ has simple poles at $z = \underline{m}\cdot\underline{\omega}$ and $\psi_r^{(n)}(z|\underline{\omega}) := \frac{d^n}{dz^n}\psi_r(z|\underline{\omega})$ has poles of order n at $z = \underline{m}\cdot\underline{\omega}$, where $\underline{m} \in (\mathbf{Z}_{\leq 0})^r$.

For the following argument, we introduce several notations. We denote $\underline{\omega}_{\leq j}$ by

$$\underline{\omega}_{\leq j} := (\omega_1, \omega_2, \cdots, \omega_j),$$

and define $\psi_j(z) := \psi_j(z|\underline{\omega}_{\leq j})$, and $\psi_j^{(n)}(z) := \frac{d^n}{dz^n}\psi_j(z)$.

3 Generalized Hölder's Theorem

In this section, we give a proof of generalized Hölder's theorem.

Theorem 3 (i) *There exists no $n+1$-variable polynomial*

$$g(t_0, t_1, \cdots, t_n) \in \mathbf{C}(z)[t_0, t_1, \cdots, t_n]$$

such that

$$g(\psi_r(z), \psi_r^{(1)}(z), \cdots \psi_r^{(n)}(z)) = 0. \tag{4}$$

(ii) *There exists no $n+1$-variable polynomial*

$$f(t_0, t_1, \cdots, t_n) \in \mathbf{C}(z)[t_0, t_1, \cdots t_n]$$

such that

$$f(\Gamma_r(z|\underline{\omega}), \Gamma_r^{(1)}(z|\underline{\omega}), \cdots \Gamma_r^{(n)}(z|\underline{\omega})) = 0.$$

In other words, Barnes' multiple gamma function satisfies no algebraic differential equation. ▌

We know the fact that if some function satisfies an algebraic differential equation, then its logarithmic derivative also satisfies an algebraic differential equation. (cf. Komatsu[7], Pastro[17]). Thus, it is sufficient to show the following theorem since Theorem 3 (i) corresponds to the case when $p = r$:

Theorem 4 *For $p \in \{1, \cdots r\}$ and arbitrary fixed $n_l \in \mathbf{Z}_{\geq 0}$ ($l := (l_{p+1}, l_{p+2}, \cdots l_r) \in \mathbf{Z}^{r-p}$), there is no polynomial*

$$f(T) \in \mathbf{C}(z)\left[T = (t_{l,k})\middle| l = (l_{p+1}, l_{p+2}, \cdots, l_r) \in \mathbf{Z}^{r-p},\ 0 \leq k \leq n_l,\right]$$

such that it vanishes when we substitute $\psi_p(z + \sum_{i=p+1}^{r} l_i\omega_i)$ for $t_{l,k}$

$$f\left(\left(\psi_p^{(j)}((z + \sum_{i=p+1}^{r} l_i\omega_i)\right)\right) \equiv 0. \tag{5}$$

Proof: : We prove this theorem by induction. In the case when $r = 1$, we can prove it in a similar way to the original Hölder's theorem for Euler's gamma function. We suppose that this claim holds for $p-1$. First, we fix some notations. We define $\Xi_p(z)$ as the ring generated by z and $\psi_{p-1}^{(j)}(z+\sum_{i=p}^{r} l_i\omega_r)$, $(j \in \mathbf{Z}_{\geq 0},\ l_i \in \mathbf{Z}\ )$

$$\Xi_p(z) := \mathbf{C}\left[z, \psi_{p-1}^{(j)}(z + \sum_{i=p}^{r} l_i\omega_i) \;\middle|\; j \in \mathbf{Z}_{\geq 0},\ \ l_i \in \mathbf{Z} \right].$$

and $\Lambda_p(z)$ as the field generated by z and $\psi_{p-1}^{(j)}(z + \sum_{i=p}^{r} l_i\omega_i)$, $(j \in \mathbf{Z}_{\geq 0}$, $l_i \in \mathbf{Z}\ )$

$$\Lambda_p(z) := \mathbf{C}\left(z, \psi_{p-1}^{(j)}(z + \sum_{i=p}^{r} l_i\omega_r) \;\middle|\; j \in \mathbf{Z}_{\geq 0},\ l_i \in \mathbf{Z} \right)$$

and $\Lambda_r(z) := \mathbf{C}(z)$. That is, any element of $\Lambda_p(z)$ is represented as $Q_1(z)/Q_2(z)$, where $Q_1(z),\ Q_2(z) \in \Xi_p(z)$ and $Q_2(z) \neq 0$. Key of proof is to show the following lemma:

Lemma 5 *If there exist some polynomial*

$$E(T) \in \Lambda_p(z)\left[T = (t_{l,j}) \;\middle|\; 0 \leq j \leq m_l,\ l \in \mathbf{Z}^{r-p} \right]$$

such that

$$E\left(\left(\psi_p^{(j)}(z + \sum_{i=p+1}^{r} l_i\omega_i) \right) \right) = 0 \tag{6}$$

then, we have

$$\psi_p(z + \sum_{i=p+1}^{r} l_i\omega_i) \in \Lambda_p(z)$$

for some $l \in \mathbf{Z}^{r-p}$. In other words, for some $l \in \mathbf{Z}^{r-p}$, we can represent the element $\psi_p(z + \sum_{i=p+1}^{r} l_i\omega_i)$ as

$$A(z)\psi_p(z + \sum_{i=p+1}^{r} l_i\omega_i) = B(z), \tag{7}$$

where $A(z),\ B(z) \in \Xi_p(z)$.

Proof of Lemma: We prove this proposition by induction. In the case when $p = 1$, we can prove the claim by the same argument as Hölder's theorem for Euler's gamma function. We suppose that there is no algebraic relation between $\psi_{p-1}^{(j)}(z + \sum_{i=p}^{r} l_i\omega_i)$ $(j \in \mathbf{Z}_{\geq 0},\ l \in \mathbf{Z}^{r-p+1})$. We introduce weight wt for element of $\Lambda_p(z)[T]$. For a monomial $\prod_l \prod_{j=0}^{n_l} t_{l,j}^{\mu_{l,j}}$, we define $wt(\prod_l \prod_{j=0}^{n_l} t_{l,j}^{\mu_{l,j}})$ as

$$wt(\prod_l \prod_{j=0}^{n_l} t_{l,j}^{\mu_{l,j}}) := \sum_l \sum_{k=0}^{n_l} (k+1)\mu_{j,k}.$$

For polynomials, we define its weight by the maximum of the weights of monomials included in the polynomial. We often abbreviates indeterminates $(\psi_p^{(j)}(z + \sum_{i=p+1}^{r} l_i\omega_i))$ by $\Psi_p(z)$. We suppose that there exists a polynomial $E_z(T)$ satisfying

$$E_z(\Psi_p(z)) = 0. \tag{8}$$

Without loss of generality, we can suppose that $E_z(T)$ has the lowest weight s among such polynomials. We denote $k_\nu + 1$ by the number of monomials of weight ν $(0 \leq \nu \leq s)$. For $0 \leq i \leq k_\nu$, we define $Q_{\nu,i}(T)$ by the following formula:

$$Q_{\nu,i}(T) := \prod_l \prod_{j=0}^{n_l} t_{l,j}^{\mu_{l,j}(\nu,i)} \qquad \text{for} \qquad 0 \leq i \leq k_\nu.$$

We can suppose that $i \neq j \Rightarrow Q_{l,i} \neq Q_{l,j}$. We define Δ_s as the sum of terms whose weight is lower than s.

By using these notation, we can see that $E_z(T)$ has the following representation:

$$E_z(T) := \sum_{i=0}^{k_s} R_{s,i}(z) Q_{s,i}(T) + \Delta_s,$$

where $R_{s,i}(z) \in \Lambda_p(z)$. Without loss of generality, we can suppose that one of nonzero coefficients, say $R_{s,0}(z)$, is equal to 1. We remark that $E_{z+\omega_p}(T)$ satisfies the relation

$$E_{z+\omega_p}(\Psi_p(z+\omega_p)) \equiv 0,$$

and that $wt(E_{z+\omega_p}(T)) = s$. From relation (3), it follows that

$$\begin{aligned} E_{z+\omega_p}(\Psi_p(z+\omega_p)) &= E_{z+\omega_p}(\Psi_p(z) - \Psi_{p-1}(z)) \\ &= \sum_{i=0}^{k_s} R_{s,i}(z+\omega_p) Q_{s,i}(\Psi_p(z)) + \Delta_s, \end{aligned}$$

where

$$\Psi_p(z) - \Psi_{p-1}(z) := \left(\psi_p^{(j)}(z + \sum_{i=p+1}^{r} l_i\omega_i) - \psi_{p-1}^{(j)}(z + \sum_{i=p+1}^{r} l_i\omega_i) \right).$$

Furthermore, we can regard an element of $\Lambda_p(z)\,[\Psi_p(z+\omega_p)]$ as an element of $\Lambda_p(z)[\Psi_p(z)]$ by using (3). In this sense, we may regard that $E_{z+\omega_p}(\Psi_p(z+\omega_p))$ as an element of $\Lambda_p(z)[\Psi_p(z)]$. We define it as $\widetilde{E}_{z+\omega_p}(\Psi_p(z))$. Let us put $F_z(T) = E_z(T) - \widetilde{E}_{z+\omega_p}(T) \in \Lambda_p(z)[T]$. We can see that

$$\begin{aligned} F_z(T) &:= E_z(T) - \widetilde{E}_{z+\omega_p}(T) \\ &= \sum_{i=0}^{k_s} (R_{s,i}(z+\omega_p) - R_{s,i}(z))\, Q_{s,i}(T) + \Delta_s. \end{aligned}$$

From the fact that $R_{s,0}(z+\omega_p) \equiv R_{s,0}(z) \equiv 1$, it follows that the number of monomial with weight s in $F_z(T)$ is k_s. Thus, $F_z(T) \equiv 0$ because the minimal number of monomial of weight s is $k_s + 1$. Therefore, we have that

$$R_{s,i}(z+\omega_p) - R_{s,i}(z) \equiv 0 \qquad \text{for} \qquad 1 \le i \le k_s. \tag{9}$$

However, if (9) have a nontrivial solution, then there exist an algebraic relation between $\psi_{p-1}^{(j)}(z + \sum_{i=p}^{r} l_i\omega_i)$ ($l \in \mathbf{Z}^{r-p+1}$, $0 \le j \le n_l$). This contradict to the assumption of the induction. Thus, $R_{s,i}(z) := a_{s,i} \in \mathbf{C}$ and $F_z(T)$ can be represented by the formula

$$F_z(\Psi_p(z)) = \sum_{i=1}^{k_s} a_{s,i}\,(Q_{s,i}(\Psi_p(z)) - Q_{s,i}(\Psi_p(z+\omega_p))) + \Delta_s.$$

From the definition of weight, it follows that

$$\begin{aligned} &Q_{s,i}(\Psi_p(z)) - Q_{s,i}(\Psi_p(z+\omega_p)) \\ &\quad = \delta_{s,i}\psi_{p-1}(z + \sum_{i=p+1}^{r} l_i\omega_i) Q_{s,i}(\Psi_p(z))/\psi_p(z + \sum_{i=p+1}^{r} l_i\omega_i) + \Delta_{s-1}, \end{aligned}$$

for some $l \in \mathbf{Z}^{r-p}$, where $\delta_{s,i}$ is a constant. Therefore, $F_z(\Psi_p(z))$ is represented by the formula,

$$\begin{aligned} F_z(\Psi_p(z)) = \sum_{i=0}^{k_{s-1}} &\left(\delta_{i,j}^{(s,s-1)}\psi_{p-1}(z + \sum_{i=p+1}^{r} l_i\omega_i) + R_{s-1,i}(z) - R_{s-1,i}(z+\omega_p) \right) \\ &\times Q_{s-1,i}(\Psi_p(z)) + \Delta_{s-1}, \end{aligned} \tag{10}$$

where $\delta_{i,j}^{(s,s-1)}$ is such a constant that it is equal to zero if there exists no j $(1 \leq j \leq k_s)$ satisfying

$$Q_{s,i}(\Psi_p(z)) - Q_{s,i}(\Psi_p(z+\omega_p)) = Q_{s-1,j}(\Psi_p(z)) + \Delta_{s-1}.$$

If (10) is a nontrivial representation of $F_z(\Psi_p(z))$, then there exists such a polynomial in $g(T) \in \Lambda_p(z)[T]\backslash\{0\}$ with weight $s-1$ that $g(\Psi_p(z)) \equiv 0$. It is contradiction to the assumption of the induction. Thus, all coefficients on the right hand side of (10) vanish

$$R_{s-1,i}(z+\omega_p) - R_{s-1,i}(z) = \delta_{i,j}^{(s,s-1)} \psi_{p-1}(z + \sum_{i=p+1}^{r} l_i\omega_i)$$

for $l \in \mathbf{Z}^{r-p}$. Therefore, from the assumption of induction, the following facts hold:

(i) $\delta_{i,j}^{(s,s-1)} = 0$ for $j = 0, 1, \cdots k_s$ and $i = 0, 1, \cdots, k_{s-1}$,

(ii) $R_{s-1,i}(z) = a_{s-1,i} \in \mathbf{C}$.

If we define $\eta_j(s,i)$ as the sum of all $\mu_{l,j}(s,i)$ over $l \in \mathbf{Z}^{r-p}$

$$\eta_j(s,j) := \sum_l \mu_{l,j}(s,i).$$

It follows from (i) that

$$\eta_0(s,i) = 0 \quad \text{for} \quad i = 0, 1, \cdots, k_s.$$

Repeating this procedure successively, we can prove that

$$\eta_j(v,w) = 0 \quad \text{for} \quad 0 \leq j \leq \max_l n_l, \ v = 1, \cdots, s, \ w = 0, \cdots, k_v.$$

Thus, $\mu_{l,j}(s,i) = 0$ for $\forall l \in \mathbf{Z}^{r-p}$ since $\mu_{l,j}(s,i)$ are nonnegative integers. It means that $wt(E_z(T)) = 1$. Therefore, we conclude that

$$A(z)\psi_p(z + \sum_{i=p+1}^{r} l_i\omega_i) = B(z),$$

for some $l \in \mathbf{Z}^{r-p}$ where $A(z)$, $B(z) \in \Xi_p(z)$. ∎

Once, we have this lemma, we can prove Theorem 4 easily. If there is no polynomial

$$E(T) \in \Lambda_p(z)\left[T = (t_{l,j}) \middle| l \in \mathbf{Z}^{r-p}, 0 \leq j \leq n_l\right]$$

such that

$$E\left(\Psi_p(z)\right) \equiv 0$$

then, of course, there is no differential relation like formula (5). Thus, it is sufficient to consider the claim in the case when there exist such differential relations. If there exists a differential relation (6), then, from Lemma 5, it follows that the relation is transformed to a formula represented as

$$A(z)\psi_p(z + \sum_{i=p+1}^{r} l_i\omega_i) = B(z) \quad \text{for some} \quad l \in \mathbf{Z}^{r-p}.$$

By using it, we can reduce any algebraic relation over $\mathbf{C}(z)$ between $\psi_p^{(j)}(z + \sum_{i=p+1}^{r} l_i\omega_i)$ ($l \in \mathbf{Z}^{r-p}$, $0 \leq j \leq n_l$) to some algebraic relation over $\mathbf{C}(z)$ between $\psi_{p-1}^{(j)}(z + \sum_{i=p}^{r} l_i\omega_i)$ ($l \in \mathbf{Z}^{r-p+1}$, $j \in \mathbf{Z}_{\geq 0}$). This is a contradiction to the assumption of the induction. Therefore, there is no algebraic relation over $\mathbf{C}(z)$ between $\psi_p^{(j)}(z + \sum_{i=p+1}^{r} l_i\omega_i)$ ($l \in \mathbf{Z}^{r-p}$, $0 \leq j \leq n_l$). ▮

We remark a generalization to its q-analogue. The author[14] introduced multiple q-gamma function $G_r(z;q)$ which satisfies a q-analogue of (2)

$$G_r(z+1;q) = G_{r-1}(z;q)G_r(z;q), \qquad G_r(1;q) = 1,$$

$$\frac{d}{dz}\log G_r(z+1;q) \geq 0 \quad \text{for} \quad z \geq 0, \qquad G_0(z;q) = \frac{1-q^z}{1-q}.$$

On the other hand, Pastro[17] proved that q-gamma function satisfies no algebraic differential equation over $\mathbf{C}(z, q^z)$. By using the same argument as above, we can prove the following theorem:

Theorem 6 *When $0 < q < 1$, the multiple q-gamma function $G_r(z;q)$ satisfies no algebraic differential equation. In other words, there exists no $n+1$-variable polynomial*

$$f(t_0, t_1, \cdots, t_n) \in \mathbf{C}(z, q^z)[t_0, t_1, \cdots t_n]$$

such that

$$f(G_r(z;q), G_r^{(1)}(z;q), \cdots G_r^{(n)}(z;q)) = 0. \qquad ▮$$

4 Differential Relation in a Special Case

As we have seen in the previous section, there is no algebraic differential equation for a multiple gamma function. However, it is not always true that there is no algebraic relation between several multiple gamma functions with different r. For example, we can derive a differential relation between $\Gamma_r(z|(\omega_1,\cdots,\omega_r))$ and $\Gamma_{r+1}(z|(\omega_1,\cdots,\omega_{r+1}))$ in the case when

$$\omega_1=\omega_2=\cdots=\omega_r=\omega_{r+1}=\omega. \tag{11}$$

Let $\widetilde{\psi}_r(z)$ be the infinite sum part of $\psi_r(z)$

$$\widetilde{\psi}_r(z):=-\sum_{k=0}^{\infty}\left\{\frac{1}{z+k\omega}\binom{k}{r-1}+\phi_r(z,k|\omega)\right\},$$

where $\phi_r(z,k|\omega)$ is a convergence factor

$$\phi_r(z,k|\omega)\ :=\ \sum_{m_1+\cdots+m_r=k}\ \sum_{l=1}^{r}\left.\frac{(-z)^{l-1}}{(\underline{m}\cdot\underline{\omega})^l}\right|_{\omega_1=\omega_2=\cdots=\omega_r=\omega}.$$

Then, we have the following relation:

Lemma 7

$$\frac{z+(r-1)\omega}{r\omega}\widetilde{\psi}_r(z)=-\widetilde{\psi}_{r+1}(z).$$

Proof: First, we notice the following identity:

$$\frac{z+(r-1)\omega}{r\omega(z+k\omega)}\binom{k}{r-1}=-\frac{1}{z+k\omega}\binom{k}{r}+\frac{1}{r\omega}\binom{k}{r-1}$$

for $r,k\in\mathbf{Z}_{>0}$. We have

$$\frac{z+(r-1)\omega}{r\omega}\widetilde{\psi}_r(z)=$$

$$=\sum_{k=1}^{\infty}\left\{\frac{1}{z+k\omega}\binom{k}{r}-\frac{1}{r\omega}\binom{k}{r-1}+\frac{z+(r-1)\omega}{r\omega}\phi_r(z,k|\omega)\right\}.$$

From the absolute convergence of $\widetilde{\psi}_r(z)$, it follows that the sum of the right hand side is absolutely convergent. The convergent factor of this sum have to be equal to $\phi_{r+1}(z,k|\omega)$. In fact, the factor is uniquely determined when its degree of k is larger than -2. ▮

Remark 8 *We can compute $\phi_r(z,k|\omega)$ explicitly. We have the following formula by the above argument:*

$$\phi_{r+1}(z,k|\omega)=\frac{1}{r\omega}\binom{k}{r-1}-\frac{z+(r-1)\omega}{r\omega}\phi_r(z,k|\omega).$$

We can calculate the convergent factor by using this formula as a recurrence formula. (*cf.* Nishizawa[15]). ∎

From Lemma 7, we have relations between logarithmic derivatives.

Proposition 9 (i) *$\psi_r(z)$ satisfies a recurrence formula*

$$\psi_{r+1}(z)=-\frac{z+(r-1)\omega}{r\omega}\psi_r(z)+p_r(z),$$

where $p_r(z)$ is a polynomial with degree less than equal r.

(ii) *$\psi_{r+1}(z)$ is transformed by the following formula:*

$$\psi_{r+1}(z)=(-1)^r\binom{\frac{z}{\omega}+r-1}{r}\psi_1(z)+P_r(z),$$

where $P_r(z)$ is a polynomial with degree less than equal r. ∎

Corollary 10 *In the case when $\omega_1=\cdots=\omega_r=\omega_{r+1}=\omega$, we have*

$$\left\{\frac{d}{dz}\Gamma_{r+1}(z)\right\}\Gamma_r(z)=-\frac{z+(r-1)\omega}{r\omega}\left\{\frac{d}{dz}\Gamma_r(z)\right\}\Gamma_{r+1}(z)+p(z)\Gamma_{r+1}(z)\Gamma_r(z),$$

where

$$\Gamma_r(z):=\Gamma_r(z|(\omega,\omega,\cdots,\omega)).$$ ∎

We remark that Proposition 9 (i) corresponds to the formula (7) in Lemma 5. Thus, in this case, we can see a differential relation between logarithmic derivatives and we can easily prove Theorem 3. If some polynomial satisfies (4), then there exists an algebraic differential equation for $\psi_r(z)$. However, from Proposition 9 (ii), it follows that there exists an algebraic differential equation for $\psi_1(z)$ over $\mathbf{C}(z)$. This contradict to Hölder's theorem for the gamma function.

Acknowledgment: The author expresses his deep gratitude to the organizing committee of International Workshop on Physics and Combinatorics 2000 and the staff of Graduate School of Mathematical Sciences of Nagoya University. He also thanks a referee for his careful reading of the manuscript and for valuable comments. The author is partially supported by Grant-in-Aid for Scientific Research from the Ministry of Education (12640046) and Waseda University Grant for Special Research Project 2000A-172.

References

1. E.W. Barnes, *The Theory of G-Function*, Quat. J. Math. **31**(1899), 264–314.
2. E.W. Barnes, *The theory of the Multiple Gamma Function*, Trans. Cambridge Phil. Soc. **19** (1904), 374–425.
3. E.W. Barnes, *On Functions Generated by Linear Difference Equations of the First Order*, Proc. London Math. Soc. (2) **2**(1905), 280–292 .
4. G.H. Hardy, *On the expression of the double zeta-function and double gamma functions*, Trans. Cambridge Phil. Soc. **20** (1905), 395–437.
5. G.H. Hardy, *On double Fourier series and especially these which represent the double zeta-function and incommensurable parameters*, Quart. J. Math. **37** (1906). 53–79
6. K. Katayama and M. Ohtsuki, *On the Multiple Gamma Functions*, Tokyo J. Math. bf 21 (1998), 159–182
7. Y. Komotsu *Special Functions (in Japanese)*, Asakura-Shoten.
8. M. Jimbo and T. Miwa, *Quantum KZ equation with* $|q| = 1$ *and correlation functions of the XXZ models in the gapless regime*, J. Phys. A: Math. Gen. **29** (1996) 2923–2958 .
9. K. Matsumoto, *Asymptotic series for the double zeta, double gamma and Hecke L-functions*, Math. Proc. Camb. Phil. Soc., **123** (1998), 385–405
10. T. Miwa and Y. Takeyama, *Determinant Formula for the Solutions of the Quantum Kniznik-Zamolodchikov Equation with* $|q| = 1$, RIMS preprint 1220.
11. T. Miwa and Y.Takeyama, *The integral formula for the solutions of the quantum Kniznik-Zamolodchikov equation associated with* $U_q(\hat{sl}_n)$ with $|q| = 1$, RIMS preprint 1226.
12. T. Miwa, Y. Takeyama and V. Tarasov, *Determinant Formula for Solutions of the* $U_q(\hat{sl}_n)$ *qKZ Equation with* $|q| = 1$, RIMS preprint 1232.
13. M. Nishizawa, *On a q-Analogue of the Multiple Gamma Function*, Lett.

Math. Phys. **37** (1996), 201–208

14. M. Nishizawa, *On a Solution of q-Difference Analogue of Lauricella's Hypergeometric Equation with* $|q| = 1$, Publ. RIMS. Kyoto Univ. **34** (1998), 277–290.
15. M. Nishizawa, *Generalized Hölder's Theorem for Vignéras' Multiple Gamma Function, To appear in* Tokyo J. Math.
16. M. Nishizawa and K. Ueno, *Integral Solution of q-Difference equation of the Hypergeometric Type with* $|q| = 1$, IIAS Report No.1997-001, "Infinite Analysis - Integral Systems and Representation Theory", 247–255.
17. P.I. Pastro, *The q-Analogue of Hölder's Theorem for the Gamma Function*, Rend Sem. Mat. Univ. Padova, **79** (1983), 47–53.
18. S.N.M. Ruijsenaars, First order difference equations and integrable quantum systems, *J. Math. Phys.* **38** (1997) 1069–1146.
19. S.N.M. Ruijsenaars, *A Generalized Hypergeometric Function Satisfying Four Analytic Difference Equations of Askey-Wilson Type*, Comm. Math. Phys. **206** (1999), 639–690.
20. T. Shintani, *On a Kronecker limit formula for real quadratic fields*, J. Fac. Sci. Univ. Tokyo, **24** (1977), 167–199.
21. T. Shintani, *A poof of classical Kronecker limit formula*, Tokyo J. Math. **3**(1980), 191–199.
22. Y. Takeyama, *The q-twisted Cohomology and The q-hypergeometric Function at* $|q| = 1$, RIMS preprint 1278.
23. K. Ueno and M. Nishizawa, *Multiple Gamma Functions and Multiple q-Gamma Functions,* Publ. RIMS. Kyoto Univ. **33** (1997), 813–838.
24. I.Vardi, *Determinants of Laplacians and multiple gamma functions*, SIAM. J. Math. Anal 19 (1988), 493–507.
25. M.F.Vignéras, *L'équation fonctionalie de la fonction zeta de Selberg de groupe modulaire* $PSL(2, Z)$, Asterisque. 61 (1979), 235–249.
26. A.Voros, *Spectral functions, Special functions and the Selberg zeta functions*, Comm. Math. Phys. 110 (1987), 431–465
27. E.T. Whittacker and G.N. Watson *A Course of Modern Analysis*, Fourth Edition, Camb. Univ. Press, 1927

QUANTUM CALOGERO–MOSER MODELS: COMPLETE INTEGRABILITY FOR ALL THE ROOT SYSTEMS

RYU SASAKI

Yukawa Institute for Theoretical Physics, Kyoto University
Kyoto 606-8502, Japan
E-mail: ryu@yukawa.kyoto-u.ac.jp

Complete integrability of Calogero–Moser models is discussed from various angles. Calogero–Moser models and Toda models are best known examples of solvable many-particle dynamics on a line which are based on root systems. At the classical level, the former (C–M models) is integrable for elliptic potentials (Weierstraß $\wp$ function) and their various degenerate limits. The latter (Toda) has exponential potentials, which is obtained from the former as a special limit of the elliptic potential. Here we discuss quantum Calogero–Moser models based on any root system. For the models with degenerate potentials, *i.e.* the rational with/without the harmonic confining force, the hyperbolic and the trigonometric, we demonstrate the following for all the root systems: (i) Construction of a complete set of quantum conserved quantities in terms of a Lax pair. (ii) Liouville integrability. (iii) Triangularity of the quantum Hamiltonian and the entire discrete spectrum. (v) Algebraic construction of all excited states in terms of creation operators. These are mainly generalizations of the results known for the models based on the A series, *i.e.* $su(N)$ type, root systems, which are treated rather explicitly in the Introduction. The independence of the conserved quantities (necessary for the Liouville integrability) obtained from the universal Lax pair is discussed in some detail.

1 Introduction

Calogero–Moser models [1,2,3] are one-dimensional multi–particle dynamical systems with long range interactions. They are integrable at both classical and quantum levels. In this paper we consider quantum models. The simplest example is described by the following Hamiltonian with r degrees of freedom:

$$\mathcal{H} = \frac{1}{2}\sum_{j=1}^{r} p_j^2 + \frac{1}{2}g(g-1)\sum_{j\neq k}^{r}\frac{1}{(q_j-q_k)^2}, \tag{1.1}$$

in which g is a real coupling constant (assumed to be $g \geq 1$ for simplicity) and $q = (q_1, \ldots, q_r)$ are the coordinates and $p = (p_1, \ldots, p_r)$ are the conjugate canonical momenta obeying the canonical commutation relations:

$$[q_j, p_k] = i\delta_{jk}, \qquad [q_j, q_k] = [p_j, p_k] = 0, \quad j, k = 1, \ldots, r. \tag{1.2}$$

The Heisenberg equations of motion read

$$\dot{q}_j = i[\mathcal{H}, q_j] = p_j,$$
$$\ddot{q}_j = \dot{p}_j = i[\mathcal{H}, p_j] = 2g(g-1)\sum_{k\neq j}\frac{1}{(q_j - q_k)^3}. \tag{1.3}$$

The repulsive $1/(\text{distance})^2$ potential cannot be surmounted classically and the relative position of the particles on the line is not changed during the time evolution. It is remarkable that these equations of motion can be expressed in a matrix form (Lax pair)

$$i[\mathcal{H}, L] = \frac{dL}{dt} = LM - ML = [L, M],$$
$$\Updownarrow$$
$$\dot{q}_j = p_j, \quad \dot{p}_j = 2g(g-1)\sum_{k\neq j}\frac{1}{(q_j - q_k)^3}, \tag{1.4}$$

in which L and M are defined by

$$L = \begin{pmatrix} p_1 & \frac{ig}{q_1 - q_2} & \cdots & \frac{ig}{q_1 - q_r} \\ \frac{ig}{q_2 - q_1} & p_2 & \cdots & \frac{ig}{q_2 - q_r} \\ \vdots & \vdots & \ddots & \vdots \\ \frac{ig}{q_r - q_1} & \frac{ig}{q_r - q_2} & \cdots & p_r \end{pmatrix} \tag{1.5}$$

and

$$M = \begin{pmatrix} m_1 & -\frac{ig}{(q_1 - q_2)^2} & \cdots & -\frac{ig}{(q_1 - q_r)^2} \\ -\frac{ig}{(q_2 - q_1)^2} & m_2 & \cdots & -\frac{ig}{(q_2 - q_r)^2} \\ \vdots & \vdots & \ddots & \vdots \\ -\frac{ig}{(q_r - q_1)^2} & -\frac{ig}{(q_r - q_2)^2} & \cdots & m_r \end{pmatrix}. \tag{1.6}$$

The diagonal element m_j of M is given by

$$m_j = ig\sum_{k\neq j}\frac{1}{(q_j - q_k)^2}. \tag{1.7}$$

The matrix M has a special property

$$\sum_{j=1}^{r} M_{jk} = \sum_{k=1}^{r} M_{jk} = 0. \tag{1.8}$$

It is straightforward to derive *quantum conserved quantities* from the *total sum* of powers of Lax matrix L:

$$[\mathcal{H}, Q_n] = 0, \quad Q_n \equiv \mathrm{Ts}(L^n) = \sum_{j,k}(L^n)_{jk}, \quad n = 1, \ldots, \tag{1.9}$$

since

$$\begin{aligned} \frac{d}{dt}\sum_{j,k}(L^n)_{jk} &= i\sum_{j,k}[\mathcal{H}, (L^n)_{jk}] = \sum_{j,k}[L^n, M]_{jk} \\ &= \sum_{j,k,\ell}\left((L^n)_{j\ell}M_{\ell k} - M_{j\ell}(L^n)_{\ell k}\right) = 0, \end{aligned} \tag{1.10}$$

in which (1.8) is used at the last equality. It is elementary to verify that the Hamiltonian is equivalent to Q_2,

$$\mathcal{H} \propto Q_2 + const.$$

In other words, the Lax matrix L is a kind of "square root" of the Hamiltonian. These are fundamental results for the integrability of the model (1.1). Similar integrable quantum many–particle dynamics are obtained by replacing the potential in (1.1) by

$$\frac{1}{(q_j - q_k)^2} \rightarrow \begin{cases} \frac{1}{\sinh^2(q_j - q_k)} \\ \frac{1}{\sin^2(q_j - q_k)} \end{cases} \quad \text{in } \mathcal{H}. \tag{1.11}$$

The quantum equations of motion are again expressed by Lax pairs if the following replacements are made:

$$\frac{1}{(q_j - q_k)} \rightarrow \begin{cases} \coth(q_j - q_k) \\ \cot(q_j - q_k) \end{cases} \text{in } L, \quad \frac{1}{(q_j - q_k)^2} \rightarrow \begin{cases} \frac{1}{\sinh^2(q_j - q_k)} \\ \frac{1}{\sin^2(q_j - q_k)} \end{cases} \text{in } M. \tag{1.12}$$

Since the property (1.8) holds also in these cases, we obtain quantum conserved quantities in the same manner as above (1.9) for the models with the trigonometric and hyperbolic interactions.

The main goal is to find all the eigenvalues $\{\lambda\}$ and eigenfunctions $\{\psi\}$ of the above Hamiltonian:

$$\mathcal{H}\psi = \lambda\psi.$$

For the rational model Hamiltonian (1.1), it reads

$$\left[-\frac{1}{2}\sum_{j=1}^{r}\frac{\partial^2}{\partial q_j^2}+\frac{1}{2}g(g-1)\sum_{j\neq k}^{r}\frac{1}{(q_j-q_k)^2}\right]\psi=\lambda\psi, \tag{1.13}$$

which is a second order Fuchsian differential equation for each variable $\{q_j\}$ with regular singularity at each hyperplane $q_j = q_k$ whose exponents are g, $1-g$. Any solution ψ of (1.13) is regular at all points except those on the union of hyperplanes $q_j = q_k$. Since the structure of the singularity is the same for the hyperbolic and trigonometric potentials (1.11), the same assertion for the regularity and singularity of the solution ψ holds for these cases. For the trigonometric case there are other singularities at $q_j - q_k = \ell\pi$, $\ell \in \mathbf{Z}$, due to the periodicity of the potential.

The integrability or more precisely the triangularity of the quantum Hamiltonian was first discovered by Calogero [1] for the model with inverse square potential plus a confining harmonic force and by Sutherland [2] for the particles on a circle with the trigonometric potential. Later, classical integrability of the models in terms of Lax pairs was proved by Moser [3]. Olshanetsky and Perelomov [4] showed that these models were based on A_{r-1} root systems, i.e. $q_j - q_k = \alpha \cdot q$ and α is one of the root vectors of A_{r-1} root system (A.6). They also introduced generalizations of the models [5] based on other root systems including the non–crystallographic ones.

In this paper we discuss quantum Calogero–Moser models with degenerate potentials, that is the rational with/without harmonic force, the hyperbolic and the trigonometric potentials based on any root system. We demonstrate that various results known for the quantum A_r models can be generalized universally to the models based on any other root systems as well [6,7,8,9,10]. They are (i) Construction of a complete set of quantum conserved quantities in terms of quantum Lax pairs and other methods. (ii) Universal proof of quantum Liouville integrability for the rational, hyperbolic and trigonometric potential models. Namely, these quantum conserved quantities commute among themselves. (iii) Triangularity of the quantum Hamiltonian for all the models. In other words, the Hamiltonian is demonstrated, in certain bases, to be decomposed into a sum of finite dimensional triangular matrices. Thus any eigenvalue equation can be solved by finite steps of linear algebraic processes only. This also gives the entire discrete spectrum of the models. (iv) Equivalence of the quantum Lax pair method and that of so–called differential–reflection (Dunkl) operators [11]. (v) Algebraic construction of all excited states in terms of creation (annihilation) operators for rational models with harmonic confining force. For the classical Liouville integrability see Ref.[12]. Solution methods of quantum Calogero–Moser models are expected to give some clues for understanding dynamical symmetries, non–perturbative methods for field theories, etc.

For the A_r models, the Lax pairs, conserved quantities and their involution were discussed by many authors with varied degrees of completeness and rigor, see for example Refs.[5,13–22]. The point (iv) was shown by Wadati and collaborators [20] and point (v) was initiated by Perelomov [14] and developed by Brink and collaborators [19] and Wadati and collaborators [20]. A rather different approach by Heckman and Opdam [23,24] to Calogero–Moser models with degenerate potentials based on any root systems should also be mentioned in this connection. For physical applications of the Calogero–Moser models with various potentials to lower dimensional physics, ranging from solid state to particle physics and supersymmetric Yang–Mills theory, we refer to recent papers [7,8,9] and references therein.

This paper is organized as follows. In section two, quantum Calogero–Moser Hamiltonian with degenerate potentials is introduced as a factorized form (2.5). Connection with root systems and the Coxeter invariance is emphasized. Some rudimentary facts of the root systems and reflections are summarized in Appendix A. A universal Coxeter invariant ground state wavefunction and the ground state energy are derived as simple consequences of the factorized Hamiltonian. In section three we show that all the excited states are also Coxeter invariant and that the Hamiltonian is *triangular* in certain bases. Complete sets of quantum conserved quantities are derived from quantum Lax operator L in section four. Instead of the trace, the *total sum* of L^n is conserved. That is $\mathrm{Ts}(L^n)=\sum_{\mu,\nu\in\mathcal{R}}(L^n)_{\mu\nu}$, in which $\mathcal{R}$ is a set of $\mathbf{R}^r$ vectors invariant under the action of the Coxeter group. They form a single Coxeter orbit. Independence (and thus completeness) of the conserved quantities is explained in some detail. A sufficient choice of the set $\mathcal{R}$ for each root system for the completeness of the conserved quantities is given in Appendix B. In section five the creation and annihilation operators for the rational models with harmonic force are derived. In section six, the equivalence of the Lax pair operator formalism and the so–called differential–reflection (Dunkl) operators is demonstrated. Another form of the quantum conserved quantities are given in terms of the differential–reflection (Dunkl) operators. In section seven an algebraic construction of excited states in terms of the differential–reflection (Dunkl) operators for rational models with harmonic force is presented. The complete sets of explicit eigenfunctions for the rank two models are derived. Section eight gives a universal proof of the Liouville integrability for models with rational (without the confining force), hyperbolic and trigonometric potentials. The final section is for summary, comments and outlook.

2 Quantum Calogero–Moser Models

A Calogero–Moser model is a Hamiltonian system associated with a root system Δ of rank r, which is a set of vectors in $\mathbf{R}^r$ with its standard inner product. A brief review of the properties of the root systems and the associated reflections together with explicit realization of all the root systems will be found in the Appendix A.

2.1 Factorized Hamiltonian

The dynamical variables of the quantum Calogero–Moser model are the coordinates $\{q_j\}$ and their canonically conjugate momenta $\{p_j\}$, with the canonical commutation relations:

$$[q_j, p_k] = i\delta_{jk}, \qquad [q_j, q_k] = [p_j, p_k] = 0, \quad j, k = 1, \ldots, r. \tag{2.1}$$

These will be denoted by vectors in $\mathbf{R}^r$

$$q = (q_1, \ldots, q_r), \qquad p = (p_1, \ldots, p_r). \tag{2.2}$$

The momentum operator p_j acts as differential operators

$$p_j = -i\frac{\partial}{\partial q_j}, \quad j = 1, \ldots, r.$$

As for the interactions we consider only the degenerate potentials, that is the rational (with/without harmonic force), hyperbolic and trigonometric potentials:

$$V(\rho \cdot q) = \begin{cases} 1/(\rho \cdot q)^2, & \text{type I}, \\ a^2/\sinh^2 a(\rho \cdot q), & \text{type II}, \\ a^2/\sin^2 a(\rho \cdot q), & \text{type III}, \end{cases} \qquad \rho \in \Delta, \tag{2.3}$$

in which a is an arbitrary real positive constant, determining the period of the trigonometric potentials. They imply integrability for all of the Calogero–Moser models based on the crystallographic root systems. Those models based on the non–crystallographic root systems, the dihedral group $I_2(m)$, H_3, and H_4, are integrable only for the rational potential. The rational potential models are also integrable if a confining harmonic potential

$$\frac{1}{2}\omega^2 q^2, \quad \omega > 0, \qquad \text{type V}, \tag{2.4}$$

is added to the Hamiltonian. Since we will discuss the universal properties and solutions applicable to all the interaction types as well as those for specific interaction potentials, let us adopt the conventional nomenclature for them. We call the models with rational, hyperbolic, trigonometric and rational with harmonic force the type I, II, III and V models, respectively. (Type IV models have elliptic potentials which will not be discussed in this paper.)

The Hamiltonian for the *quantum* Calogero–Moser model can be written in a 'factorized form':

$$\mathcal{H} = \frac{1}{2}\sum_{j=1}^{r}\left(p_j - i\frac{\partial W}{\partial q_j}\right)\left(p_j + i\frac{\partial W}{\partial q_j}\right), \tag{2.5}$$

$$= \frac{1}{2}\sum_{j=1}^{r}\left(p_j^2 + \left(\frac{\partial W}{\partial q_j}\right)^2\right) + \frac{1}{2}\sum_{j=1}^{r}\frac{\partial^2 W}{\partial q_j^2}, \tag{2.6}$$

$$= \frac{1}{2}p^2 + \frac{1}{2}\sum_{\rho\in\Delta_+} g_{|\rho|}(g_{|\rho|}-1)|\rho|^2\, V(\rho\cdot q) + (\frac{\omega^2}{2}q^2) - \mathcal{E}_0. \tag{2.7}$$

It should be noted that the above factorized Hamiltonian (2.7) consists of an operator part $\hat{\mathcal{H}}$, which is the Hamiltonian in the usual definition, and a constant $\mathcal{E}_0$ which is the ground state energy to be discussed later:

$$\mathcal{H} = \hat{\mathcal{H}} - \mathcal{E}_0, \tag{2.8}$$

$$\hat{\mathcal{H}} = \frac{1}{2}p^2 + \frac{1}{2}\sum_{\rho\in\Delta_+} g_{|\rho|}(g_{|\rho|}-1)|\rho|^2\, V(\rho\cdot q) + (\frac{\omega^2}{2}q^2). \tag{2.9}$$

The real *positive* coupling constants $g_{|\rho|}$ are defined on orbits of the corresponding Coxeter group, *i.e.* they are identical for roots in the same orbit. That is, for the simple Lie algebra cases one coupling constant $g_{|\rho|} = g$ for all roots in simply–laced models and two independent coupling constants, $g_{|\rho|} = g_L$ for long roots and $g_{|\rho|} = g_S$ for short roots in non–simply laced models. For the $I_2(m)$ models, there is one coupling if m is odd, and two independent ones if m is even. Let us call them g_e for even roots and g_o for odd roots. The H_3 and H_4 models have one coupling constant $g_{|\rho|} = g$, since these root systems are simply–laced. It should be noted that the operator part of the Hamiltonian $\hat{\mathcal{H}}$ is strictly positive for $g_{|\rho|} \geq 1$, which we will assume throughout this paper for simplicity of the arguments. Throughout this paper we consider the coupling constants at generic values.

The simplest way to introduce the factorized form is through supersymmetry [10,25], in which function W is called a superpotential:

$$W(q) = \sum_{\rho\in\Delta_+} g_{|\rho|} \ln |w(\rho \cdot q)| + (-\frac{\omega}{2}q^2), \quad g_{|\rho|} \geq 1, \quad \omega > 0. \tag{2.10}$$

The potential $V(u)$ (2.3) and the function $w(u)$ are related by

$$y(u) \equiv \frac{d}{du}x(u), \quad \frac{dw(u)}{du}/w(u) \equiv x(u), \tag{2.11}$$

$$V(u) = -y(u) = x^2(u) + a^2 \times \begin{cases} 0 \text{ rational,} \\ -1 \text{ hyperbolic,} \\ 1 \text{ trigonometric.} \end{cases} \tag{2.12}$$

The following Table I gives these functions for each potential:

potential	type	$w(u)$	$x(u)$	$y(u)$
rational	I & V	u	$1/u$	$-1/u^2$
hyperbolic	II	$\sinh au$	$a\coth au$	$-a^2/\sinh^2 au$
trigonometric	III	$\sin au$	$a\cot au$	$-a^2/\sin^2 au$

Table I: Functions appearing in the Lax pair and superpotential.

For proofs that the factorized Hamiltonian (2.5) actually gives the quantum Hamiltonian (2.7) for *any root system and potential* see Refs.[5,13,10]. It is easy to verify that for any potential $V(u)$, the Hamiltonian is invariant under reflection of the phase space variables in the hyperplane perpendicular to any root

$$\mathcal{H}(s_\alpha(p), s_\alpha(q)) = \mathcal{H}(p, q), \quad \forall\alpha \in \Delta \tag{2.13}$$

with s_α defined by (A.2).

The main problem is to find all the eigenvalues $\{\lambda\}$ and eigenfunctions $\{\psi\}$ of the above Hamiltonian:

$$\mathcal{H}\psi = \lambda\psi. \tag{2.14}$$

Some remarks are in order. For any root system and for any choice of potential (2.3), the Calogero–Moser model has a hard repulsive potential $\sim 1/(\alpha \cdot q)^2$ near the reflection hyperplane $H_\alpha = \{q \in \mathbf{R}^r, \alpha \cdot q = 0\}$. The strength of the singularity is given by the coupling constant $g_{|\alpha|}(g_{|\alpha|} - 1)$ which is *independent* of the choice of the normalization of the roots. (Thus for rational models with/without harmonic force there is equivalence, $A_2 \cong I_2(3)$, $B_2 \cong I_2(4)$, $G_2 \cong I_2(6)$, $B_r \cong C_r \cong BC_r$.) In other words, (2.14) is a second order Fuchsian differential equation with regular singularity at each reflection hyperplane H_α. That is any solution of (2.14) is regular at all points except those on the union of reflection hyperplanes $\cup_{\alpha\in\Delta_+} H_\alpha$ (and possibly at infinity for type V models). Near the reflection hyperplane H_α, the solution behaves

$$\psi \sim (\alpha \cdot q)^{g_{|\alpha|}}(1 + \text{regular terms}), \quad \text{or} \quad \psi \sim (\alpha \cdot q)^{1-g_{|\alpha|}}(1 + \text{regular terms}).$$

The former solution is chosen for the square integrability. This determines the form of the ground state wavefunction, as we will see in subsection 2.2. This repulsive potential is classically insurmountable. Thus the motion is always confined within one Weyl chamber. This feature allows us to constrain the configuration space to the principal Weyl chamber (Π: set of simple roots, see Appendix A)

$$PW = \{q \in \mathbf{R}^r |\ \alpha \cdot q > 0, \quad \alpha \in \Pi\}, \tag{2.15}$$

without loss of generality. In the case of the trigonometric potential, the configuration space is further limited due to the periodicity of the potential to

$$PW_T = \{q \in \mathbf{R}^r |\ \alpha \cdot q > 0, \quad \alpha \in \Pi, \quad \alpha_h \cdot q < \pi/a\}, \tag{2.16}$$

where α_h is the highest root.

The fact that the classical motions are confined in the Weyl chamber (alcove) $PW(PW_T)$ does not mean that the corresponding quantum wavefunctions vanish identically outside of the region [a]. On the contrary, as we will see soon, the ground state wavefunction (subsection 2.2) and all the other excited states wavefunctions (section 3) are Coxeter invariant, reflecting the Coxeter invariance of the Hamiltonian (2.13). In the early years of Calogero–Moser models in which those based on the A_r root system were mainly discussed, these Coxeter invariant solutions were considered as totally symmetric states of bosonic systems. We will not, however, adopt this interpretation, for in the models based on the other root systems the reflection is not the same as particle interchange. The quantum theory we are discussing is the so–called first quantized theory. That is, the notions of identical particles and the associated statistics are non–existent.

[a] Remember the quantum mechanical system defined in a finite interval on a line with the periodic boundary condition. It is described by a Fourier series which does not vanish outside of the interval.

2.2 Ground state wavefunction and energy

One straightforward outcome of the factorized Hamiltonian (2.5) is the *universal* ground state wavefunction which is given by

$$\Phi_0(q) = e^{W(q)} = \prod_{\rho\in\Delta_+} |w(\rho\cdot q)|^{g_{|\rho|}}\, e^{-\frac{\omega}{2}q^2}. \tag{2.17}$$

The exponential factor $e^{-\frac{\omega}{2}q^2}$ exists only for the rational potential case with the harmonic confining force. It is easy to see that it is an eigenstate of the Hamiltonian (2.5) with *zero* eigenvalue:

$$\mathcal{H}\Phi_0(q) = \frac{1}{2}\sum_{j=1}^{r}\left(p_j - i\frac{\partial W}{\partial q_j}\right)\left(p_j + i\frac{\partial W}{\partial q_j}\right)\Phi_0(q) = 0, \tag{2.18}$$

since it satisfies

$$\left(p_j + i\frac{\partial W}{\partial q_j}\right) e^{W(q)} = 0, \quad j = 1,\ldots,r. \tag{2.19}$$

By using the decomposition of the factorized Hamiltonian into the operator Hamiltonian (2.9) and a constant, we obtain

$$\hat{\mathcal{H}}\, e^W \equiv \left(\frac{1}{2}p^2 + \frac{1}{2}\sum_{\rho\in\Delta_+} g_{|\rho|}(g_{|\rho|}-1)|\rho|^2\, V(\rho\cdot q) + (\frac{\omega^2}{2}q^2)\right) e^W = \mathcal{E}_0\, e^W. \tag{2.20}$$

In other words, the above solution (2.17) provides an eigenstate of the Hamiltonian operator $\hat{\mathcal{H}}$ with energy $\mathcal{E}_0$. The fact that it is a ground state (for type I, III and V) can be easily shown within the framework of the supersymmetric model [10] thanks to the positivity of the supersymmetric Hamiltonian. It should be stressed that $\mathcal{E}_0$ is determined purely algebraically [10], without really applying the operator on the left hand side of (2.20) to the wavefunction. This type of ground states has been known for some time. It is derived by various methods, see for example Refs.[3,5], and also by using supersymmetric quantum mechanics for the models based on classical root systems [25,18].

The ground state energy for the rational potential cases are

$$\mathcal{E}_0 = \begin{cases} 0 & \text{type I,} \\ \omega\left(\frac{r}{2} + \sum_{\rho\in\Delta_+} g_{|\rho|}\right) & \text{type V.} \end{cases} \tag{2.21}$$

The same for the hyperbolic and trigonometric potential cases are

$$\mathcal{E}_0 = 2a^2\varrho^2 \times \begin{cases} -1 & \text{hyperbolic,} \\ 1 & \text{trigonometric,} \end{cases} \tag{2.22}$$

in which

$$\varrho = \frac{1}{2}\sum_{\rho\in\Delta_+} g_{|\rho|}\rho \tag{2.23}$$

can be considered as a 'deformed Weyl vector' [5,24]. Again these formulas are universal. That is they apply to all of the Calogero–Moser models based on any root system. A negative $\mathcal{E}_0$ for the obviously positive Hamiltonian of the hyperbolic potential model indicates that the interpretation of e^W as an eigenfunction is not correct. This function diverges as $|q| \to \infty$ for the hyperbolic and the rational potential cases, destroying the hermiticity of the Hamiltonian. Obviously we have

$$\int_{PW\,(PW_T)} e^{2W(q)}\,dq = \begin{cases} \infty & : \text{type I and II}, \\ \text{finite} & : \text{type III and V}, \end{cases} \tag{2.24}$$

in which PW and PW_T denote that the integration is over the regions defined in (2.15) and (2.16). Naturally, most existing results in quantum Calogero–Moser models are for the models with trigonometric potential and the rational potential with harmonic force which have square integrable states and discrete spectra.

It should be remarked that the universal ground state wavefunction Φ_0 and W are characterized as *Coxeter invariant*:

$$\check{s}_\rho\Phi_0 = \Phi_0, \quad \check{s}_\rho W = W, \qquad \forall\rho\in\Delta, \tag{2.25}$$

in which $\check{s}_\rho$ is the representation of the reflection in the function space. For an arbitrary function f of $q\in\mathbf{R}^r$, its action is defined by

$$(\check{s}_\rho f)(q) = f(s_\rho(q)). \tag{2.26}$$

This definition can be generalized to the entire Coxeter group G_Δ: for an arbitrary element g of G_Δ, $\check{g}$ is defined by:

$$(\check{g}f)(q) = f(g^{-1}(q)), \qquad \forall g\in G_\Delta. \tag{2.27}$$

In the rest of this paper we mainly discuss the type III and V models which have normalisable states and discrete spectra.

3 Coxeter invariant excited states, triangularity and spectrum

In this section we show that all the excited states wavefunctions are Coxeter invariant, too. In other words, the Fock space consists of Coxeter invariant functions only. With the knowledge of the ground state wavefunction e^W, the other states of the Calogero–Moser models can be easily obtained as eigenfunctions of a differential operator $\tilde{\mathcal{H}}$ obtained from $\mathcal{H}$ by a similarity transformation:

$$\begin{aligned}\tilde{\mathcal{H}} &= e^{-W}\,\mathcal{H}\,e^{W}\\ &= e^{-W}\left(\frac{1}{2}p^2 + \frac{1}{2}\sum_{\rho\in\Delta_+} g_{|\rho|}(g_{|\rho|}-1)|\rho|^2 V(\rho\cdot q) + (\frac{\omega^2}{2}q^2) - \mathcal{E}_0\right)e^{W}, \quad (3.1)\end{aligned}$$

$$\tilde{\mathcal{H}}\Psi_\lambda = \lambda\Psi_\lambda \quad \Longleftrightarrow \quad \mathcal{H}\,\Psi_\lambda\,e^{W} = \lambda\Psi_\lambda\,e^{W}. \tag{3.2}$$

Since all the singularities of the Fuchsian differential equation (2.14) are contained in the ground state wavefunction e^W the function Ψ_λ above must be regular at finite q including all the reflection boundaries.

Thanks to the factorized form of the Hamiltonian $\mathcal{H}$ (2.5), (2.6), the transformed Hamiltonian $\tilde{\mathcal{H}}$ takes a simple form:

$$\tilde{\mathcal{H}} = -\frac{1}{2}\sum_{j=1}^{r}\left(\frac{\partial^2}{\partial q_j^2} + 2\frac{\partial W}{\partial q_j}\frac{\partial}{\partial q_j}\right). \tag{3.3}$$

The Coxeter invariance of W implies those of $\mathcal{H}$ and $\tilde{\mathcal{H}}$:

$$\check{s}_\rho\mathcal{H}\check{s}_\rho = \mathcal{H}, \quad \check{s}_\rho\tilde{\mathcal{H}}\check{s}_\rho = \tilde{\mathcal{H}}, \qquad \forall\rho\in\Delta. \tag{3.4}$$

For type V and III models we introduce proper bases of Fock space consisting of Coxeter invariant functions and show that the above Hamiltonian $\tilde{\mathcal{H}}$ (3.3) is *triangular* in these bases. This establishes the integrability of the type V and III models universally [b] and also gives the entire spectrum of the Hamiltonian, see (3.12), (3.13) and (3.42).

3.1 Rational potential with harmonic force

First, let us determine the structure of the set of eigenfunctions of the transformed Hamiltonian $\tilde{\mathcal{H}}$, for the type V models:

$$\tilde{\mathcal{H}} = \omega\,q\cdot\frac{\partial}{\partial q} - \frac{1}{2}\sum_{j=1}^{r}\frac{\partial^2}{\partial q_j^2} - \sum_{\rho\in\Delta_+}\frac{g_{|\rho|}}{\rho\cdot q}\rho\cdot\frac{\partial}{\partial q}. \tag{3.5}$$

As remarked above, the eigenfunctions of $\tilde{\mathcal{H}}$ have no singularities at finite q. Thus we look for polynomial (in q) eigenfunctions. The situation is very

---

[b] Triangularity of the A_r type V and III Hamiltonians was noted in the original papers of Calogero [1] and Sutherland [2]. That of rank two models in the Coxeter invariant bases was shown in Refs.[24,26].

similar to that of the harmonic oscillators to which the type V Calogero–Moser models reduce in the free limit $g_{|\rho|} \to 0$. Obviously a constant and a quadratic polynomial $\omega q^2 - \mathcal{E}_0/\omega$ are its eigenfunctions with eigenvalue 0 and 2ω, respectively. Let us suppose that a polynomial $P(q)$ is an eigenfunction of $\tilde{\mathcal{H}}$:

$$\tilde{\mathcal{H}}P(q) = \lambda P(q). \tag{3.6}$$

Due to the Coxeter invariance of $\tilde{\mathcal{H}}$ (3.4), we know that $\check{s}_\rho P$ together with the difference

$$Q = (1 - \check{s}_\rho)P$$

are also eigenfunctions with the same eigenvalue:

$$\tilde{\mathcal{H}}Q(q) = \lambda Q(q), \tag{3.7}$$

if the latter is not identically zero. Since Q is a polynomial which is odd under reflection $\check{s}_\rho$

$$\check{s}_\rho Q(q) = -Q(q),$$

it can be factorized as

$$Q(q) = (\rho \cdot q)^{2n+1}\tilde{Q}(q), \quad \tilde{Q}\Big|_{\rho\cdot q=0} \neq 0, \tag{3.8}$$

with a non–negative integer n and a polynomial $\tilde{Q}$. By substituting (3.8) into (3.7) and using the explicit form of $\tilde{\mathcal{H}}$ near the reflection hyperplane $\rho \cdot q = 0$, we obtain

$$-(\rho \cdot q)^{2n-1}(2n+1)(n+g_{|\rho|})\rho^2\tilde{Q} + \mathcal{O}[(\rho \cdot q)^{2n}] = \lambda(\rho \cdot q)^{2n+1}\tilde{Q}, \tag{3.9}$$

which would imply the vanishing of $\tilde{Q}$ on the reflection hyperplane

$$\tilde{Q}\Big|_{\rho\cdot q=0} = 0,$$

an obvious contradiction. Thus we are led to the conclusion that *the eigenfunctions are Coxeter invariant polynomials* and that the Hamiltonian $\tilde{\mathcal{H}}$ (3.5) maps a Coxeter invariant polynomial to another.

An obvious basis in the space of Coxeter invariant polynomials is the homogeneous polynomials of various degrees. This basis has a natural order given by the degree. For a given degree the space of homogeneous Coxeter invariant polynomials is finite–dimensional. The explicit form of $\tilde{\mathcal{H}}$ (3.5) shows that it is *lower triangular* in this basis and the diagonal elements are $\omega \times$*degree* as given by the first term. Independent Coxeter invariant polynomials exist at the degrees f_j listed in Table II:

$$f_j = 1 + e_j, \quad j = 1, \ldots, r, \tag{3.10}$$

in which $\{e_j\}$, $j = 1, \ldots, r$, are the *exponents* of Δ. Let us denote them by

$$z_1(q), \ldots, z_r(q); \quad z_j(\kappa q) = \kappa^{f_j} z_j(q). \tag{3.11}$$

Δ	$f_j = 1 + e_j$	Δ	$f_j = 1 + e_j$
A_r	$2, 3, 4, \ldots, r+1$	E_8	$2, 8, 12, 14, 18, 20, 24, 30$
B_r	$2, 4, 6, \ldots, 2r$	F_4	$2, 6, 8, 12$
C_r	$2, 4, 6, \ldots, 2r$	G_2	$2, 6$
D_r	$2, 4, \ldots, 2r-2; r$	$I_2(m)$	$2, m$
E_6	$2, 5, 6, 8, 9, 12$	H_3	$2, 6, 10$
E_7	$2, 6, 8, 10, 12, 14, 18$	H_4	$2, 12, 20, 30$

Table II: The degrees f_j in which independent Coxeter invariant polynomials exist.

Thus we arrive at:
the quantum Calogero–Moser models with the rational potential and the harmonic confining force is algebraically solvable for any (crystallographic and non–crystallographic) root system Δ. (See Ref.[27] for algebraic linearisability theorem of the classical Calogero–Moser system.) The spectrum of the operator Hamiltonian $\hat{\mathcal{H}}$ is

$$\omega N + \mathcal{E}_0, \tag{3.12}$$

with a non–negative integer N which can be expressed as

$$N = \sum_{j=1}^{r} n_j f_j, \quad n_j \in \mathbf{Z}_+, \tag{3.13}$$

and the degeneracy of the above eigenvalue (3.12) is the number of different solutions of (3.13) for given N. This is generalization of Calogero's original argument for the A_r model [1] to the models based on any root system. Now let us denote by $\vec{N}$ the set of non–negative integers in (3.13):

$$\vec{N} = (n_1, n_2, \ldots, n_r), \tag{3.14}$$

and by $\phi_{\vec{N}}(q)$ the homogeneous Coxeter invariant polynomial determined by $\vec{N}$ and the above basis $\{z_j\}$ (3.11):

$$\phi_{\vec{N}}(q) = \prod_{j=1}^{r} z_j^{n_j}(q). \tag{3.15}$$

As shown above, there exists a unique eigenstate $\psi_{\vec{N}}(q)$ for each $\phi_{\vec{N}}(q)$:

$$\psi_{\vec{N}}(q) = \phi_{\vec{N}}(q) + \sum_{\vec{N}'<\vec{N}} d_{\vec{N}'}\phi_{\vec{N}'}(q), \quad d_{\vec{N}'} : \; const, \tag{3.16}$$

$$\tilde{\mathcal{H}}\psi_{\vec{N}}(q) = \omega N \psi_{\vec{N}}(q). \tag{3.17}$$

It satisfies the orthogonality relation

$$(\psi_{\vec{N}}, \phi_{\vec{N}'}) = 0, \quad \vec{N}' < \vec{N}, \tag{3.18}$$

with respect to the inner product in PW:

$$(\psi, \varphi) = \int_{PW} \psi^*(q)\varphi(q)\, e^{2W(q)}\, dq. \tag{3.19}$$

These polynomials $\{\psi_{\vec{N}}(q)\}$ are generalizations of the multivariable Laguerre (Hermite) polynomials [16] known for the A_r (B_r, D_r) root systems to any root system.

Some remarks are in order.

1. There is no Coxeter invariant linear function in q. The quadratic invariant polynomial $q^2 = q \cdot q$ exists in all the root systems. This corresponds to the universal fact that $f_1 = 2, (e_1 = 1)$ for all the root systems. Moreover, this is related to the fact that a special sub–series of the excited states with $N = 2n_1 = n_1 f_1$, $n_j = 0$, $j \geq 2$, can be expressed universally in terms of Laguerre polynomials in q^2. This will be discussed at the end of section 5 and in subsection 7.2.

2. The other Coxeter invariants corresponding to the degrees $f_2, \ldots, f_r$ could be interpreted as special 'angular' variables of a unit sphere S^{r-1} ($q^2 = 1$), with the first Coxeter invariant $\sqrt{q^2}$ being the radial coordinate. These would provide proper variables for describing solutions. Solutions in terms of separation of variables are in general possible only for the simplest cases, that is the rank two models, which will be demonstrated in subsection 7.3.

3. For $\Delta = A_1$, the simplest root system of rank one, the Hamiltonian $\tilde{\mathcal{H}}$ can be rewritten in terms of a Coxeter invariant variable $u = \omega q^2$ as:

$$\tilde{\mathcal{H}} = \omega q \frac{d}{dq} - \frac{1}{2}\frac{d^2}{dq^2} - \frac{g}{q}\frac{d}{dq} = -2\omega \left\{ u\frac{d^2}{du^2} + (g + \frac{1}{2} - u)\frac{d}{du} \right\}. \tag{3.20}$$

The Laguerre polynomial satisfying the differential equation

$$\left\{ u\frac{d^2}{du^2} + (g + \frac{1}{2} - u)\frac{d}{du} + n \right\} L_n^{(g-\frac{1}{2})}(u) = 0, \tag{3.21}$$

provides an eigenfunction with eigenvalue $2\omega n$, which corresponds to the eigenvalue $2\omega n + \mathcal{E}_0$ of $\hat{\mathcal{H}}$. This is a well–known result.

3.2 Trigonometric potential

Here we consider those root systems associated with Lie algebras. In order to determine the excited states of the type III models, we have to consider the periodicity. The superpotential W and the Hamiltonian $\mathcal{H}$ are invariant under the following translation:

$$W(q') = W(q), \quad \mathcal{H}(p, q') = \mathcal{H}(p, q), \quad q' = q + l^\vee \pi/a, \tag{3.22}$$

in which $l^\vee$ is an element of the dual weight lattice, that is

$$l^\vee = \sum_{j=1}^{r} l_j \frac{2}{\alpha_j^2} \lambda_j, \quad l_j \in \mathbf{Z}, \quad \alpha_j \in \Pi, \quad \alpha_j^\vee \cdot \lambda_k = \delta_{jk}. \tag{3.23}$$

As is well known in quantum mechanics with periodic potentials, the (Bloch) wavefunctions diagonalizing the translation operators are expressed as

$$e^{2ia\mu\cdot q} \left(\sum_{\alpha \in L(\Delta)} b_\alpha e^{2ia\alpha\cdot q} \right) e^W, \quad b_\alpha : const, \quad L(\Delta) : \text{root lattice}, \tag{3.24}$$

in which a vector $\mu \in \mathbf{R}^r$ is as yet unspecified. In other words, up to the overall phase factor $e^{2ia\mu\cdot q}$, this is a Fourier expansion in terms of the simple roots.

Let $P_T(q)$ be a polynomial in $e^{\pm 2ia\alpha_j\cdot q}$, $\alpha_j \in \Pi$ and suppose that a function $\phi(q)$

$$\phi(q) = e^{2ia\mu\cdot q} P_T(q), \quad \mu \in \mathbf{R}^r, \tag{3.25}$$

is an eigenfunction of $\tilde{\mathcal{H}}$:

$$\tilde{\mathcal{H}}\phi(q) = \lambda\phi(q), \tag{3.26}$$

in which the explicit form of $\tilde{\mathcal{H}}$ is given by

$$\tilde{\mathcal{H}} = -\frac{1}{2}\sum_{j=1}^{r} \frac{\partial^2}{\partial q_j^2} - a \sum_{\rho \in \Delta_+} g_{|\rho|} \cot(a\rho\cdot q)\rho \cdot \frac{\partial}{\partial q}. \tag{3.27}$$

As above, due to the Coxeter invariance of $\tilde{\mathcal{H}}$ (3.4), we know that $\check{s}_\rho \phi$ together with the difference

$$\varphi = (1 - \check{s}_\rho)\phi$$

are also eigenfunctions with the same eigenvalue:

$$\tilde{\mathcal{H}}\varphi(q) = \lambda\varphi(q).$$

Since φ is odd under reflection $\check{s}_\rho$

$$\check{s}_\rho\varphi(q) = -\varphi(q),$$

it can be expressed as

$$\varphi(q) = (\rho\cdot q)^{2n+1}\tilde{\varphi}(q) + \mathcal{O}[(\rho\cdot q)^{2n+3}], \quad \tilde{\varphi}|_{\rho\cdot q=0} \neq 0,$$

in a neighbourhood of the reflection hyperplane $\rho\cdot q = 0$. In this neighbourhood, the singularity structure of $\tilde{\mathcal{H}}$ for the trigonometric potential is the same as that of $\tilde{\mathcal{H}}$ for the rational potential discussed in the previous subsection. Thus we obtain, as before, a contradiction $\tilde{\varphi}|_{\rho\cdot q=0} = 0$. In other words, the eigenfunction ϕ (3.25) must be *Coxeter invariant*. This in turn requires that the unspecified vector μ in (3.25) to be an element of the weight lattice

$$\mu \in \Lambda(\Delta). \tag{3.28}$$

Since

$$\check{s}_\rho\phi(q) = e^{2ia\, s_\rho(\mu)\cdot q}\check{s}_\rho P_T(q) = e^{2ia(\mu\cdot q-\rho^\vee\cdot\mu\rho\cdot q)}\check{s}_\rho P_T(q),$$

the following condition is necessary, but not sufficient,

$$\rho^\vee\cdot\mu \in \mathbf{Z}, \quad \forall\rho\in\Delta \tag{3.29}$$

for the Coxeter invariance of ϕ. Thus we arrive at (3.28).

Let us introduce a basis for the Coxeter invariant functions of the form (3.25). Let λ be a dominant weight

$$\lambda = \sum_{j=1}^{r} m_j\lambda_j, \quad m_j \in \mathbf{Z}_+, \tag{3.30}$$

and W_λ be the orbit of λ by the action of the Weyl group:

$$W_\lambda = \{\mu\in\Lambda(\Delta)|\ \mu = g(\lambda),\ \forall g\in G_\Delta\}. \tag{3.31}$$

We define

$$\phi_\lambda(q) \equiv \sum_{\mu\in W_\lambda} e^{2ia\mu\cdot q}, \tag{3.32}$$

which is Coxeter invariant. The set of functions $\{\phi_\lambda\}$ has an order $>$:

$$|\lambda|^2 > |\lambda'|^2 \quad \Rightarrow \quad \phi_\lambda > \phi_{\lambda'}. \tag{3.33}$$

Next we show that $\tilde{\mathcal{H}}$ is *lower triangular* in this basis. By using (3.27) we obtain

$$\tilde{\mathcal{H}}\phi_\lambda = 2a^2\lambda^2\phi_\lambda - 2ia^2 \sum_{\rho\in\Delta_+}\sum_{\mu\in W_\lambda} g_{|\rho|}\cot(a\rho\cdot q)(\rho\cdot\mu)e^{2ia\mu\cdot q}. \tag{3.34}$$

First let us fix one positive root ρ and a weight μ in W_λ such that $\rho\cdot\mu\neq 0$. Then

$$\mu' \equiv s_\rho(\mu) = \mu - (\rho^\vee\cdot\mu)\rho \in W_\lambda, \quad \rho\cdot\mu' = -\rho\cdot\mu. \tag{3.35}$$

Without loss of generality we may assume

$$\rho^\vee\cdot\mu = k > 0, \quad k\in\mathbf{Z}. \tag{3.36}$$

The contribution of the pair (μ,μ') in the summation of (3.34) reads

$$\begin{aligned} &|\rho\cdot\mu|e^{2ai\mu\cdot q}(1-e^{-2aik\rho\cdot q})\cot(a\rho\cdot q) \\ &= i|\rho\cdot\mu|\left(e^{2ai\mu\cdot q}+e^{2ai\mu'\cdot q}+2\sum_{j=1}^{k-1}e^{2ai(\mu-j\rho)\cdot q}\right), \end{aligned} \tag{3.37}$$

which is the generalization of Sutherland's fundamental identity eq(15) in Ref.[2] to any root system. The summation in the expression correspond to $\phi_{\lambda'}$ with λ' being lower than λ. Thus (3.34) reads

$$\tilde{\mathcal{H}}\phi_\lambda = 2a^2\lambda^2\phi_\lambda + 2a^2\sum_{\rho\in\Delta_+}\sum_{\mu\in W_\lambda} g_{|\rho|}|\rho\cdot\mu|e^{2ia\mu\cdot q} + \sum_{|\lambda'|<|\lambda|} c_{\lambda'}\phi_{\lambda'}, \tag{3.38}$$

in which $\{c_{\lambda'}\}$'s are constants. It is easy to see that ($\mu = g(\lambda)$, $\exists g\in G_\Delta$)

$$\sum_{\rho\in\Delta_+} g_{|\rho|}|\rho\cdot\mu| = \sum_{\rho\in\Delta_+} g_{|\rho|}|g(\rho)\cdot\lambda| = (\sum_{\rho\in\Delta_+} g_{|\rho|}\rho)\cdot\lambda = 2\varrho\cdot\lambda, \tag{3.39}$$

which is independent of μ. Thus we have demonstrated the triangularity of $\tilde{\mathcal{H}}$:

$$\tilde{\mathcal{H}}\phi_\lambda = 2a^2(\lambda^2+2\varrho\cdot\lambda)\phi_\lambda + \sum_{|\lambda'|<|\lambda|} c_{\lambda'}\phi_{\lambda'}, \tag{3.40}$$

or that of $\hat{\mathcal{H}}$

$$\hat{\mathcal{H}}\phi_\lambda e^W = 2a^2(\lambda+\varrho)^2\phi_\lambda e^W + \sum_{|\lambda'|<|\lambda|} c_{\lambda'}\phi_{\lambda'}e^W, \tag{3.41}$$

with the eigenvalue

$$2a^2(\lambda+\varrho)^2. \tag{3.42}$$

In other words, for each dominant weight λ there exists an eigenstate of $\tilde{\mathcal{H}}$ with eigenvalue proportional to $\lambda(\lambda + 2\varrho)$. Let us denote this eigenfunction by $\psi_\lambda(q)$:

$$\psi_\lambda(q) = \phi_\lambda(q) + \sum_{|\lambda|'<|\lambda|} d_{\lambda'}\phi_{\lambda'}(q), \quad d_{\lambda'} : \mathit{const}, \tag{3.43}$$

$$\tilde{\mathcal{H}}\psi_\lambda(q) = 2a^2\lambda(\lambda + 2\varrho)\psi_\lambda(q), \tag{3.44}$$

and call it a *generalized Jack polynomial* [28-31]. It satisfies the orthogonality relation.

$$(\psi_\lambda, \phi_{\lambda'}) = 0, \quad |\lambda|' < |\lambda|, \tag{3.45}$$

with respect to the inner product in PW_T:

$$(\psi, \varphi) = \int_{PW_T} \psi^*(q)\varphi(q)\, e^{2W(q)}\, dq. \tag{3.46}$$

In the A_r model, specifying a dominant weight λ is the same as giving a Young diagram which designates a Jack polynomial. It should be emphasized, however, that $\{\psi_\lambda\}$ are not identical to the Jack polynomials even for the A_r root systems, because of different treatments of the center of mass coordinates.

Thus we arrive at:
the quantum Calogero–Moser models with the trigonometric potential are algebraically solvable for any crystallographic root system Δ. The spectrum of the Hamiltonian $\hat{\mathcal{H}}$ is given by (3.42) in which λ is an arbitrary dominant weight. This is generalization of Sutherland's original argument [2] to the models based on any root system.

Some remarks are in order.

1. The weights μ appearing in the lower order terms $\{\phi_{\lambda'}\}$'s are those weights contained in the Lie algebra representation belonging to the highest weight λ.

2. As a simple corollary we find that for a minimal weight λ,

$$\psi_\lambda(q) = \phi_\lambda(q) = \sum_{\mu \in W_\lambda} e^{2ia\mu\cdot q}$$

is an eigenfunction of $\tilde{\mathcal{H}}$. A minimal representation [7] consists of a single Weyl orbit and all of its weights μ satisfy $\rho^\vee \cdot \mu = 0, \pm 1, \forall \rho \in \Delta$.

3. If $\lambda = \alpha_h$, the highest root of a simply–laced root system, W_λ is the set of roots itself. Then the lower order terms are constants only. We find that

$$\psi_{\alpha_h}(q) = 2 \sum_{\rho\in\Delta_+} \left(\cos(2a\rho\cdot q) + g\rho^2/\alpha_h\cdot(\alpha_h + 2\varrho)\right)$$

is an eigenstate of $\tilde{\mathcal{H}}$.

4. If $\lambda = \alpha_{Sh}$, the highest short root of a non–simply laced root system, W_λ is the set of short roots itself. The lower order terms are constants, too. Similarly as above, we find that

$$\begin{aligned}\psi_{\alpha_{Sh}}(q) = 2 &\sum_{\rho\in\Delta_{L+}} \cos(2a\rho\cdot q)\\ +2 &\sum_{\rho\in\Delta_{S+}} \left(\cos(2a\rho\cdot q) + g_S\rho_S^2/\alpha_{Sh}\cdot(\alpha_{Sh} + 2\varrho)\right)\end{aligned}$$

is an eigenstate of $\tilde{\mathcal{H}}$. Here $\Delta_{L(S)}$ is the set of long (short) roots.

5. If $-\lambda \notin W_\lambda$ then there is another set of functions containing the weight $-\lambda$ which belongs to the same eigenvalue.

6. The Coxeter invariant trigonometric polynomials specified by the fundamental weights $\{\lambda_j\}$

$$\phi_{\lambda_j}(q) = \sum_{\mu\in W_{\lambda_j}} e^{2ia\mu\cdot q}, \quad \lambda_j : \text{ fundamental weight}, \quad j = 1,\ldots,r \tag{3.47}$$

are expected to play the role of the fundamental variables [24,26].

7. Let us consider the well–known case $\Delta = A_1$, the simplest root system of rank one. By rewriting the Hamiltonian $\tilde{\mathcal{H}}$ in terms of the Coxeter invariant variable $z = \cos(a\rho q)$, we obtain

$$\hat{\mathcal{H}} = -\frac{1}{2}\frac{d^2}{dq^2} - ag\rho\cot(a\rho q)\frac{d}{dq} = -\frac{a^2|\rho|^2}{2}\left\{(1-z^2)\frac{d^2}{dz^2} - (1+2g)z\frac{d}{dz}\right\}. \tag{3.48}$$

The Gegenbauer polynomials [5], a special case of Jacobi polynomials $P_n^{(\alpha,\beta)}$ provide eigenfunctions:

$$P_n^{(g-\frac{1}{2},g-\frac{1}{2})}(\cos(a\rho q)), \quad \mathcal{E} = a^2|\rho|^2(n+g)^2/2, \quad n\in\mathbf{Z}_+. \tag{3.49}$$

The Jacobi polynomial $P_n^{(\alpha,\beta)}(z)$ satisfies a differential equation

$$\left\{(1-z^2)\frac{d^2}{dz^2}+(\beta-\alpha-(2+\alpha+\beta)z)\frac{d}{dz}\right.$$
$$\left.+n(n+\alpha+\beta+1)\right\}P_n^{(\alpha,\beta)}(z)=0. \qquad (3.50)$$

Here we follow the notation of Ref.[32]. They form orthogonal polynomials with weight $e^{2W}=|\sin(a\rho q)|^{2g}$ in the interval $q\in[0,\pi/a\rho]$, (2.16).

8. Triangularity of type II models follows from the same algebraic reasoning.

4 Quantum Lax pair and quantum conserved quantities

Historically, Lax pairs for Calogero–Moser models were presented in terms of Lie algebra representations [3,5], in particular, the vector representation of the A_r models as shown in Introduction. However, the invariance of Calogero–Moser models is that of Coxeter group but not that of the associated Lie algebra, which does not exist for the non–crystallographic root systems. Thus the universal and Coxeter covariant Lax pairs are given in terms of the representations of the Coxeter group.

4.1 General case

Here we recapitulate the essence of the quantum Lax pair operators for the Calogero–Moser models with degenerate potentials and without spectral parameter. The quantum Lax pair in this subsection applies to all the degenerate potential cases except for the case of the rational potential with the harmonic force, which will be treated separately in subsection 4.2. For details and a full exposition, see Ref.[10]. The Lax operators without spectral parameter are

$$L(p,q)=p\cdot\hat{H}+X(q), \qquad X(q)=i\sum_{\rho\in\Delta_+}g_{|\rho|}\,(\rho\cdot\hat{H})x(\rho\cdot q)\hat{s}_\rho, \quad (4.1)$$

$$M(q)=\frac{i}{2}\sum_{\rho\in\Delta_+}g_{|\rho|}|\rho|^2\,y(\rho\cdot q)\,(\hat{s}_\rho-I), \qquad (4.2)$$

in which I is the identity operator and $\{\hat{s}_\alpha,\ \alpha\in\Delta\}$ are the reflection operators of the root system. In contrast to the $\{\check{s}_\alpha\}$ operators (2.26) which act in function space, $\{\hat{s}_\alpha\}$ act on a set of $\mathbf{R}^r$ vectors $\mathcal{R}=\{\mu^{(k)}\in\mathbf{R}^r,\ k=1,\ldots,d\}$, permuting them under the action of the reflection group. The vectors in $\mathcal{R}$

form a basis for the representation space $\mathbf{V}$ of dimension d. The operator M satisfies the relation

$$\sum_{\mu\in\mathcal{R}} M_{\mu\nu} = \sum_{\nu\in\mathcal{R}} M_{\mu\nu} = 0, \tag{4.3}$$

which is essential for deriving quantum conserved quantities. The matrix elements of the operators $\{\hat{s}_\alpha,\, \alpha\in\Delta\}$ and $\{\hat{H}_j,\, j=1,\ldots,r\}$ are defined as follows:

$$(\hat{s}_\rho)_{\mu\nu} = \delta_{\mu,s_\rho(\nu)} = \delta_{\nu,s_\rho(\mu)}, \quad (\hat{H}_j)_{\mu\nu} = \mu_j\delta_{\mu\nu}, \quad \rho\in\Delta, \quad \mu,\nu\in\mathcal{R}. \tag{4.4}$$

The form of the function x depends on the chosen potential, and the function y are defined by (2.11), (2.12). Note that these relations are only valid for the degenerate potentials (2.3).

The underlying idea of the Lax operator L, (4.1), is quite simple. As seen from (4.10), L is a "square root" of the Hamiltonian. Thus one part of L contains p which is not associated with roots and another part contains $x(\rho\cdot q)$, a "square root" of the potential $V(\rho\cdot q)$, which being associated with a root ρ is therefore accompanied by the reflection operator $\hat{s}_\rho$. Another explanation is the factorized Hamiltonian $\mathcal{H}$ (2.5). We obtain, roughly speaking, $L\sim\sqrt{\mathcal{H}}\sim p+i\frac{\partial W}{\partial q}\hat{s}$ and the property of reflection $\hat{s}^2=1$ explains the sign change in the first term in (2.5).

It is straightforward to show that the quantum Lax equation

$$\frac{d}{dt}L = i[\mathcal{H},L] = [L,M], \tag{4.5}$$

is equivalent to the quantum equations of motion derived from the Hamiltonian (2.7). From this it follows:

$$\begin{aligned}\frac{d}{dt}(L^n)_{\mu\nu} &= i[\mathcal{H},(L^n)_{\mu\nu}] = [L^n,M]_{\mu\nu}\\ &= \sum_{\lambda\in\mathcal{R}}\Big((L^n)_{\mu\lambda}M_{\lambda\nu} - M_{\mu\lambda}(L^n)_{\lambda\nu}\Big), \quad n=1,\ldots.\end{aligned} \tag{4.6}$$

Thanks to the property of the M operator (4.3):

$$\sum_{\mu\in\mathcal{R}} M_{\mu\nu} = \sum_{\nu\in\mathcal{R}} M_{\mu\nu} = 0,$$

we obtain *quantum conserved quantities* as the *total sum* (Ts) of all the matrix elements of L^n [c]:

$$Q_n = \mathrm{Ts}(L^n) \equiv \sum_{\mu,\nu\in\mathcal{R}} (L^n)_{\mu\nu}, \qquad [\mathcal{H}, Q_n] = 0, \quad n = 1, \ldots. \tag{4.7}$$

The Lax pair is Coxeter covariant:

$$L\left(s_\rho(p), s_\rho(q)\right)_{\mu\nu} = L\left(p, q\right)_{\mu'\nu'}, \; M\left(s_\rho(q)\right)_{\mu\nu} = M(q)_{\mu'\nu'},$$
$$\mu' \equiv s_\rho(\mu), \; \nu' \equiv s_\rho(\nu), \tag{4.8}$$

which ensures the Coxeter invariance of the conserved quantities. Independent conserved quantities appear at such power n that

$$n = 1 + exponent \tag{4.9}$$

of each root system. These are the degrees at which independent Coxeter invariant polynomials exist. There are r exponents for each root system Δ of rank r. Thus we have r independent conserved quantities in Calogero–Moser models. We list in Table II these powers for each root system. In particular, the power 2 is universal to all the root systems and the quantum Hamiltonian (2.7) is given by

$$\mathcal{H} = \frac{1}{2C_{\mathcal{R}}}\mathrm{Ts}(L^2) + const, \tag{4.10}$$

where the constant $C_{\mathcal{R}}$ is the quadratic Casimir invariant, which depends on the representation. It is defined by

$$\mathrm{Tr}(\hat{H}_j\hat{H}_k) \equiv \sum_{\mu\in\mathcal{R}} (\hat{H}_j\hat{H}_k)_{\mu\mu} = \sum_{\mu\in\mathcal{R}} \mu_j\mu_k = C_{\mathcal{R}}\,\delta_{jk}. \tag{4.11}$$

Some remarks are in order.

1. Lax pairs can be written down in various representations and the quantum conserved quantities Q_n do depend on the representations, in general. If necessary, we denote by $Q_n^{\mathcal{R}}$ the explicit representation dependence.

2. The availability of plural representations of the Lax pair and the conserved quantities is essential for the completeness of the set of conserved quantities as polynomials of the momentum operators. For example, let us consider the case of D_r with even r, which has two independent

[c] This type of conserved quantities is known for A_r models [18,20].

conserved quantities at power r, see Table II. At least two different representations of the Lax pair are necessary in order to represent them in the form of (4.7). Those based on the vector, and the (anti)–spinor representations give two independent conserved quantities. For D_4 case, we obtain

$$Q_4^v = 2\sum_{j=1}^{4} p_j^4, \quad Q_4^s - Q_4^a = 24\prod_{j=1}^{4} p_j, \tag{4.12}$$

in which v, s and a stand for the vector, spinor and anti–spinor representations and we have set $g = 0$ for simplicity. If two conserved quantities are independent for zero coupling constants, surely they are so at non–vanishing couplings.

3. If a representation $\mathcal{R}$ contains a vector μ and its negative $-\mu$ at the same time, then we have $\mathrm{Ts}(L^{odd}) = 0$. In such a case the corresponding Lie algebra representations are called real. In order to construct the odd power conserved quantities appearing in A_r for all r, D_r for odd r, E_6 and $I(m)$ for odd m, we need a Lax pair in non–real representations. For A_r all the fundamental representations corresponding to the fundamental weights λ_j, $j = 1, \ldots, r$ except for the middle one $\lambda_{(r+1)/2}$ for odd r are non–real. For D_r with odd r, the spinor and anti–spinor representations are non–real. For E_6 the **27** and $\overline{\mathbf{27}}$ are non–real. $I_2(m)$ is the symmetry group of a regular m–sided polygon. The set of m vectors R_m corresponding to the vertices of the regular m–gon (B.1) provides a non–real representation when m is odd.

4. The independence of the conserved quantities can be easily verified by considering the free limit, *i.e.* $g_{|\rho|} = 0$. In this limit we have the conserved quantities

$$Q_n(p,q) \to Q_n^0(p) = \mathrm{Ts}(p\cdot\hat{H})^n = \sum_{\mu\in\mathcal{R}} (p\cdot\mu)^n, \quad n:\ \textit{degree}, \tag{4.13}$$

which is essentially the same result of the standard Lax pair formulation. In the latter formulation based on the weights of a Lie algebra representation, sometimes the zero weights $\mu \equiv 0$ are contained on top of those other weights (roots) involved in the present formulation. However, it is obvious that they give zero contribution and the results (4.13) are the same in both formulations. Since $\{Q_n^0(p)\}$ are obviously Coxeter invariant polynomials in $\{p\}$, the problem that the totality of $\{Q_n^0\}$ (4.13) provides the complete (total of r *independent*) set of free conserved quantities can be rephrased that if and how the basis of Coxeter invariant polynomials can be obtained as *power sums*. The latter problem was addressed in [33] about thirty years ago. If the Jacobian

$$\frac{\partial(Q^0_{f_1},\ldots,Q^0_{f_r})}{\partial(p_1,\ldots,p_r)}, \quad f_j:\ \mathit{degree}, \tag{4.14}$$

is non–vanishing the map $(p_1,\ldots,p_r)\to(Q^0_{f_1},\ldots,Q^0_{f_r})$ is invertible and the basis is complete. If the map is invertible at one point, it should be so in the same Weyl chamber and the invertibility breaks down at the boundary. By simple comparison of the powers in $\{p\}$ we arrive at

$$\frac{\partial(Q^0_{f_1},\ldots,Q^0_{f_r})}{\partial(p_1,\ldots,p_r)} = \lambda \prod_{\rho\in\Delta_+}(p\cdot\rho), \quad \lambda\in\mathbf{R}. \tag{4.15}$$

For lower rank cases, G_2, $I_2(m)$, H_3 and F_4 one can show (4.15) by direct calculation for the representations given in Appendix B. In [34] the affirmative answers are given for the classical root systems for the choices given in Appendix B. For E_6, E_7 and E_8 one could invoke the theorems in [35] for demonstrating the non–vanishing of the Jacobian for those $\mathcal{R}$ listed in Appendix B. We have verified the independence of the free conserved quantities (4.15) for all the cases (except for the classical ones) listed in Appendix B with the aid of formula manipulation programs. The verification is exact in which left hand side of (4.15) is evaluated for integer (fraction) values of $\{p_j\}$.

5. In Appendix B we list for each root system how the complete set of independent conserved quantities are obtained by choosing proper representations of the Lax pair.

4.2 Rational potential with harmonic force

The quantum Lax pair for the type V models needs a separate formulation. The explicit form of the Hamiltonian is

$$\mathcal{H} = \frac{1}{2}p^2 + \frac{1}{2}\omega^2 q^2 + \frac{1}{2}\sum_{\rho\in\Delta_+} g_{|\rho|}(g_{|\rho|}-1)\frac{|\rho|^2}{(\rho\cdot q)^2} - \mathcal{E}_0. \tag{4.16}$$

The canonical equations of motion are equivalent to the following Lax equations for $L^\pm$:

$$\frac{d}{dt}L^\pm = i[\mathcal{H},L^\pm] = [L^\pm, M] \pm i\omega L^\pm, \tag{4.17}$$

in which (see section 4 of Ref.[8]) M is the same as before (4.2), and $L^{\pm}$ and Q are defined by

$$L^{\pm} = L \pm i\omega Q, \quad Q = q \cdot \hat{H}, \tag{4.18}$$

with L, $\hat{H}$ as earlier (4.1), (4.4). If we define hermitian operators $\mathcal{L}_1$ and $\mathcal{L}_2$ by

$$\mathcal{L}_1 = L^+ L^-, \quad \mathcal{L}_2 = L^- L^+, \tag{4.19}$$

they satisfy Lax–like equations

$$\dot{\mathcal{L}}_k = i[\mathcal{H}, \mathcal{L}_k] = [\mathcal{L}_k, M], \quad k = 1, 2. \tag{4.20}$$

From these we can construct conserved quantities

$$\mathrm{Ts}(\mathcal{L}_j^n), \quad j = 1, 2, \quad n = 1, 2, \ldots, \tag{4.21}$$

as before. Such quantum conserved quantities have been previously reported for models based on A_r root systems [18,20]. It should be remarked that $\mathrm{Ts}(\mathcal{L}_2^n)$ is no longer the same as $\mathrm{Ts}(\mathcal{L}_1^n)$ due to quantum corrections. It is elementary to check that the first conserved quantities give the Hamiltonian (4.16)

$$\mathcal{H} \propto \mathrm{Ts}(\mathcal{L}_1) = \mathrm{Ts}(\mathcal{L}_2) + const. \tag{4.22}$$

This then completes the presentation of the quantum Lax pairs and quantum conserved quantities for all of the quantum Calogero–Moser models with non–elliptic potentials.

5 Algebraic construction of excited states I

In this section we show that all the excited states of the type V Calogero–Moser models can be constructed algebraically. Later in section 7 we show the same results in terms of the ℓ operators to be introduced in section 6. The main result is surprisingly simple and can be stated universally:

Corresponding to each partition of an integer N which specify the energy level (3.12) into the sum of the degrees of Coxeter invariant polynomials (3.13), we have an eigenstate of the Hamiltonian $\hat{\mathcal{H}}$ with eigenvalue $\omega N + \mathcal{E}_0$:

$$\prod_{j=1}^{r} (B_{f_j}^{+})^{n_j}\, e^W, \quad N = \sum_{j=1}^{r} n_j f_j, \quad n_j \in \mathbf{Z}_+, \tag{5.1}$$

in which the integers $\{f_j\}$, $j = 1, \ldots, r$ are listed in Table II. They exhaust all the excited states. In other words the above states give the complete basis of the Fock space. The creation operators $B_{f_j}^{+}$ and the corresponding annihilation operators[d] $B^-_{f_j}$ are defined in terms of the Lax operators $L^\pm$ (4.18) as follows:

$$B^\pm_{f_j} = \mathrm{Ts}(L^\pm)^{f_j}, \quad j = 1, \ldots, r. \tag{5.2}$$

They are hermitian conjugate to each other

$$(B^\pm_{f_j})^\dagger = B^\mp_{f_j} \tag{5.3}$$

with respect to the standard hermitian inner product of the states defined in PW:

$$(\psi, \varphi) = \int_{PW} \psi^*(q)\varphi(q)\, dq. \tag{5.4}$$

We will show later in section 6, (6.15) that the creation (annihilation) operators commute among themselves:

$$[B^+_k, B^+_l] = [B^-_k, B^-_l] = 0, \quad k, l \in \{f_j |\, j = 1, \ldots, r\}, \tag{5.5}$$

so that the state (5.1) does not depend on the order of the creation.

The proof is very simple. By using (4.17) we obtain

$$\frac{d}{dt}(L^\pm)^n - i[\mathcal{H}, (L^+)^n] = [(L^\pm)^n, M] \pm in\omega(L^\pm)^n, \tag{5.6}$$

from which

$$[\mathcal{H}, B^\pm_n] = \pm n\omega B^\pm_n, \tag{5.7}$$

follows after taking the total sum. This simply says that $B^\pm_n$ creates (annihilates) a state having energy $n\omega$. In other words we have

$$\hat{\mathcal{H}} \prod_{j=1}^{r} (B^+_{f_j})^{n_j}\, e^W = \left(\mathcal{E}_0 + \omega \sum_{k=1}^{r} n_k f_k\right) \prod_{j=1}^{r} (B^+_{f_j})^{n_j}\, e^W.$$

Moreover, it is trivial to show that

$$\sum_{\nu\in\mathcal{R}} (L^-)_{\mu\nu}\, e^W = \left(p\cdot\mu - i\omega q\cdot\mu + i \sum_{\rho\in\Delta_+} \frac{\rho\cdot\mu}{\rho\cdot q}\right) e^W = \mu\cdot\left(p + i\frac{\partial W}{\partial q}\right) e^W = 0, \tag{5.8}$$

which implies that the ground state is annihilated by all the annihilation operators

$$B^-_{f_j}\, e^W = 0, \quad j = 1, \ldots, r. \tag{5.9}$$

Some remarks are in order.

---

[d]We adopt the notation by Olshanetsky and Perelomov [5,14].

1. In most cases the energy levels are highly degenerate. The above basis is neither orthogonal nor normalized.

2. The independence of the creation–annihilation operators can also be shown in a similar way to that of the conserved quantities. As with the conserved quantities, plural representations are necessary to define the full set of creation–annihilation operators in some models. This aspect will be discussed in later sections in connection with the ℓ operators.

3. Reflecting the universality of the first exponent, $f_1 = 2$, the creation and annihilation operators of the least quanta, 2ω, exist in all the models. They form an $sl(2,\mathbf{R})$ algebra together with the Hamiltonian $\hat{\mathcal{H}}$:

$$[\hat{\mathcal{H}}, b_2^{\pm}] = \pm 2\omega b_2^{\pm}, \quad [b_2^+, b_2^-] = -\omega^{-1}\hat{\mathcal{H}}, \tag{5.10}$$

in which $b_2^{\pm}$ are normalized forms of $B_2^{\pm}$:

$$b_2^{\pm} = \sum_{\mu,\nu\in\mathcal{R}} (L^{\pm})^2_{\mu\nu}/(4\omega C_{\mathcal{R}}). \tag{5.11}$$

The $sl(2,\mathbf{R})$ algebra was discussed by many authors (see, for example [14,36,19,23], and others) in connection with the models based on classical root systems. We will show later in subsection 7.2 that the states created by B_2^+ (b_2^+) only can be expressed by the Laguerre polynomial:

$$(b_2^+)^n\, e^W = n!L_n^{(\tilde{\mathcal{E}}_0-1)}(\omega q^2)\, e^W, \quad \tilde{\mathcal{E}}_0 \equiv \mathcal{E}_0/\omega. \tag{5.12}$$

It is trivial to verify that $L_n^{(\tilde{\mathcal{E}}_0-1)}(\omega q^2)$ is an eigenfunction of $\tilde{\mathcal{H}}$ (3.5)

$$\tilde{\mathcal{H}} L_n^{(\tilde{\mathcal{E}}_0-1)}(\omega q^2) = 2n\omega L_n^{(\tilde{\mathcal{E}}_0-1)}(\omega q^2). \tag{5.13}$$

The normalisation of the state

$$\|(b_2^+)^n\, e^W\|^2 = n!\mathcal{N}_0/\Gamma(n+\tilde{\mathcal{E}}_0), \quad \mathcal{N}_0 \equiv \|e^W\|^2\Gamma(\tilde{\mathcal{E}}_0), \tag{5.14}$$

is also dictated by the $sl(2,\mathbf{R})$ relations. The Laguerre polynomial wavefunctions appear as 'radial' wavefunctions in all the cases [37]. This will be shown explicitly for for the rank two models given in subsection 7.3.

4. As is emphasized by Perelomov [14] and Gambardella [36] the $sl(2,\mathbf{R})$ algebra and the corresponding Laguerre wavefunctions are more universal than Calogero–Moser models. They arise when the potentials are homogeneous functions in q of degree -2 with the confining harmonic force.

5. The operators $\{Q_n\}$ and $\{B_n^{\pm}\}$ do not form a Lie algebra. They satisfy interesting non–linear relations, for example,

$$[[B_n^+, b_2^-], b_2^+] = nB_n^+, \quad [[B_n^-, b_2^+], b_2^-] = nB_n^-. \tag{5.15}$$

This tells, for example, that although B_n^+ and b_2^+ create different units of quanta n and 2, they are not independent

$$[B_n^+, b_2^-] \neq 0 \neq [B_n^-, b_2^+].$$

6 ℓ operators

In this section we will show the equivalence of the quantum conserved quantities obtained in the Lax operator formalism of section 4 and those derived in the 'commuting differential operators' formalism initiated by Dunkl [11] and followed by many authors. Again the equivalence is universal, applicable to the models based on any root system. We propose to call the operators in the latter approach simply 'ℓ operators', since they are essentially the same as the L operator in the Lax pair formalism and that they are not mutually commuting, as we will show presently, when the interaction potentials are trigonometric (hyperbolic), (6.14). Although these two formalisms are formally equivalent, the ℓ operator formalism has many advantages over the Lax pair one. Roughly speaking, the 'vector–like' objects ℓ_μ's are easier to handle than the matrix $L_{\mu\nu}$.

Let us fix a representation $\mathcal{R}$ of the Coxeter group G_Δ and define for each element $\mu \in \mathcal{R}$ the following differential–reflection operator

$$\ell_\mu = \ell \cdot \mu = p \cdot \mu + i \sum_{\rho \in \Delta_+} g_{|\rho|} \, (\rho \cdot \mu) \, x(\rho \cdot q) \check{s}_\rho, \quad \mu \in \mathcal{R}. \tag{6.1}$$

It is linear in μ and Coxeter covariant

$$\ell_{\mu+\nu} = \ell_\mu + \ell_\nu, \qquad \check{s}_\rho \ell_\mu \check{s}_\rho = \ell_{s_\rho(\mu)}, \quad \forall \rho \in \Delta. \tag{6.2}$$

They are hermitian operators, $\ell_\mu^\dagger = \ell_\mu$, with respect to the standard inner product for the states (5.4).

It is straightforward to show that the quantum conserved quantities Q_n derived in the previous section (4.7) can be expressed as polynomials in the ℓ operators as follows:

$$Q_n \psi = \sum_{\mu,\nu \in \mathcal{R}} (L^n)_{\mu\nu} \psi = (\sum_{\mu \in \mathcal{R}} \ell_\mu^n) \psi, \tag{6.3}$$

in which ψ is an arbitrary Coxeter invariant state, $\check{s}_\rho\psi = \psi$. This also illustrates the Coxeter invariance of Q_n clearly, since $\check{s}_\rho(\sum_{\mu\in\mathcal{R}} \ell_\mu^n)\check{s}_\rho = \sum_{\mu\in\mathcal{R}} \ell_{s_\rho(\mu)}^n = \sum_{\mu\in\mathcal{R}} \ell_\mu^n$. For $n = 1$ it is trivial, since

$$\sum_{\nu\in\mathcal{R}} (L)_{\mu\nu}\psi = \left(p\cdot\mu + i\sum_{\rho\in\Delta_+} g_{|\rho|}\,(\rho\cdot\mu)\,x(\rho\cdot q)\sum_{\nu\in\mathcal{R}}(\hat{s}_\rho)_{\mu\nu}\right)\psi$$
$$= \left(p\cdot\mu + i\sum_{\rho\in\Delta_+} g_{|\rho|}\,(\rho\cdot\mu)\,x(\rho\cdot q)\check{s}_\rho\right)\psi = \ell_\mu\psi, \qquad (6.4)$$

in which $\sum_{\nu\in\mathcal{R}}(\hat{s}_\rho)_{\mu\nu} = 1$ and $\check{s}_\rho\psi = \psi$ are used. Let us assume that

$$\sum_{\nu\in\mathcal{R}} (L^n)_{\mu\nu}\psi = \ell_\mu^n\psi, \qquad (6.5)$$

is correct, then we obtain

$$\sum_{\nu\in\mathcal{R}} (L^{n+1})_{\mu\nu}\psi = \sum_{\lambda,\nu\in\mathcal{R}} L_{\mu\lambda}(L^n)_{\lambda\nu}\psi = \sum_{\lambda\in\mathcal{R}} L_{\mu\lambda}\ell_\lambda^n\psi,$$
$$= \sum_{\lambda\in\mathcal{R}}\left(p\cdot\mu\delta_{\mu\lambda} + i\sum_{\rho\in\Delta_+} g_{|\rho|}\,(\rho\cdot\mu)\,x(\rho\cdot q)(\hat{s}_\rho)_{\mu\lambda}\right)\ell_\lambda^n\psi.$$

In the second summation only such λ as $\lambda = s_\rho(\mu)$ contributes and we find

$$\ell_{s_\rho(\mu)}^n\psi = (\check{s}_\rho\ell_\mu^n\check{s}_\rho)\psi = \check{s}_\rho\ell_\mu^n\psi.$$

Thus we arrive at

$$\sum_{\nu\in\mathcal{R}} (L^{n+1})_{\mu\nu}\psi = \ell_\mu^{n+1}\psi, \qquad (6.6)$$

and the equivalence of the two expressions of the conserved quantity (6.3) is proved.

Commutation relations among ℓ operators can be evaluated in a similar manner as those appearing in the Lax pair [8,10], that is, by decomposing the roots into two–dimensional sub–root systems. We obtain

$$[\ell_\mu, \ell_\nu] = -a^2\sum_{\rho,\sigma\in\Delta_+} g_{|\rho|}g_{|\sigma|}\,(\rho\cdot\mu)\,(\sigma\cdot\nu)[\check{s}_\rho, \check{s}_\sigma]\times\begin{cases} 0 & \text{rational,}\\ -1 & \text{hyperbolic,}\\ 1 & \text{trigonometric.}\end{cases} \qquad (6.7)$$

One important use of the ℓ operators is the proof of involution of quantum conserved quantities. For type I models Heckman [23] gave a universal proof based on the commutation relation (6.7):

$$[Q_n, Q_m]\psi = \sum_{\mu,\nu\in\mathcal{R}} [\ell_\mu^n, \ell_\nu^m]\psi = 0, \quad \text{rational model.} \tag{6.8}$$

This was the motivation for the introduction of the commuting differential–reflection operators by Dunkl [11]. In fact, Dunkl's and Heckman's operators were the similarity transformation of ℓ_μ by the ground state wavefunction e^W:

$$\tilde{\ell}_\mu = e^{-W}\ell_\mu e^W = p\cdot\mu + i\sum_{\rho\in\Delta_+} g_{|\rho|}\frac{(\rho\cdot\mu)}{(\rho\cdot q)}(\check{s}_\rho - I). \tag{6.9}$$

As for type V models, we define $\ell^\pm$ corresponding to $L^\pm$ (4.18):

$$\ell_\mu^\pm = \ell^\pm\cdot\mu = p\cdot\mu \pm i\omega(q\cdot\mu) + i\sum_{\rho\in\Delta_+} g_{|\rho|}\frac{(\rho\cdot\mu)}{(\rho\cdot q)}\check{s}_\rho, \quad \mu\in\mathcal{R}. \tag{6.10}$$

They are linear in μ, Coxeter covariant and hermitian conjugate of each other with respect to the standard inner product (5.4):

$$\check{s}_\rho\ell_\mu^\pm\check{s}_\rho = \ell_{s_\rho(\mu)}^\pm, \qquad (\ell_\mu^\pm)^\dagger = \ell_\mu^\mp. \tag{6.11}$$

The conserved quantities are expressed as polynomials in $\ell^\pm$ operators:

$$\begin{aligned}\mathrm{Ts}(\mathcal{L}_1^n)\psi &= \sum_{\mu,\nu\in\mathcal{R}}(L^+L^-)_{\mu\nu}^n\psi = \sum_{\mu\in\mathcal{R}}(\ell_\mu^+\ell_\mu^-)^n\psi,\\ \mathrm{Ts}(\mathcal{L}_2^n)\psi &= \sum_{\mu,\nu\in\mathcal{R}}(L^-L^+)_{\mu\nu}^n\psi = \sum_{\mu\in\mathcal{R}}(\ell_\mu^-\ell_\mu^+)^n\psi.\end{aligned} \tag{6.12}$$

Likewise the creation and annihilation operators $B_n^\pm$ (5.2) are expressed as

$$B_n^\pm\psi = \mathrm{Ts}(L^\pm)^n\psi = \sum_{\mu,\nu\in\mathcal{R}}(L^\pm)_{\mu\nu}\psi = \sum_{\mu\in\mathcal{R}}(\ell_\mu^\pm)^n\psi. \tag{6.13}$$

The commutation relations among $\ell^\pm$ operators are easy to evaluate, since ℓ operators commute in the rational potential models (6.7):

$$[\ell_\mu^+, \ell_\nu^+] = [\ell_\mu^-, \ell_\nu^-] = 0, \quad [\ell_\mu^-, \ell_\nu^+] = 2\omega\left(\mu\cdot\nu + \sum_{\rho\in\Delta_+} g_{|\rho|}\,(\rho\cdot\mu)\,(\rho^\vee\cdot\nu)\check{s}_\rho\right). \tag{6.14}$$

From these it follows that the creation (annihilation) operators $B_n^\pm$ do commute among themselves:

$$[B_n^+, B_m^+]\psi = [B_n^-, B_m^-]\psi = 0. \tag{6.15}$$

It is also clear that $\ell_\mu^\pm/\sqrt{2\omega}$ are the 'deformation' of the creation (annihilation) operators of the ordinary multicomponent harmonic oscillators. In fact we have

$$\ell_\mu^+ \, e^W = 2i\omega(\mu\cdot q)\, e^W \quad \text{and} \quad \ell_\mu^- \, e^W = 0. \tag{6.16}$$

In the next section we present an alternative scheme of algebraic construction of excited states of type V models by pursuing the analogy that $\ell^\pm$ are the creation and annihilation operators of the unit quantum. This method was applied to the A_r models by Brink et. al and others [19,17,20].

7 Algebraic construction of excited states II

7.1 Operator solution of the triangular Hamiltonian

In subsection 3.1, we have shown that an eigenfunction of $\mathcal{H}$ with eigenvalue $N\omega$ is given by

$$\left(P_N(q) + \tilde{P}_{N-2}(q)\right) e^W, \tag{7.1}$$

in which $P_N(q)$ is a Coxeter invariant polynomial in q of homogeneous degree N and $\tilde{P}_{N-2}(q)$ is a Coxeter invariant polynomial in q of degree $N-2$ and lower. The non–leading polynomial $\tilde{P}_{N-2}(q)$ is completely determined by the leading one $P_N(q)$ due to the triangularity. This solution can be written in an operator form as follows.

Suppose $P_N(q)$ is expressed as

$$P_N(q) = \sum_{\{\mu\}} c_{\{\mu\}}(q\cdot\mu_1)\cdots(q\cdot\mu_N), \quad \mu_j \in \mathcal{R}, \quad c_{\{\mu\}} : \ const. \tag{7.2}$$

We obtain a Coxeter invariant polynomial in the creation operators ℓ^+ by replacing $q\cdot\mu$ by $\ell_\mu^+/(2i\omega)$:

$$P_N(q) \Rightarrow \frac{1}{(2i\omega)^N} P_N(\ell^+).$$

This creates the above eigenfunction of $\mathcal{H}$ from the ground state:

$$\frac{1}{(2i\omega)^N} P_N(\ell^+)\, e^W = \left(P_N(q) + \tilde{P}_{N-2}(q)\right) e^W. \tag{7.3}$$

The proof is again elementary. By using the commutation relations among $\ell^\pm$ operators it is straightforward to derive the explicit expression of the Hamiltonian in terms of $\ell^\pm$:

$$\mathcal{H} = \frac{1}{2C_\mathcal{R}} \sum_{\mu\in\mathcal{R}} \ell_\mu^+ \ell_\mu^- + \sum_{\rho\in\Delta_+} g_{|\rho|}\left(\omega + \frac{1}{2}\frac{|\rho|^2}{(\rho\cdot q)^2}\right)(\check{s}_\rho - I), \tag{7.4}$$

in which the second term vanishes upon acting on a Coxeter invariant state. Next we obtain

$$\frac{1}{2\mathcal{C}_{\mathcal{R}}}\sum_{\mu\in\mathcal{R}}[\ell_\mu^+\ell_\mu^-,\ell_\nu^\pm]=[\ell_\nu^\pm,S]\pm\omega\ell_\nu^\pm,\quad S\equiv\sum_{\rho\in\Delta_+}g_{|\rho|}\check{s}_\rho,\tag{7.5}$$

which is an ℓ operator version of (4.17). Since a commutator is a derivation, we obtain

$$\frac{1}{2\mathcal{C}_{\mathcal{R}}}\sum_{\mu\in\mathcal{R}}[\ell_\mu^+\ell_\mu^-,P_N(\ell^+)]=[P_N(\ell^+),S]+N\omega P_N(\ell^+),\tag{7.6}$$

in which the first term in r.h.s. vanishes due to the Coxeter invariance of P_N. Thus we arrive at the desired commutation relation

$$[\mathcal{H},P_N(\ell^+)]=N\omega P_N(\ell^+)+\sum_{\rho\in\Delta_+}g_{|\rho|}\left[\frac{1}{2}\frac{|\rho|^2}{(\rho\cdot q)^2},P_N(\ell^+)\right](\check{s}_\rho-I),\tag{7.7}$$

and the eigenvalue equation

$$\mathcal{H}\,P_N(\ell^+)\,e^W=N\omega P_N(\ell^+)\,e^W.\tag{7.8}$$

Since the action of the creation operators on the ground state is

$$\ell_{\mu_1}^+\cdots\ell_{\mu_N}^+\,e^W=[(2i\omega)^N(q\cdot\mu_1)\cdots(q\cdot\mu_N)+\text{lower powers of }q]\,e^W,\tag{7.9}$$

our assertion (7.3) is proved. It should be stressed that in this formalism the Coxeter invariance of the polynomial P is important but not how it is obtained.

Like the above Hamiltonian (7.4), the ℓ operator formulas of higher conserved quantities (6.12) contain extra terms:

$$\mathrm{Ts}(\mathcal{L}_1^n)=\sum_{\mu,\nu\in\mathcal{R}}(L^+L^-)^n_{\mu\nu}=\sum_{\mu\in\mathcal{R}}(\ell_\mu^+\ell_\mu^-)^n+VT.\tag{7.10}$$

Here VT stands for vanishing terms when they act on a Coxeter invariant state. The same is true for most formulas derived in section 6.

7.2 States Created by B_2^+

Here we derive the explicit forms of the subseries of eigenstates obtained by multiple applications of the least quanta creation operator B_2^+ (5.2), or its normalized form b_2^+ (5.11). It is convenient to work with the similarity transformed operator

$$\tilde{b}_2^+=e^{-W}b_2^+\,e^W=\frac{1}{4\omega C_{\mathcal{R}}}\sum_{\mu\in\mathcal{R}}(\tilde{\ell}_\mu^+)^2+VT,\tag{7.11}$$

in which

$$\tilde{\ell}^+_\mu = p \cdot \mu + 2i\omega(q \cdot \mu) + i \sum_{\rho \in \Delta_+} \frac{\rho \cdot \mu}{\rho \cdot q}(\check{s}_\rho - I). \tag{7.12}$$

Let $f(u)$ be an arbitrary function of $u \equiv \omega q^2$, then it is Coxeter invariant. We find

$$\tilde{\ell}^+_\mu f(u) = 2i\omega(q \cdot \mu)(1 - \frac{d}{du})f(u), \quad u \equiv \omega q^2, \tag{7.13}$$

and

$$\tilde{b}^+_2 f(u) = -\left[u\left(1 - \frac{d}{du}\right)^2 - \tilde{\mathcal{E}}_0\left(1 - \frac{d}{du}\right)\right] f(u). \tag{7.14}$$

Since $\tilde{b}^+_2 1 = \tilde{\mathcal{E}}_0 - u = L_1^{(\tilde{\mathcal{E}}_0 - 1)}(u)$, we assume

$$(\tilde{b}^+_2)^n 1 = n! L_n^{(\tilde{\mathcal{E}}_0 - 1)}(u). \tag{7.15}$$

By using the Laguerre differential equation (3.21) and the recurrence formulas of the Laguerre polynomial $L_n^{(\alpha)}(u)$,

$$u\frac{d}{du}L_n^{(\alpha)}(u) = nL_n^{(\alpha)}(u) - (n+\alpha)L_{n-1}^{(\alpha)}(u), \tag{7.16}$$

$$nL_n^{(\alpha)}(u) + (u - 2n - \alpha + 1)L_{n-1}^{(\alpha)}(u) + (n + \alpha - 1)L_{n-2}^{(\alpha)}(u) = 0, \tag{7.17}$$

we can show

$$-\left[u(1 - \frac{d}{du})^2 - \tilde{\mathcal{E}}_0(1 - \frac{d}{du})\right] L_n^{(\tilde{\mathcal{E}}_0 - 1)}(u) = (n+1)L_{n+1}^{(\tilde{\mathcal{E}}_0 - 1)}(u). \tag{7.18}$$

Thus the induction is proved and we arrive at (5.12). The orthogonality of the states

$$((B^+_2)^n e^W, (B^+_2)^m e^W) = 0, \quad n \neq m \tag{7.19}$$

can be easily understood as the du part of the measure

$$e^{2W} d^r q = e^{-u} u^{\tilde{\mathcal{E}}_0 - 1} du d\Omega, \quad d\Omega : \text{angular part},$$

is the proper weight function for the Laguerre polynomial $L_n^{(\tilde{\mathcal{E}}_0 - 1)}(u)$.

7.3 Explicit solutions of the rank two models

For rank two models, the Liouville integrability, or the involution of conserved quantities is automatically satisfied since the second conserved quantity is already obtained. For rank two type V models, the complete set of orthogonal wavefunctions can be written down explicitly in terms of separation of variables by using the Coxeter invariant polynomials. These are based on the dihedral root systems $I_2(m)$, with $A_2 \cong I_2(3)$ [37], $B_2 \cong I_2(4)$ and $G_2 \cong I_2(6)$ [38]. The Coxeter invariant polynomials exist at degree 2, *i.e.* q^2 and m which is

$$\prod_{j=1}^{m}(v_j \cdot q), \tag{7.20}$$

where $\{v_j\}$ is a set of vectors given in (B.2). If we introduce the two–dimensional polar coordinates system[e] for q

$$q = r(\sin\theta, \cos\theta), \tag{7.21}$$

then the principal Weyl chamber is

$$PW: \ 0 < r^2 < \infty, \quad 0 < \theta < \pi/m. \tag{7.22}$$

The two Coxeter invariant variables read:

$$q^2 = r^2, \quad \prod_{j=1}^{m}(v_j \cdot q) = 2(\frac{r}{2})^m \cos m\theta, \tag{7.23}$$

and the latter variable varies the full range, $-1 < \cos m\theta < 1$ in the PW. Thus solving the eigenvalue equation for $\tilde{\mathcal{H}}$ (3.6) by separation of variables in the polar coordinate system is compatible with Coxeter invariance. We adopt as two independent variables

$$u \equiv \omega r^2, \quad z \equiv \cos m\theta. \tag{7.24}$$

The solutions consist of a Gegenbauer (Jacobi) polynomial in $\cos m\theta$ times a Laguerre polynomial in ωr^2. The former we have encountered in the A_1 Sutherland problem, subsection 3.2 and the latter in the A_1 Calogero problem subsections 3.1 and 7.2.

Let us demonstrate this for odd m with a single coupling constant and for even m with two independent coupling constants, in parallel. In terms of the

---

[e] We believe no confusion arises here, between the radial coordinate variable r and the rank of the root system r, which in this case is 2 of $I_2(m)$.

Coxeter invariant variables (7.24) the $I_2(m)$ Hamiltonians take surprisingly simple forms:

$$\tilde{\mathcal{H}} = \omega r\frac{\partial}{\partial r} - \frac{1}{2r}\frac{\partial}{\partial r}\left(r\frac{\partial}{\partial r}\right) - \frac{m}{r}\begin{Bmatrix} g \\ \frac{1}{2}(g_o+g_e)\end{Bmatrix}\frac{\partial}{\partial r}$$
$$-\frac{1}{2r^2}\left[\frac{\partial^2}{\partial\theta^2} + m\begin{Bmatrix} 2g\cot m\theta \\ -g_o\tan\frac{m\theta}{2} + g_e\cot\frac{m\theta}{2}\end{Bmatrix}\frac{\partial}{\partial\theta}\right] \tag{7.25}$$

$$=-2\omega\left[u\frac{\partial^2}{\partial u^2} + (\tilde{\mathcal{E}}_0 - u)\frac{\partial}{\partial u}\right]$$
$$-\frac{\omega m^2}{2u}\left[(1-z^2)\frac{\partial^2}{\partial z^2} + \left(\begin{Bmatrix} 0 \\ g_o - g_e\end{Bmatrix} - \begin{Bmatrix} 1+2g \\ 1+g_e+g_o\end{Bmatrix}z\right)\frac{\partial}{\partial z}\right].$$

The z part admits polynomial solutions

$$\left[(1-z^2)\frac{d^2}{dz^2} + \left(\begin{Bmatrix} 0 \\ g_o - g_e\end{Bmatrix} - \begin{Bmatrix} 1+2g \\ 1+g_e+g_o\end{Bmatrix}z\right)\frac{d}{dz}\right]P_l^{\left\{\substack{(g-\frac{1}{2},g-\frac{1}{2}) \\ (g_o-\frac{1}{2},g_e-\frac{1}{2})}\right\}}(z)$$

$$= -l\left(l + \begin{Bmatrix} 2g \\ g_e+g_o\end{Bmatrix}\right)P_l^{\left\{\substack{(g-\frac{1}{2},g-\frac{1}{2}) \\ (g_o-\frac{1}{2},g_e-\frac{1}{2})}\right\}}(z), \tag{7.26}$$

in which l is the degree of the polynomial. After substituting them, the radial part of the Hamiltonian $\tilde{\mathcal{H}}_r$ reads

$$\tilde{\mathcal{H}}_r = -2\omega\left[u\frac{d^2}{du^2} + (\tilde{\mathcal{E}}_0 - u)\frac{d}{du} - \frac{m^2}{4u}l\left(l + \begin{Bmatrix} 2g \\ g_e+g_o\end{Bmatrix}\right)\right]. \tag{7.27}$$

By similarity transformation in terms of $u^{ml/2} \propto r^{ml}$, which is the radial part of the highest term of the polynomial $P_l^{(\alpha,\beta)}(r^m\cos m\theta)$, it reads

$$u^{-ml/2}\tilde{\mathcal{H}}_r u^{ml/2} = -2\omega\left[u\frac{d^2}{du^2} + \left(ml + \tilde{\mathcal{E}}_0 - u\right)\frac{d}{du} - \frac{ml}{2}\right]. \tag{7.28}$$

This is the main part of the differential equation for the Laguerre polynomial (3.21):

$$\left[u\frac{d^2}{du^2} + \left(ml + \tilde{\mathcal{E}}_0 - u\right)\frac{d}{du} - n\right]L_n^{(ml+\tilde{\mathcal{E}}_0-1)}(u) = 0.$$

Thus the eigenstates of the Hamiltonian are obtained:

$$\tilde{\mathcal{H}}_r u^{ml/2}L_n^{(ml+\tilde{\mathcal{E}}_0-1)}(u) = \omega(2n+ml)u^{ml/2}L_n^{(ml+\tilde{\mathcal{E}}_0-1)}(u), \tag{7.29}$$

$$\tilde{\mathcal{H}} u^{ml/2} L_n^{(ml+\tilde{\mathcal{E}}_0-1)}(u)\, P_l^{\left\{ {(g-\frac{1}{2},g-\frac{1}{2}) \atop (g_o-\frac{1}{2},g_e-\frac{1}{2})} \right\}}(z) \tag{7.30}$$

$$= (\omega(2n+ml)+\mathcal{E}_o)u^{ml/2} L_n^{(ml+\tilde{\mathcal{E}}_0-1)}(u) P_l^{\left\{ {(g-\frac{1}{2},g-\frac{1}{2}) \atop (g_o-\frac{1}{2},g_e-\frac{1}{2})} \right\}}(z).$$

It is instructive to note that the Hamiltonians $\hat{\mathcal{H}}$ look also simple:

$$\hat{\mathcal{H}} = \frac{1}{2}\omega^2 r^2 - \frac{1}{2r}\frac{\partial}{\partial r}\left(r\frac{\partial}{\partial r}\right) - \frac{1}{2r^2}\frac{\partial^2}{\partial \theta^2} + \frac{m^2}{2r^2}\left\{ \begin{array}{c} \frac{g(g-1)}{\sin^2 m\theta} \\ \frac{g_o(g_o-1)}{4\cos^2\frac{m\theta}{2}} + \frac{g_e(g_e-1)}{4\sin^2\frac{m\theta}{2}} \end{array} \right\}. \tag{7.31}$$

Olshanetsky and Perelomov [5] obtained the above solutions starting from these formulas.

8 Universal proof of involution of quantum conserved quantities for type I, II and III models

Here we present a proof of involution of quantum conserved quantities $\{Q_n\}$ derived from the universal Lax pair in subsection 4.1 for type I, II and III models. The proof is applicable to all models based on any root system. Though a universal proof of involution for type I models is given by Heckman [23] as recapitulated in section 6, we believe the universal proof applicable to type II and III models as well is new. It depends on a theorem by Olshanetsky and Perelomov [13]. Our own contribution is that we have provided a universal Lax pair and conserved quantities satisfying all the requirements of the theorem.

Liouville's theorem states the complete integrability as the existence of an involutive set of conserved quantities as many as the degrees of freedom. We have already given conserved quantities $\{Q_n\}$ (4.7) independent and as many as the degrees of freedom (see Appendix B). They have the following properties:

1. Coxeter invariance

$$Q_n(s_\rho(p), s_\rho(q)) = Q_n(p,q), \quad \forall \rho \in \Delta. \tag{8.1}$$

2. $Q_n(p,q)$ is a homogeneous polynomial of degree n in variables $(p_1,\ldots,p_r,x(\rho\cdot q))$.

3. Scaling property for those of type I models:

$${}^I Q_n(\kappa^{-1}p, \kappa q) = \kappa^{-n}\, {}^I Q_n(p,q) \tag{8.2}$$

as a consequence of the above point.

4. For type II and III models, the asymptotic behaviour near the origin:

$$Q_n(p,q) = {}^I Q_n(p,q)(1+\mathcal{O}(|q|)), \quad \text{for} \quad |q| \to 0. \tag{8.3}$$

We need to show the vanishing of

$$J_{lm} \equiv [Q_l, Q_m], \tag{8.4}$$

which is a polynomial in $\{p\}$ of degree s

$$s < l + m. \tag{8.5}$$

Let us decompose J_{lm} into the leading part and the rest:

$$J_{lm} = J^0_{lm} + J^{rest}_{lm}, \quad J^0_{lm} = \sum c^{j_1,\ldots,j_s}(q) p_{j_1} \cdots p_{j_s} \tag{8.6}$$

and J^{rest}_{lm} is a polynomial in $\{p\}$ of degree less than s. From Jacobi identity and conservation $[\mathcal{H}, Q_{l(m)}] = 0$, we obtain

$$[\mathcal{H}, J_{lm}] = 0. \tag{8.7}$$

Considering the explicit form of the Hamiltonian (2.7) ($\omega = 0$), the leading (*i.e.* of degree $s+1$ in $\{p\}$) part of $[\mathcal{H}, J_{lm}]$ comes only from the free part

$$[p^2, J^0_{lm}]$$

and it vanishes if the following conditions are satisfied:

$$\sum_\sigma \frac{\partial}{\partial q_t} c^{k_1,\ldots,k_s}(q) = 0, \tag{8.8}$$

where the sum is taken over all permutations of indices $\sigma(t, k_1, \ldots, k_s) = (j_1, \ldots, j_{s+1})$. In Ref.[39]. it is proved (Lemma 2.5, p. 407) that the system (8.8) has only polynomial solutions. Then Olshanetsky and Perelomov argue that for type I models the scaling property tells that $c^{k_1,\ldots,k_s}(\kappa q) = \kappa^{s-l-m} c^{k_1,\ldots,k_s}(q)$. Since $s < l+m$ (8.5), it follows that the only polynomial solution satisfying the condition is the null polynomial. Thus we obtain $c^{j_1,\ldots,j_s}(q) = 0 \Rightarrow J^0_{lm} = 0$ and $J_{lm} = 0$. The same results follow for type II and III models by considering the asymptotic behaviour for $|q| \to 0$. Thus the involution of all the conserved quantities $\{Q_n\}$ is proved. This result also implies the involution of classical conserved quantities by taking the classical limit ($\hbar \to 0$). See Ref.[12] for the classical Liouville integrability of the most general Calogero–Moser models with elliptic potentials.

9 Summary, comments and outlook

Various issues related to quantum integrability of Calogero–Moser models based on any root system are presented. These are construction of quantum conserved quantities and a unified proof of their involution, the relationship between the Lax pair and the differential–reflection (Dunkl) operators formalisms, construction of excited states by creation operators, etc. The independence (completeness) of the set of conserved quantities is discussed in some detail. They are mainly generalizations of the results known for the models based on A_r root systems. Integrability of the models based on other classical root systems and the exceptional ones including the non–crystallographic models are also discussed in Refs. [38,41–44].

There are still many interesting problems to be addressed to: The structure and properties of the eigenfunctions of the trigonometric potential models, which are generalizations of the Jack polynomials [28–31]. Comprehensive treatment of Liouville integrability of rational models with harmonic force. Understanding the roles of supersymmetry and shape invariance in Calogero–Moser models [45,44]. Formulation of various aspects of quantum Calogero–Moser models with elliptic potentials; Lax pair, the differential–reflection operators [46,47], conserved quantities, supersymmetry and excited states wavefunctions.

Acknowledgements

I thank K. Saito, J. Sekiguchi and T. Yano for bringing [33,35] to our attention. This work is partially supported by the Grant–in–aid from the Ministry of Education, Science and Culture, priority area (#707) "Supersymmetry and unified theory of elementary particles".

Appendix A: Root Systems

In this Appendix we recapitulate the rudimentary facts of the root systems and reflections to be used in the main text. The set of roots Δ is invariant under reflections in the hyperplane perpendicular to each vector in Δ. In other words,

$$s_\alpha(\beta) \in \Delta, \quad \forall \alpha, \beta \in \Delta, \tag{A.1}$$

where

$$s_\alpha(\beta) = \beta - (\alpha^\vee \cdot \beta)\alpha, \quad \alpha^\vee \equiv 2\alpha/|\alpha|^2. \tag{A.2}$$

The set of reflections $\{s_\alpha,\ \alpha \in \Delta\}$ generates a group G_Δ, known as a Coxeter group, or finite reflection group. The orbit of $\beta \in \Delta$ is the set of root vectors resulting from the action of the Coxeter group on it. The set of positive roots Δ_+ may be defined in terms of a vector $U \in \mathbf{R}^r$, with $\alpha \cdot U \neq 0$, $\forall \alpha \in \Delta$, as those roots $\alpha \in \Delta$ such that $\alpha \cdot U > 0$. Given Δ_+, there is a unique set of r simple roots $\Pi = \{\alpha_j,\ j = 1, \ldots, r\}$ defined such that they span the root space and the coefficients $\{a_j\}$ in $\beta = \sum_{j=1}^r a_j \alpha_j$ for $\beta \in \Delta_+$ are all non–negative. The highest root α_h, for which $\sum_{j=1}^r a_j$ is maximal, is then also determined uniquely. The subset of reflections $\{s_\alpha,\ \alpha \in \Pi\}$ in fact generates the Coxeter group G_Δ. The products of s_α, with $\alpha \in \Pi$, are subject solely to the relations

$$(s_\alpha s_\beta)^{m(\alpha,\beta)} = 1, \qquad \alpha, \beta \in \Pi. \tag{A.3}$$

The interpretation is that $s_\alpha s_\beta$ is a rotation in some plane by $2\pi/m(\alpha, \beta)$. The set of positive integers $m(\alpha, \beta)$ (with $m(\alpha, \alpha) = 1$, $\forall \alpha \in \Pi$) uniquely specify the Coxeter group. The weight lattice $\Lambda(\Delta)$ is defined as the $\mathbf{Z}$–span of the fundamental weights $\{\lambda_j\}$, $j = 1, \ldots, r$, defined by

$$\alpha_j^\vee \cdot \lambda_k = \delta_{jk}, \quad \alpha_j \in \Pi. \tag{A.4}$$

The root systems for finite reflection groups may be divided into two types: crystallographic and non–crystallographic. Crystallographic root systems satisfy the additional condition

$$\alpha^\vee \cdot \beta \in \mathbf{Z}, \quad \forall \alpha, \beta \in \Delta, \tag{A.5}$$

which implies that the $\mathbf{Z}$–span of Π is a lattice in $\mathbf{R}^r$ and contains all roots in Δ. We call this the root lattice, which is denoted by $L(\Delta)$. These root systems are associated with simple Lie algebras: $\{A_r,\ r \geq 1\}$, $\{B_r,\ r \geq 2\}$, $\{C_r,\ r \geq 2\}$, $\{D_r,\ r \geq 4\}$, E_6, E_7, E_8, F_4 and G_2. The Coxeter groups for these root systems are called Weyl groups. The remaining non–crystallographic root systems are H_3, H_4, whose Coxeter groups are the symmetry groups of the icosahedron and four–dimensional 600–cell, respectively, and the dihedral group of order $2m$, $\{I_2(m),\ m \geq 4\}$.

Here we give the explicit examples of root systems. In all cases but the A_r, $\{e_j\}$ denotes an orthonormal basis in $\mathbf{R}^r$, $e_j \in \mathbf{R}^r, e_j \cdot e_k = \delta_{jk}$.

1. A_r: This root system is related with the Lie algebra $su(r+1)$. It is convenient to have the $r+1$ dimensional realization:

$$\begin{aligned} &\Delta = \{\ \pm(e_j - e_k),\ \ j \neq k = 1, \ldots, r+1 | e_j \in \mathbf{R}^{r+1}, e_j \cdot e_k = \delta_{jk}\}, \\ &\Pi = \{e_j - e_{j+1},\ j = 1, \ldots, r\}. \end{aligned} \tag{A.6}$$

2. B_r: This root system is associated with Lie algebra $so(2r+1)$. The long roots have $(\text{length})^2 = 2$ and short roots have $(\text{length})^2 = 1$:

$$\begin{aligned}\Delta &= \{\pm e_j \pm e_k,\ \pm e_j,\ j \neq k = 1,\ldots,r\},\\ \Pi &= \{e_j - e_{j+1},\ j = 1,\ldots,r-1\} \cup \{e_r\}.\end{aligned} \tag{A.7}$$

3. C_r: This root system is associated with Lie algebra $sp(2r)$. The long roots have $(\text{length})^2 = 4$ and short roots have $(\text{length})^2 = 2$:

$$\begin{aligned}\Delta &= \{\pm e_j \pm e_k,\ \pm 2e_j,\ j,k = 1,\ldots,r\},\\ \Pi &= \{e_j - e_{j+1},\ j = 1,\ldots,r-1\} \cup \{2e_r\}.\end{aligned} \tag{A.8}$$

4. D_r: This root system is associated with Lie algebra $so(2r)$:

$$\begin{aligned}\Delta &= \{\pm e_j \pm e_k,\ j \neq k = 1,\ldots,r\},\\ \Pi &= \{e_j - e_{j+1},\ j = 1,\ldots,r-1\} \cup \{e_{r-1} + e_r\}.\end{aligned} \tag{A.9}$$

5. E_6: All the roots have the same $(\text{length})^2$, which is chosen to be 2:

$$\begin{aligned}\Delta &= \{\pm e_j \pm e_k,\ j \neq k = 1,\ldots,5\} \cup \{\frac{1}{2}(\pm e_1 \cdots \pm e_5 \pm \sqrt{3}e_6),\ (\text{even }+)\},\\ \Pi &= \{\ \frac{1}{2}(e_1 - e_2 - e_3 - e_4 + e_5 - \sqrt{3}e_6),\ e_4 - e_5,\ e_3 - e_4,\ e_4 + e_5,\\ &\quad \frac{1}{2}(e_1 - e_2 - e_3 - e_4 - e_5 + \sqrt{3}e_6),\ e_2 - e_3\}.\end{aligned} \tag{A.10}$$

6. E_7: All the roots have the same $(\text{length})^2$, which is chosen to be 2:

$$\begin{aligned}\Delta &= \{\pm e_j \pm e_k,\ j \neq k = 1,\ldots,6\} \cup \{\pm\sqrt{2}e_7\}\\ &\quad \cup\{\frac{1}{2}(\pm e_1 \cdots \pm e_6 \pm \sqrt{2}e_7),\ (\text{even }+)\},\\ \Pi &= \{\ e_2 - e_3,\ e_3 - e_4,\ e_4 - e_5,\ e_5 - e_6,\\ &\quad \frac{1}{2}(e_1 - e_2 - e_3 - e_4 - e_5 + e_6 - \sqrt{2}e_7),\ \sqrt{2}e_7,\ e_5 + e_6\}.\end{aligned} \tag{A.11}$$

7. E_8: All the roots have the same $(\text{length})^2$, which is chosen to be 2:

$$\begin{aligned}\Delta &= \{\pm e_j \pm e_k,\ j \neq k = 1,\ldots,8\} \cup \{\frac{1}{2}(\pm e_1 \cdots \pm e_8),\ (\text{even }+)\},\\ \Pi &= \{\frac{1}{2}(e_1 - e_2 - e_3 - e_4 - e_5 - e_6 - e_7 + e_8),\ e_7 + e_8\}\\ &\quad \cup\{e_j - e_{j+1},\quad j = 2,\ldots,7\}.\end{aligned} \tag{A.12}$$

8. F_4: The long roots ((length)$^2 = 2$) are those of D_4 and the short roots ((length)$^2 = 1$) are the union of the vector, spinor and anti–spinor weights of D_4:

$$\Delta = \{\pm \boldsymbol{e}_j \pm \boldsymbol{e}_k,\ \pm \boldsymbol{e}_j, \frac{1}{2}(\pm \boldsymbol{e}_1 \cdots \pm \boldsymbol{e}_4),\ j \neq k = 1, \ldots, 4\},$$
$$\Pi = \{\boldsymbol{e}_2 - \boldsymbol{e}_3,\ \boldsymbol{e}_3 - \boldsymbol{e}_4,\ \boldsymbol{e}_4,\ \frac{1}{2}(\boldsymbol{e}_1 - \boldsymbol{e}_2 - \boldsymbol{e}_3 - \boldsymbol{e}_4)\}. \qquad (A.13)$$

9. G_2: The G_2 root system consists of six long roots and six short roots, and the sets of long and short roots have the same structure as the A_2 roots, scaled $|\alpha_S|^2/|\alpha_L|^2 = 1/3$ and rotated by $\pi/6$. They are

$$\Delta = \{(\pm\sqrt{2}, 0),\ (\pm\sqrt{\frac{3}{2}}, \pm\frac{1}{\sqrt{2}}),\ (0, \pm\sqrt{\frac{2}{3}}),\ (\pm\frac{1}{\sqrt{2}}, \pm\frac{1}{\sqrt{6}})\},$$
$$\Pi = \{(\sqrt{2}, 0),\ (-\frac{1}{\sqrt{2}}, \frac{1}{\sqrt{6}})\} \qquad (A.14)$$

10. $I_2(m)$: This is a symmetry group of a regular m–gon. For odd m Δ consists of a single orbit, whereas for even m it has two orbits. In both cases we have a representation in which all the roots have length unity

$$\Delta = \{(\cos((j-1)\pi/m), \sin((j-1)\pi/m)),\ j = 1, \ldots, m\},$$
$$\Pi = \{(1, 0),\ (\cos((m-1)\pi/m), \sin((m-1)\pi/m)))\} \qquad (A.15)$$

11. H_4: Define $a \equiv \cos\pi/5 = (1+\sqrt{5})/4$, $b \equiv \cos 2\pi/5 = (-1+\sqrt{5})/4$. Then the H_4 roots are generated by the following simple roots [34]:

$$\alpha_1 = \left(a, -\frac{1}{2}, b, 0\right), \qquad \alpha_2 = \left(-a, \frac{1}{2}, b, 0\right), \qquad (A.16)$$
$$\alpha_3 = \left(\frac{1}{2}, b, -a, 0\right), \qquad \alpha_4 = \left(-\frac{1}{2}, -a, 0, b\right).$$

The full set of roots of H_4 in this basis may be obtained from $(1,0,0,0)$, $(\frac{1}{2}, \frac{1}{2}, \frac{1}{2}, \frac{1}{2})$, and $(a, \frac{1}{2}, b, 0)$ by even permutations and arbitrary sign changes of coordinates. These 120 roots form a single orbit.

12. H_3: A subset of (A.16), $\{\alpha_1, \alpha_2, \alpha_3\}$ is a choice of simple roots for the H_3 root system. In this basis, the full set of roots for H_3 results from even permutations and arbitrary sign changes of $(1,0,0)$ and $(a, \frac{1}{2}, b)$. These 30 roots also form a single orbit.

Appendix B: Conserved quantities

Here we list for each root system how the complete set of independent conserved quantities is obtained by choosing proper representations $\{\mathcal{R}\}$ of the Lax pair. We choose those of the lowest dimensionality for the convenience of practical calculation. Of course there are many other choices of representations giving equally good sets of conserved quantities.

1. A_r: For all powers, the vector representation ($r+1$ dimensions) is enough.

2. B_r: For all powers, the representation consisting of short roots $\{\pm \boldsymbol{e}_j : j = 1, \ldots, r\}$, ($2r$ dimensions) is enough.

3. C_r: For all powers, the representation consisting of long roots $\{\pm 2\boldsymbol{e}_j : j = 1, \ldots, r\}$, ($2r$ dimensions) is enough.

4. D_r: For all even powers, the vector representation ($2r$ dimensions) is enough. For the additional one occurring at power r, the (anti)–spinor representation (2^{r-1} dimensions) would be necessary. They are minimal representations.

5. E_6: For all powers, the weights of the **27** (or $\overline{\mathbf{27}}$) dimensional representation of the Lie algebra is enough. They are minimal representations.

6. E_7: For all powers, the weights of the **56** dimensional representation of the Lie algebra is enough. This is a minimal representation.

7. E_8: For all powers, the 240 dimensional representation consisting of all the roots is enough. This is not the same as the adjoint representation of the Lie algebra.

8. F_4: For all powers, either of the 24 dimensional representation consisting of all the long roots or the short roots is enough. These are not Lie algebra representations.

9. G_2: For all powers, either of the 6 dimensional representations consisting of all the long roots or the short roots is enough. These are not Lie algebra representations.

10. $I_2(m)$: For both powers 2 and m, the representation consisting of the vertices R_m of the regular m–gon is enough:

$$R_m = \{(\cos(2k/m + t_0), \sin(2k/m + t_0)) \in \mathbf{R}^2 |\ k = 1, \ldots, m\}, \quad \text{(B.1)}$$

in which $t_0 = 0$ $(1/2m)$ for m even (odd). Another set V_m is used in 7.3. Here, V_m is the set of vectors with 'half' angles of the roots (see (A.15)) given by

$$V_m = \{v_j = (\cos((2j-1)\pi/2m), \sin((2j-1)\pi/2m)) \in \mathbf{R}^2 |\, j = 1, \ldots, m\}. \tag{B.2}$$

11. H_3: For all powers, the representation consisting of all the 30 roots is enough.

12. H_4: For all powers, the representation consisting of all the 120 roots is enough.

References

1. F. Calogero, *Solution of the one-dimensional N-body problem with quadratic and/or inversely quadratic pair potentials*, J. Math. Phys. **12** (1971) 419-436.
2. B. Sutherland, *Exact results for a quantum many-body problem in one-dimension, II*, Phys. Rev. **A5** (1972) 1372-1376.
3. J. Moser, *Three integrable Hamiltonian systems connected with isospectral deformations*, Adv. Math. **16** (1975) 197-220; J. Moser, *Integrable systems of non-linear evolution equations*, in *Dynamical Systems, Theory and Applications*; J. Moser, ed., Lecture Notes in Physics **38** (1975), Springer–Verlag; F. Calogero, C. Marchioro and O. Ragnisco, *Exact solution of the classical and quantal one-dimensional many body problems with the two body potential $V_a(x) = g^2a^2/\sinh^2 ax$*, Lett. Nuovo Cim. **13** (1975) 383-387; F. Calogero, *Exactly solvable one-dimensional many body problems*, Lett. Nuovo Cim. **13** (1975) 411-416.
4. M. A. Olshanetsky and A. M. Perelomov, *Completely integrable Hamiltonian systems connected with semisimple Lie algebras*, Inventions Math. **37** (1976), 93-108.
5. M. A. Olshanetsky and A. M. Perelomov, *Quantum integrable systems related to Lie algebras*, Phys. Rep. **94** (1983) 313-404.
6. S. P. Khastgir, A. J. Pocklington and R. Sasaki, *Quantum Calogero–Moser Models: Integrability for all Root Systems*, J. Phys. **A33** (2000) 9033-9064, hep-th/0005277.
7. A. J. Bordner, E. Corrigan and R. Sasaki, *Calogero–Moser models I: a new formulation*, Prog. Theor. Phys. **100** (1998) 1107-1129, hep-th/9805106; A. J. Bordner, R. Sasaki and K. Takasaki, *Calogero–Moser models II: symmetries and foldings*, Prog. Theor. Phys. **101**

(1999) 487-518, hep-th/9809068; A. J. Bordner and R. Sasaki, *Calogero–Moser models III: elliptic potentials and twisting*, Prog. Theor. Phys. **101** (1999) 799-829, hep-th/9812232; S. P. Khastgir, R. Sasaki and K. Takasaki, *Calogero–Moser Models IV: Limits to Toda theory*, Prog. Theor. Phys. **102** (1999), 749-776, hep-th/9907102.

8. A. J. Bordner, E. Corrigan and R. Sasaki, *Generalised Calogero–Moser models and universal Lax pair operators*, Prog. Theor. Phys. **102** (1999) 499-529, hep-th/9905011.
9. E. D'Hoker and D. H. Phong, *Calogero–Moser Lax pairs with spectral parameter for general Lie algebras*, Nucl. Phys. **B530** (1998) 537-610, hep-th/9804124.
10. A. J. Bordner, N. S. Manton and R. Sasaki, *Calogero–Moser Models V: Supersymmetry and Quantum Lax Pair*, Prog. Theor. Phys. **103** (2000) 463-487, hep-th/9910033.
11. C. F. Dunkl, *Differential–difference operators associated to reflection groups*, Trans. Amer. Math. Soc. **311** (1989) 167-183; V. M. Buchstaber, G. Felder and A. P. Veselov, *Elliptic Dunkl operators, root systems and functional equations*, Duke Math. J. **76** (1994) 885-911.
12. S. P. Khastgir and R. Sasaki, Kyoto preprint, *Liouville Integrability of Classical Calogero Moser Models*, YITP 00-31, hep-th/0005278, May 2000, to be published in Phys. Lett. A; J. C. Hurtubise and E. Markman, *Calogero–Moser systems and Hitchin systems*, (1999) math.AG/9912161.
13. M. A. Olshanetsky and A. M. Perelomov, *Quantum systems related to root systems, and radial parts of Laplace operators*, Funct. Anal. Appl. **12** (1977) 121-128.
14. A. M. Perelomov, *Algebraic approach to the solution of a one-dimensional model of interacting particles*, Theor. Math. Phys. **6** (1971) 263-282.
15. S. Wojciechowski, *Involutive set of integrals for completely integrable many–body problems with pair interaction*, Lett. Nuouv. Cim. **18** (1976) 103-107.
16. M. Lassalle, "Polynômes de Jacobi, généralisés", "Polynômes de Laguerre généralisés", "Polynômes de Hermite généralisés", C. R. Acad. Sci. Paris, t. Sér. I Math. **312** (1991) 425-428, 725-728, **313** (1991) 579-582.
17. A. P. Polykronakos, *Exchange operator formalism for integrable systems of particles*, Phys. Rev. Lett. **69** (1992) 703–705.
18. B. S. Shastry and B. Sutherland, *Superlax pairs and infinite symmetries in the* $1/r^2$ *system*, Phys. Rev. Lett. **70** (1993) 4029-4033.
19. L. Brink, T. H. Hansson and M. A. Vasiliev, *Explicit solution to the N body Calogero problem*, Phys. Lett. **B286** (1992) 109-

111, hep-th/9206049; L. Brink, T.H. Hansson, S. Konstein and M.A. Vasiliev, *The Calogero model: anyonic representation, fermionic extension and supersymmetry*, Nucl. Phys. **B401** (1993) 591-612, hep-th/9302023; L. Brink, A. Turbiner and N. Wyllard, *Hidden algebras of the (super) Calogero and Sutherland models*, J. Math. Phys. **39** (1998) 1285-1315, hep-th/9705219.

20. H. Ujino M. Wadati and K. Hikami, *The quantum Calogero–Moser model: algebraic structures*, J. Phys. Soc. Jpn. **62** (1993) 3035–3043; H. Ujino and M. Wadati, *Rodrigues formula for Hi–Jack symmetric polynomials associated with the quantum Calogero model*, J. Phys. Soc. Jpn. **65** (1996) 2423–2439, cond-mat/9609041; H. Ujino, *Orthogonal symmetric polynomials associated with the Calogero model*, J. Phys. Soc. Jpn **64** (1995) 2703-2706, cond-mat/9706133; A. Nishino, H. Ujino and M. Wadati, *Symmetric Fock space and orthogonal symmetric polynomials associated with the Calogero model*, Chaos Solitons Fractals **11** (2000) 657-674, cond-mat/9803284.
21. K. Sogo, *A simple derivation of multivariable Hermite and Legendre polynomials*, J. Phys. Soc. Jpn **65** (1996) 3097-3099; N. Gurappa and P. K. Panigrahi, *Equivalence of the Calogero–Sutherland model to free harmonic oscillators*, Phys. Rev. **B59** (1999) R2490-R2493, cond-mat/9710035.
22. S. N. M. Ruijsenaars, *Sytems of Calogero–Moser type*, CRM Series in Math. Phys. **1** 251-352, Springer, 1999.
23. G. J. Heckman, *A remark on the Dunkl differential–difference operators*, in W. Barker and P. Sally (eds.) *Harmonic analysis on reductive groups*, Birkhäuser, Basel (1991); *An elementary approach to the hypergeometric shift operators of Opdam*, Inv. math. **103** (1991) 341–350.
24. G. J. Heckman and E. M. Opdam, *Root systems and hypergeometric functions I*, Comp. Math. **64** (1987), 329–352; G. J. Heckman, *Root systems and hypergeometric functions II*, Comp. Math. **64** (1987), 353–373; E. M. Opdam, *Root systems and hypergeometric functions III*, Comp. Math. **67** (1988), 21–49; *Root systems and hypergeometric functions IV*, Comp. Math. **67** (1988), 191–209.
25. D. Z. Freedman and P. F. Mende, *An exactly solvable N–particle system in supersymmetric quantum mechanics*, Nucl. Phys. **344** (1990) 317-343.
26. W. Rühl and A. Turbiner, *Exact solvability of the Calogero and Sutherland models*, Mod. Phys. Lett. **A10** (1995) 2213-2222, hep-th/9506105; O. Haschke and W. Rühl, *Is it possible to construct exactly solvable models?*, Lecture Notes in Physics **539** 118-140, Springer, Berlin 2000, hep-th/9809152.

27. R. Caseiro, J. P. Françoise and R. Sasaki, *Algebraic linearization of dynamics of Calogero type for any Coxeter group*, J. Math. Phys. **41** (2000) 4679-4689, hep-th/0001074.
28. R. Stanley, *Some combinatorial properties of Jack symmetric function*, Adv. Math. **77** (1989) 76-115; I. G. Macdonald, *Symmetric functions and Hall polynomials*, second edition, Oxford University Press, 1995.
29. L. Lapointe and L. Vinet, *Rodrigues formulas for the Macdonald polynomials*, Adv. Math. **130** (1997) 261-279, q-alg/9607025; it Exact operator solution of the Calogero–Sutherland model, Commun. Math. Phys. **178** (1996) 425-452, q-alg/9509003.
30. T. H. Baker and P. J. Forrester, *The Calogero–Sutherland model and generalized classical polynomials*, Commun. Math. Phys. **188** (1997) 175-216, solv-int/9608004.
31. H. Awata, Y. Matsuo, S. Odake and J. Shiraishi, *Excited states of Calogero–Sutherland model and singular vectors of W_N algebra*, Nucl. Phys. **B449** (1995) 347-374, hep-th/9503043.
32. A. Erdélyi et al. *Higher Transcendental Functions III*, McGraw–Hill, New York, 1955.
33. L. Flatto and M. M. Wiener, *Invariants of finite reflection groups and mean value problems*, Amer. J. Math. **91** (1969) 591-598; L. Flatto, *Invariants of finite reflection groups and mean value problems II*, Amer. J. Math. **92** (1970) 552-561.
34. J. E. Humphreys, *Reflection groups and Coxeter groups*, Cambridge Univ. Press, Cambridge 1990.
35. T. Shioda, *Construction of elliptic curves with high rank via the invariants of the Weyl groups*, J. Math. Soc. Japan **43** (1991) 673–719.
36. P. J. Gambardella, *Exact results in quantum many–body systems of interacting particles in many dimensions with $\overline{SU(1,1)}$ as the dynamical group*, J. Math. Phys. **16** (1975) 1172-1187.
37. F. Calogero, *Solution of a three body problem in one dimension*, J. Math. Phys. **10** (1969) 2191-2196; *Ground state of a one–dimensional N–body problem,* J. Math. Phys. **10** (1969) 2197-2200.
38. J. Wolfes, *On the three–body linear problem with three–body interaction*, J. Math. Phys. **15** (1974) 1420-1424; F. Calogero and C. Marchioro, *Exact solution of a one–dimensional three–body scattering problem with two–body and/or three–body inverse–square potential*, J. Math. Phys. **15** (1974) 1425–1430.
39. F. A. Berezin, *Laplace operators on semisimple Lie groups*, Tr. Mosk. Mat. Ob-va, **6** (1957) 371-463.
40. I. M. Krichever, *Elliptic solutions of the Kadomtsev–Petviashvili equation*

and integrable systems of particles, Funct. Anal. Appl. **14** (1980) 282-289.

41. M. Rosenbaum, A. Turbiner and A. Capella, *Solvability of the g_2 integrable system*, Int. J. Mod. Phys. **A13** (1998) 3885-3904, solv-int/9707005; N. Gurappa, A. Khare and P. K. Panigrahi, *Connection between Calogero–Marchioro–Wolfes type few–body models and free oscillators*, cond-mat/9804207.
42. O. Haschke and W. Rühl, *An exactly solvable model of the Calogero type for the icosahedral group*, Mod. Phys. Lett. **A13** (1998) 3109-3122, hep-th/9811011; *The construction of trigonometric invariants for Weyl groups and the derivation of corresponding exactly solvable Sutherland models,* Modern Phys. Lett. **A14** (1999) 937-949, math-ph/9904002.
43. S. Kakei, *Common algebraic structure for the Calogero–Sutherland models*, J. Phys. **A29** (1996) L619–L624; *An orthogonal basis for the B_N–type Calogero model*, J. Phys. **A30** (1997) L535–L541.
44. P. K. Ghosh, A. Khare and M. Sivakumar, *Supersymmetry, shape invariance and solvability of A_{N-1} and BC_N Calogero–Sutherland model*, Phys. Rev. **A58** (1998) 821-825, cond-mat/9710206; U. Sukhatme and A. Khare, *Comment on self–isospectral periodic potentials and supersymmetric quantum mechanics*, quant-ph/9902072.
45. C. Efthimiou and D. Spector, *Shape invariance in the Calogero and Calogero–Sutherland models*, Phys. Rev. **A56** (1997) 208-219, quant-ph/9702017.
46. I. V. Cherednik, *Elliptic quantum many–body problem and double affine Knizhnik–Zamolodchikov equation*, Comm. Math. Phys. **169** (1995) 441-461.
47. T. Oshima and H. Sekiguchi, *Commuting families of differential operators invariant under the action of a Weyl group*, J. Math. Sci. Univ. Tokyo **2** (1995) 1-75.

GREEN FUNCTIONS ASSOCIATED TO COMPLEX REFLECTION GROUPS $G(e, 1, n)$

TOSHIAKI SHOJI
Department of Mathematics, Science University of Tokyo
Noda, Chiba 278-8510, Japan
E-mail: shoji@ma.noda.sut.ac.jp

Green functions of classical groups are determined by the data from Weyl groups and by certain combinatorial objects called symbols. Generalizing this, we define Green functions associated to complex reflection groups $G(e, 1, n)$. We also construct Hall-Littlewood functions and Schur functions associated to $G(e, 1, n)$, and show that the above Green functions are obtained as a transition matrix between those two symmetric functions. The details of this paper will appear elsewhere.

0 Introduction

Green polynomials $Q_\lambda^\mu(q)$ of $GL_n(\mathbf{F}_q)$, where λ, μ are partitions of n, were first introduced by J.A. Green [2] in 1955 in a combinatorial framework of symmetric functions. They are obtained as the transition matrix between Schur functions and Hall–Littlewood functions. Those Green functions play a crucial role in [2] in describing irreducible characters of $GL_n(\mathbf{F}_q)$. In 1976, P. Deligne and G. Lusztig constructed in [1] Green functions Q_T^G for finite reductive groups $G(\mathbf{F}_q)$ in general, as the restriction to unipotent classes of $G(\mathbf{F}_q)$ of Deligne–Lusztig's virtual character $R_T^G(\theta)$, associated to a maximal torus T of G defined over $\mathbf{F}_q$. The Green function Q_χ, for an irreducible character χ of the Weyl group W of G, is defined as a certain linear combination of Q_T^G for various T. In the case of $GL_n(\mathbf{F}_q)$, let χ^λ be the irreducible character of the symmetric group $W = \mathfrak{S}_n$ corresponding to λ, and u_μ the unipotent class corresponding to μ. Then $Q_{\chi^\lambda}(u_\mu)$ coincides with the modified Kostka polynomial $\widetilde{K}_{\lambda,\mu}(t)$, which is obtained as a certain linear combination of Green polynomials with some modification. Lusztig proved in [6], generalizing the result in [13], that there exists a simple algorithm of computing Green functions Q_χ. In other words, Green functions arise as a unique solution of a certain type of matrix equation

$$P\Lambda\, {}^tP = \Omega \qquad (*)$$

with unknown P, Λ. (The entries of the matrix P give Green functions $Q_\chi(u)$). Note that the matrix Ω is completely determined by the property of irreducible characters of W, while the shape of P and Λ are determined by the property of the Springer correspondence between unipotent classes of G and irreducible characters of W. It is known, by the geometric interpretation of Green functions in [8], that the entries of the matrices P and Λ are in $\mathbb{Z}[q]$.

In [3], Geck and Malle considered an analogy of the equation in (*) by making use of a–functions of irreducible characters of Weyl groups and families of unipotent characters, and showed that they have a unique solution with coefficients in $\mathbb{Q}(q)$. They verified in the case of exceptional groups that the matrices P, Λ have coefficients in $\mathbb{Z}[q]$, and also the matrix Λ provides the polynomials conjectured in [9] expressing the cardinality of special pieces in $G(\mathbf{F}_q)$. They conjecture that similar facts will hold for reductive groups in general.

In the case of classical groups, the Springer correspondence is described in terms of certain combinatorial objects called "u–symbols" introduced by Lusztig [5]. The remarkable fact is that the matrix equation (*) is completely described by the combinatorial property of such u–symbols. The notion of u–symbols was generalized by Malle [12] so that it fits to the case of complex reflection groups $G(e, 1, n)$ as a generalization of Weyl group of type C_n. Now the ingredients used to construct the equation (*) make sense even for this case, and one can define a matrix equation by using the combinatorics of symbols. By using a similar argument as in [3], it is shown that this equation has a unique solution. So it would be natural to call the solution of this equation as the Green function associated to $G(e, 1, n)$.

After Green [2], the character table of $GL_n(\mathbf{F}_q)$, especially the part corresponding to the values at unipotent elements of unipotent characters, is completely dominated by Green functions. Thanks to Lusztig [6], similar facts hold in general; the character values at unipotent elements of finite reductive groups, say, $Sp_{2n}(\mathbf{F}_q)$ or $SO_{2n+1}(\mathbf{F}_q)$, of unipotent characters are dominated by Green functions (or rather by generalized Green functions). Lusztig showed that his method of classifying unipotent characters of finite reductive groups makes sense even for other Coxeter groups W such as H_3, H_4, and determined their "degrees", even there does not exist a finite group $G(\mathbf{F}_q)$ whose Weyl group is W. Inspired by this, Malle [12] classified, in the case of $G(e, 1, n)$, the "unipotent characters" and described their degrees. One might expect that Green functions associated to $G(e, 1, n)$ will provide us the values of unipotent characters at "unipotent elements".

In this note, we try to find a combinatorial theory of symmetric functions which fits to the situation for $G(e, 1, n)$. In our theory, the role of partitions is replaced by symbols. In particular, we construct Hall–Littlewood functions associated to $G(e, 1, n)$, which are parametrized by symbols. We show that a large part of arguments used to construct Green polynomials of $GL_n(\mathbf{F}_q)$

can be generalized to our setting, and that Green functions are obtained as the transition matrix between Schur functions and Hall–Littlewood functions. Thus obtained Green functions are rational functions in $\mathbb{Q}(q)$. As some examples show, it is very likely that these Green functions are actually polynomials in q with positive integral coefficients.

The details of this note will be found in [15].

The author is grateful for G. Malle and F. Lübeck for some useful discussions, and for Malle for the computation of examples by using computers.

1 A background from finite reductive groups

1.1 Let G be a connected reductive algebraic group defined over a finite field $\mathbf{F}_q$, with Frobenius map $F : G \to G$. We assume that G is split over $\mathbf{F}_q$. We are interested in the representation theory of the finite group G^F over $\bar{\mathbf{Q}}_l$, an algebraic closure of the l-adic number field $\mathbf{Q}_l$ with $l \neq \mathrm{ch}F_q$. For an F-stable maximal torus T and a character $\theta : T^F \to \bar{\mathbf{Q}}_l^*$, Deligne–Lusztig's virtual character $R_T^G(\theta) : G^F \to \bar{\mathbf{Q}}_l$ is defined. The Green function Q_T^G is defined as the restriction of $R_T^G(\theta)$ to the set of unipotent elements in G^F, which is independent of the choice of θ. The functions $R_T^G(\theta)$ play a crucial role in the representation theory of G^F, and it is known that the computation of $R_T^G(\theta)$ is reduced to the computation of various Green functions of smaller reductive groups.

Let W be the Weyl group of G. Then the G^F-conjugacy classes of F-stable maximal tori are parametrized by the conjugacy classes of W. We denote by T_w the F-stable maximal torus corresponding to $w \in W$ (up to conjugacy). Let $\mathcal{V}_{\mathrm{uni}}$ be the space of G^F-invariant functions on the set of unipotent elements in G^F. Thus $Q_T^G \in \mathcal{V}_{\mathrm{uni}}$. Let $W^\wedge$ be the set of irreducible characters of W. For $\chi \in W^\wedge$, let us define a function Q_χ on $\mathcal{V}_{\mathrm{uni}}$ by

$$Q_\chi = |W|^{-1} \sum_{w \in W} \chi(w) Q_{T_w}^G.$$

We refer Q_χ also as Green functions. Clearly, the determination of $Q_{T_w}^G$ for all $w \in W$ is equivalent to that of Q_χ for all $\chi \in W^\wedge$.

1.2 Let I be the set of pairs (C, ρ), where C is a unipotent class in G, and ρ is an irreducible character of $A_G(u)$. (Here C is F-stable, and u is an element in C^F. $A_G(u)$ is the component group $Z_G(u)/Z_G^0(u)$). It is known, by a suitable choice of $u \in C^F$, that the set of G^F-conjugacy classes in C^F is in bijection with the set of conjugacy classes in $A_G(u)$. We denote by u_a the class in

C^F corresponding to $a \in A_G(u)$. For each $i = (C, \rho)$, we define a function $Y_i \in \mathcal{V}_{\text{uni}}$ by

$$Y_i(v) = \begin{cases} \rho(a) & \text{if } v \sim u_a \ (G^F - - \text{conjugate}), \\ 0 & \text{if } v \notin C^F. \end{cases}$$

Then $\{Y_i \mid i \in I\}$ forms a basis of the space $\mathcal{V}_{\text{uni}}$.

There exists a natural injection f, the so–called "Springer correspondence" from $W^\wedge$ to I. We denote by I_0 the image of $W^\wedge$ under the map f. If $f(\chi) = i$, we write Q_χ also as Q_i. Then it is known that Q_χ can be written as

$$Q_\chi = Q_i = \sum_{j \in I_0} p_{ij} Y_j$$

with coefficients $p_{ij} \in \bar{\mathbf{Q}}_l$. In fact, it is clear that Q_i is written as a linear combination of Y_j with $j \in I$, and it is a deep result due to Lusztig that $p_{ij} = 0$ if $j \notin I_0$. We fix a total order on I_0 which is compatible with the closure relations of unipotent classes, and consider the square matrix $P = (p_{ij})$. The determination of Q_i for various $i \in I_0$ is equivalent to the determination of the matrix P. Lusztig showed that P is determined as a unique solution of the matrix equation of the form $P\Lambda\, {}^tP = \Omega$ as in (*), where Ω is a known matrix obtained from the data on $W^\wedge$. In fact, we define an equivalence relation $\sim$ on I_0, by the condition that $i \sim i'$ if $i = (C, \rho), i' = (C', \rho')$ with $C = C'$, and consider the block matrix with respect to this equivalence relation. Then it is shown that P is a block lower triangular matrix with diagonal blocks of the form $q^{d_u} I$ if the block corresponds to $C \ni u$. Here d_u is the dimension of the variety $\mathcal{B}_u = \{B_1 \in G/B \mid u \in B_1\}$. Furthermore, the matrix Λ is a non–singular block diagonal matrix. By this condition, P and Λ are uniquely determined from the equation (*).

1.3 In the case where $G = Sp_{2n}$ or SO_{2n+1}, all of the ingredients involved in the above arguments are described in a purely combinatorial way; the set I is in bijection with certain combinatorial objects, "u–symbols". Then the subset I_0 of I corresponds to the set of u–symbols of defect 1. Furthermore, the equivalence class in I with respect to $\sim$ corresponds to the similarity class of u–symbols. The assignment $u \to d_u$ is defined as a function $a : I_0 \to \mathbb{Z}_{\geq 0}$ combinatorially.

As explained in the introduction, these formalism can be extended to the case of complex reflection groups $W = G(e, 1, n)$. In the next section, we formulate the equation corresponding to (*) for such W.

2 Green functions associated to $G(e, 1, n)$

2.1 Let $W = \mathfrak{S}_n \ltimes (\mathbb{Z}/e\mathbb{Z})^n$ be the imprimitive complex reflection group $G(e, 1, n)$, and $V = \mathbb{C}^n$ the natural reflection representation of W. Let $S(V)$ be the symmetric algebra of V, and I_+ the ideal of $S(V)$ generated by the homogeneous W–invariant vectors of strictly positive degrees. We denote by $R = \oplus R_i$ the coinvariant algebra of W, which is a graded algebra defined as the the quotient of $S(V)$ by I_+. Let t be an indeterminate. The Poincaré polynomial $P_W(t)$ is defined as $P_W(t) = \sum_{i\geq 0}(\dim_{\mathbb{C}} R_i)t^i$, which is explicitly given as

$$P_W(t) = \prod_{i=1}^{n} \frac{t^{ei} - 1}{t - 1}.$$

For any class function f on W, we define $R(f)$ by

$$R(f) = (t-1)^n P_W(t) \frac{1}{|W|} \sum_{w \in W} \frac{\det_V(w) f(w)}{\det_V(t \cdot \mathrm{id}_V - w)}, \tag{2.1.1}$$

where $\det_V$ denotes the determinant on V. If χ is an irreducible character of W, then $R(\chi)$ coincides with $\sum_i \langle \chi, R_i \rangle_W t^i$, where $\langle\ ,\ \rangle_W$ denotes the inner product of class functions. Hence $R(\chi)$ coincides with the fake degree of χ, and $R(f) \in \mathbb{Z}[t]$ for a generalized character f of W.

Let N^* be the number of complex reflections in W. Then N^* is the maximal degree in R, and the W–module R_{N^*} coincides with $\overline{\det_V}$.

2.2 An e–partition $\boldsymbol{\alpha} = (\alpha^{(0)}, \dots, \alpha^{(e-1)})$ is an e–tuple of partitions $\alpha^{(k)}$. The size $|\boldsymbol{\alpha}|$ of $\boldsymbol{\alpha}$ is defined as $|\boldsymbol{\alpha}| = \sum_{i=0}^{e-1} |\alpha^{(k)}|$. We denote by $\mathcal{P}_{n,e}$ the set of e–partitions of size n. Then the set $W^\wedge$ of irreducible characters of W is in bijection with the set $\mathcal{P}_{n,e}$. We denote by $\chi^{\boldsymbol{\alpha}}$ the irreducible character corresponding to $\boldsymbol{\alpha} \in \mathcal{P}_{n,e}$. In particular, the unit character corresponds to $\boldsymbol{\alpha} = ((n), -, \dots, -)$, and $\overline{\det_V}$ corresponds to $\boldsymbol{\alpha} = (-. \dots, -, (1^n))$.

Let us fix $\mathbf{m} = (m_0, \dots, m_{e-1})$, where m_k are positive integers such that $m_k \geq n$ for any k. We denote by $Z_n^{0,0} = Z_n^{0,0}(\mathbf{m})$ the set of e–partitions $\boldsymbol{\alpha}$ with size n, where $\alpha^{(k)}$ is regarded as an element in $\mathbb{Z}^{m_k}$ written in the form, $\alpha^{(k)} : \alpha_1^{(k)} \geq \dots \geq \alpha_{m_k}^{(k)} \geq 0$. We fix integers $r \geq s \geq 0$, and consider an e–tuple of partitions $\boldsymbol{\Lambda}^0 = \boldsymbol{\Lambda}^0(\mathbf{m}) = (\Lambda_0, \dots, \Lambda_{e-1})$ defined by

$$\begin{aligned} &\Lambda_0 : (m_0 - 1)r \geq \dots \geq 2r \geq r \geq 0, && (2.2.1)\\ &\Lambda_i : s + (m_i - 1)r \geq \dots \geq s + 2r \geq s + r \geq s \qquad \text{for } 1 \leq i \leq e-1. \end{aligned}$$

We denote by $Z_n^{r,s} = Z_n^{r,s}(\mathbf{m})$ the set of e–partitions of the form $\boldsymbol{\Lambda} = \boldsymbol{\alpha} + \boldsymbol{\Lambda}^0$, where $\boldsymbol{\alpha} \in Z_n^{0,0}$ and the sum is taken entry–wisely. We denote by $\boldsymbol{\Lambda} =$

$\boldsymbol{\Lambda}(\boldsymbol{\alpha})$ if $\boldsymbol{\Lambda}$ is as above, and call it the e–symbol of type $(\boldsymbol{r}, \boldsymbol{s})$ corresponding to $\boldsymbol{\alpha}$. Let us define a shift operation $Z_n^{r,s}(\mathbf{m}) \to Z_n^{r,s}(\mathbf{m}')$ by associating $\boldsymbol{\Lambda}' = (\Lambda'_0, \dots, \Lambda'_{e-1}) \in Z_n^{r,s}(\mathbf{m}')$ to $\boldsymbol{\Lambda} = (\Lambda_0, \dots, \Lambda_{e-1}) \in Z_n^{r,s}(\mathbf{m})$, where $\mathbf{m}' = (m_0+1, \dots, m_{e-1}+1)$ with $\Lambda'_0 = (\Lambda_0 + r) \cup \{0\}$ and $\Lambda'_k = (\Lambda_k + r) \cup \{s\}$ for $k = 1, \dots, e-1$.

Two elements $\boldsymbol{\Lambda}$ and $\boldsymbol{\Lambda}'$ in $Z_n^{r,s}$ are said to be similar if all the entries of them coincide each other with multiplicities. The set of symbols which are similar to a fixed symbol is called a similarity class in $Z_n^{r,s}$.

We shall define a function $a : Z_n^{r,s} \to \mathbb{Z}_{\geq 0}$. Take $\boldsymbol{\Lambda} \in Z_n^{r,s}$ and let $\boldsymbol{\Lambda}^0$ be as before. We put

$$a(\boldsymbol{\Lambda}) = \sum_{\lambda, \lambda' \in \boldsymbol{\Lambda}} \min(\lambda, \lambda') - \sum_{\mu, \mu' \in \boldsymbol{\Lambda}^0} \min(\mu, \mu'), \qquad (2.2.2)$$

where in the first sum, we assume that $\lambda \neq \lambda'$ if λ and λ' are contained in the same Λ_i, and similarly for the second sum (this occurs only when $r = 0$). The function a on $Z_n^{r,s}$ is invariant under the shift operation, and takes a constant value on each similarity class in $Z_n^{r,s}$.

Remark 2.3 The notion of symbols already appeared in several articles in various forms. If $e = 1$, $Z_n^{r,s}$ is in bijection with the set of partitions of n. So, we consider the case where $e = 2$. Then the set $Z_n^{1,0}$ (up to shift operation, all the same below) was used in [4] to parameterize unipotent characters of $Sp_{2n}(\mathbf{F}_q)$ or $SO_{2n+1}(\mathbf{F}_q)$. In that case, the a–function defined in (2.2.2) coincides with the original a–function, i.e., the exact power of q dividing the degree of the corresponding unipotent character. The set $Z_n^{2,1}$ (resp. $Z_n^{2,0}$) was used in [5] to describe the generalized Springer correspondence for $G = Sp_{2n}(k)$ (resp. $G = SO_{2n+1}(k)$) subject to the condition that ch $k \neq 2$. In this case, the similarity classes are in 1-1 correspondence with the unipotent classes of G, and the value $a(\boldsymbol{\Lambda})$ for such a symbol $\boldsymbol{\Lambda}$ coincides with $\dim \mathcal{B}_u$, (see, for example [7, 4.4]). While, the set $Z_n^{4,2}$ was used in [10] to describe the generalized Springer correspondence for $Sp_{2n}(k)$ with ch $k = 2$. On the other hand, for arbitrary e, the set $Z_n^{1,0}$ was used in [12] to parameterize unipotent degrees associated to the complex reflection group $G(e, 1, n)$.

2.4 Under the identification of $Z_n^{0,0}$ with $\mathcal{P}_{n,e}$, we consider the a function as the function on the set $\mathcal{P}_{n,e}$. Also the similarity classes in $\mathcal{P}_{n,e}$ are defined by inheriting the classes in $Z_n^{r,s}$. We denote by $\boldsymbol{\alpha} \sim \boldsymbol{\beta}$ if $\boldsymbol{\alpha}, \boldsymbol{\beta} \in \mathcal{P}_{n,e}$ are in the same similarity class. We choose a total order $\succ$ on $\mathcal{P}_{n,e}$ such that $a(\boldsymbol{\alpha}) \leq a(\boldsymbol{\beta})$ whenever $\boldsymbol{\alpha} \succ \boldsymbol{\beta}$, and that each similarity class in $\mathcal{P}_{n,e}$ forms an interval. In the following, we consider the matrices of degree $|\mathcal{P}_{n,e}|$ with respect to this order. For each $\boldsymbol{\alpha}, \boldsymbol{\beta} \in \mathcal{P}_{n,e}$, we define a polynomial $\omega_{\boldsymbol{\alpha},\boldsymbol{\beta}} \in \mathbb{Z}[t]$

by

$$\omega_{\boldsymbol{\alpha},\boldsymbol{\beta}} = t^{N^*} R(\chi^{\boldsymbol{\alpha}} \otimes \chi^{\boldsymbol{\beta}} \otimes \overline{\det}_V), \tag{2.4.1}$$

and put $\Omega = (\omega_{\boldsymbol{\alpha},\boldsymbol{\beta}})$, the matrix of degree $|\mathcal{P}_{n,e}|$. Let $P = (p_{\boldsymbol{\alpha},\boldsymbol{\beta}})$ and $\Lambda = (\lambda_{\boldsymbol{\alpha},\boldsymbol{\beta}})$ be matrices of degree $|\mathcal{P}_{n,e}|$. We consider the following system of equations with respect to the unknown variables $\lambda_{\boldsymbol{\alpha},\boldsymbol{\beta}}, p_{\boldsymbol{\alpha},\boldsymbol{\beta}}$.

$$\begin{aligned} \lambda_{\boldsymbol{\alpha},\boldsymbol{\beta}} &= 0 \quad \text{unless } \boldsymbol{\alpha} \sim \boldsymbol{\beta}, \\ p_{\boldsymbol{\alpha},\boldsymbol{\beta}} &= 0 \quad \text{unless either } \boldsymbol{\alpha} \succ \boldsymbol{\beta} \text{ and } \boldsymbol{\alpha} \not\sim \boldsymbol{\beta}, \text{ or } \boldsymbol{\alpha} = \boldsymbol{\beta}, \\ p_{\boldsymbol{\alpha},\boldsymbol{\alpha}} &= t^{a(\boldsymbol{\alpha})}, \\ P\Lambda\, {}^t\!P &= \Omega. \end{aligned} \tag{2.4.2}$$

2.5 There exists a system of equations which is closely related to (2.4.2). Let $\Omega' = (\omega'_{\boldsymbol{\alpha},\boldsymbol{\beta}})$ be a matrix defined by

$$\omega'_{\boldsymbol{\alpha},\boldsymbol{\beta}} = t^{N^*} R(\chi^{\boldsymbol{\alpha}} \otimes \overline{\chi}^{\boldsymbol{\beta}} \otimes \overline{\det}_V), \tag{2.5.1}$$

where $\overline{\chi}^{\boldsymbol{\beta}}$ is the complex conjugate of $\chi^{\boldsymbol{\beta}}$. We consider a matrix equation of the form

$$P'\Lambda'\, {}^t\!P'' = \Omega', \tag{2.5.2}$$

where either of $P' = (p'_{\boldsymbol{\alpha},\boldsymbol{\beta}})$ and $P'' = (p''_{\boldsymbol{\alpha},\boldsymbol{\beta}})$ satisfies similar conditions as in the second and the third one in (2.4.2), and $\Lambda' = (\lambda'_{\boldsymbol{\alpha},\boldsymbol{\beta}})$ as in the first one. Let $\sigma = (\sigma_{\boldsymbol{\alpha},\boldsymbol{\beta}})$ be the permutation matrix realizing the complex conjugation of irreducible characters of W, i.e., $\sigma_{\boldsymbol{\alpha},\boldsymbol{\beta}} = 1$ if $\chi^{\boldsymbol{\beta}} = \overline{\chi}^{\boldsymbol{\alpha}}$ and $\sigma_{\boldsymbol{\alpha},\boldsymbol{\beta}} = 0$ otherwise. Then we have $\Omega' = \Omega\sigma$, and the equation $P\Lambda^t P = \Omega$ implies that

$$P \cdot \Lambda\sigma \cdot \sigma^{-1 t} P\sigma = \Omega'.$$

It is easily checked that the equation (2.4.2) has a solution if and only if the equation (2.5.2) has a solution, and in that case, we have $P' = P, \Lambda' = \Lambda\sigma, P'' = \sigma^{-1 t} P\sigma$. In the remainder of this note, we report that the solution of (2.5.2) can be described in terms of certain symmetric functions as in the case of GL_n.

3 A combinatorial setting for $G(e, 1, n)$

3.1 For a given $\mathbf{m} = (m_0, \dots, m_{e-1})$ as in 2.2, we introduce indeterminates $x_j^{(k)}$ $(0 \le k \le e-1, 1 \le j \le m_k)$. We denote by x the whole variables $(x_j^{(k)})$, and also denote by $x^{(k)}$ the variables $x_1^{(k)}, \dots, x_{m_k}^{(k)}$. Power sum symmetric functions and Schur functions are defined as in [11], Appendix B. Let ζ be a primitive e–th root of unity in $\mathbb{C}$. For each integer $r \geq 1$ and i such that $0 \leq i \leq e-1$, put

$$p_r^{(i)}(x) = \sum_{j=0}^{e-1} \zeta^{ij} p_r(x^{(j)}),$$

where $p_r(x^{(j)})$ denotes the usual r–th power sum symmetric function with respect to the variables $x^{(j)}$. We put $p_r^{(i)}(x) = 1$ for $r = 0$. For an e–partition $\boldsymbol{\alpha} = (\alpha^{(0)}, \dots, \alpha^{(e-1)})$ with $\alpha^{(k)} : \alpha_1^{(k)} \geq \cdots \geq \alpha_{m_k}^{(k)}$, we define a function $p_{\boldsymbol{\alpha}}(x)$ by

$$p_{\boldsymbol{\alpha}}(x) = \prod_{k=0}^{e-1} \prod_{j=1}^{m_k} p^{(k)}_{\alpha_j^{(k)}}(x). \tag{3.1.1}$$

(Note that our power sum symmetric function $p_{\boldsymbol{\alpha}}(x)$ is not exactly the same as the function given in [11]. Our $p_{\boldsymbol{\alpha}}(x)$ coincides with the complex conjugate of the one in [11]).

Next, we define the Schur function $s_{\boldsymbol{\alpha}}(x)$ and the monomial symmetric function $m_{\boldsymbol{\alpha}}(x)$ by

$$s_{\boldsymbol{\alpha}}(x) = \prod_{k=0}^{e-1} s_{\alpha^{(k)}}(x^{(k)}), \qquad m_{\boldsymbol{\alpha}}(x) = \prod_{k=0}^{e-1} m_{\alpha^{(k)}}(x^{(k)}), \tag{3.1.2}$$

where $s_{\alpha^{(k)}}(x^{(k)})$ (resp. $m_{\alpha^{(k)}}(x^{(k)})$) denotes the usual Schur function (resp. monomial symmetric function) associated to the partition $\alpha^{(k)}$ with respect to the variables $x^{(k)}$.

3.2. Let t be an indeterminate as before. In what follows we regard the variables $x_i^{(k)}$ defined for $k \in \mathbb{Z}/e\mathbb{Z} \simeq \{0, 1, \dots, e-1\}$. For each $0 \leq k \leq e-1$ and an integer $r \geq 0$, we define a function $q_{r,\pm}^{(k)}(x;t)$ by

$$q_{r,\pm}^{(k)}(x;t) = \sum_{i \geq 1} (x_i^{(k)})^{r+\delta} \frac{\prod_j x_i^{(k)} - t x_j^{(k\pm1)}}{\prod_{j \neq i} x_i^{(k)} - x_j^{(k)}} \qquad (r \geq 1), \tag{3.2.1}$$

where $\delta = m_k - 1 - m_{k\pm1}$, and by $q_{r,\pm}^{(k)}(x;t) = 1$ for $r = 0$. In the product of the denominator, $x_j^{(k)}$ runs over all the variables in $x^{(k)}$ except $x_i^{(k)}$, while in the numerator, $x_j^{(k\pm1)}$ runs over all the variables in $x^{(k\pm1)}$. Note that if $x_i^{(k)} = x_i^{(k\pm1)} = x_i$ and $m_k = m_{k\pm1}$, $q_{r,\pm}^{(k)}(x;t)$ coincide with the function $q_r(x;t)$ given in [11], III, 2.9. It can be shown that $q_{r,\pm}^{(k)}(x;t)$ is a polynomial in x, t with integral coefficients, homogeneous of degree r with respect to the variable x.

For an e-partition $\boldsymbol{\alpha} = (\alpha^{(0)}, \dots, \alpha^{(e-1)}) \in \mathcal{P}_{n,e}$, we define a function $q_{\boldsymbol{\alpha},\pm}(x)$ by

$$q_{\boldsymbol{\alpha},\pm}(x;t) = \prod_{k=0}^{e-1} \prod_{j=1}^{m_k} q^{(k)}_{\alpha_j^{(k)},\pm}(x;t). \tag{3.2.2}$$

Next, for a partition $\alpha^{(k)} : \alpha_1^{(k)} \geq \cdots \geq \alpha_{l_k}^{(k)} > 0$, (here $l_k = l(\alpha^{(k)})$ is the number of parts of $\alpha^{(k)}$), we define a function $z_{\alpha^{(k)}}(t)$ by

$$z_{\alpha^{(k)}}(t) = \prod_{j=1}^{l_k} (1 - \zeta^k t^{\alpha_j^{(k)}})^{-1},$$

Then we define $z_{\boldsymbol{\alpha}}(t)$ by

$$z_{\boldsymbol{\alpha}}(t) = z_{\boldsymbol{\alpha}} \prod_{k=0}^{e-1} z_{\alpha^{(k)}}(t), \tag{3.2.3}$$

where $z_{\boldsymbol{\alpha}}$ is the order of the centralizer of $w_{\boldsymbol{\alpha}}$ in W. Explicitly, $z_{\boldsymbol{\alpha}}$ is given as follows. For $\boldsymbol{\alpha} \in \mathcal{P}_{n,e}$, put $l(\boldsymbol{\alpha}) = \sum_{i=0}^{e-1} l(\alpha^{(i)})$. For a partition $\alpha = (1^{n_1}, 2^{n_2}, \dots)$, put $z_\alpha = \prod_{i \geq 1} i^{n_i} n_i!$. Then $z_{\boldsymbol{\alpha}} = e^{l(\boldsymbol{\alpha})} \prod_{k=0}^{e-1} z_{\alpha^{(k)}}$.

The importance of the quantity $z_{\boldsymbol{\alpha}}(t)$ lies in the following relations with $\Pi(f)$ in (2.1.1);

$$\det_V(t \cdot \mathrm{id}_V - w_{\boldsymbol{\alpha}}) = \prod_{k=0}^{e-1} \prod_{j=1}^{l_k} (t^{\alpha_j^{(k)}} - \zeta^k).$$

In particular, we have $z_{\boldsymbol{\alpha}}(t^{-1}) = z_{\boldsymbol{\alpha}} t^n \det_V(t \cdot \mathrm{id}_V - w_{\boldsymbol{\alpha}})^{-1}$.

We now introduce infinitely many variables $x_i^{(k)}, y_i^{(k)}$ for $i = 1, 2, \dots$ and for $0 \leq k \leq e-1$. We may regard the above functions as functions with infinitely many variables $x_1^{(k)}, x_2^{(k)}, \dots$. Under this setting, the following proposition holds.

Proposition 3.3 *Let*

$$\Omega(x, y; t) = \prod_{k=0}^{e-1} \prod_{i,j} \frac{1 - t x_i^{(k+1)} y_j^{(k)}}{1 - x_i^{(k)} y_j^{(k)}}.$$

Then we have

$$\Omega(x,y;t) = \sum_{\boldsymbol{\alpha}} q_{\boldsymbol{\alpha},+}(x;t)m_{\boldsymbol{\alpha}}(y) = \sum_{\boldsymbol{\alpha}} m_{\boldsymbol{\alpha}}(x)q_{\boldsymbol{\alpha},-}(y;t), \tag{3.3.1}$$

$$\Omega(x,y;t) = \sum_{\boldsymbol{\alpha}} z_{\boldsymbol{\alpha}}(t)^{-1}p_{\boldsymbol{\alpha}}(x)\bar{p}_{\boldsymbol{\alpha}}(y), \tag{3.3.2}$$

where $\boldsymbol{\alpha}$ runs over all the e–partitions of any size. In (3.3.2), $\bar{p}_{\boldsymbol{\alpha}}(y)$ denotes the complex conjugate of $p_{\boldsymbol{\alpha}}(y)$.

3.4 We now introduce functions $Q^{\pm}_{\boldsymbol{\Lambda}}(x;t)$ and $P^{\pm}_{\boldsymbol{\Lambda}}(x;t)$ associated to symbols $\boldsymbol{\Lambda} \in Z^{r,s}_n$, which is in analogy with $Q_\lambda(x;t)$ and $P_\lambda(x;t)$ in [11], III. Let $\mathcal{A}$ be the subring of $\mathbb{Q}(t)$ consisting of functions which have no pole at $t=0$. Then $\mathcal{A}$ is a local ring with the unique maximal ideal $t\mathcal{A}$. Hence $\mathcal{A}^* = \mathcal{A} - t\mathcal{A}$ is the set of units in $\mathcal{A}$. In what follows, we identify the set $Z^{r,s}_n$ and $Z^{0,0}_n$ by the natural map, and we fix a total order $\prec$ on $Z^{0,0}_n$ as in 1.4. Then we have the following theorem.

Theorem 3.5 *(i) For each $\boldsymbol{\Lambda} \in Z^{r,s}_n$, there exists a unique function $P^{\pm}_{\boldsymbol{\Lambda}}(x;t)$ satisfying the following properties.*

(a) $P^{\pm}_{\boldsymbol{\Lambda}}(x;t)$ can be expressed as

$$P^{\pm}_{\boldsymbol{\Lambda}}(x;t) = \sum_{\boldsymbol{\beta}\in Z^{0,0}} c_{\boldsymbol{\alpha},\boldsymbol{\beta}}(t)q_{\boldsymbol{\beta},\pm}(x;t),$$

where $c_{\boldsymbol{\alpha},\boldsymbol{\beta}}(t) \in \mathbb{Q}(t)$ and $c_{\boldsymbol{\alpha},\boldsymbol{\beta}}(t) = 0$ unless $\boldsymbol{\beta} \succ \boldsymbol{\alpha}$ or $\boldsymbol{\beta} \sim \boldsymbol{\alpha}$.

(b) $P^{\pm}_{\boldsymbol{\Lambda}}(x;t)$ can be expressed as

$$P^{\pm}_{\boldsymbol{\Lambda}}(x;t) = s_{\boldsymbol{\alpha}}(x) + \sum_{\boldsymbol{\beta}\in Z^{0,0}} u_{\boldsymbol{\alpha},\boldsymbol{\beta}}(t)s_{\boldsymbol{\beta}}(x),$$

where $u_{\boldsymbol{\alpha},\boldsymbol{\beta}}(t) \in t\mathcal{A}$, and $u_{\boldsymbol{\alpha},\boldsymbol{\beta}}(t) = 0$ unless $\boldsymbol{\beta} \prec \boldsymbol{\alpha}$ and $\boldsymbol{\beta} \not\sim \boldsymbol{\alpha}$.

(ii) For each $\boldsymbol{\Lambda} \in Z^{r,s}_n$, there exists a unique function $Q^{\pm}_{\boldsymbol{\Lambda}}(x;t)$ satisfying the following properties.

(a) $Q^{\pm}_{\boldsymbol{\Lambda}}(x;t)$ can be expressed as

$$Q^{\pm}_{\boldsymbol{\Lambda}}(x;t) = q_{\boldsymbol{\alpha},\pm}(x;t) + \sum_{\boldsymbol{\beta}\in Z^{0,0}} d_{\boldsymbol{\alpha},\boldsymbol{\beta}}(t)q_{\boldsymbol{\beta},\pm}(x;t),$$

where $d_{\boldsymbol{\alpha},\boldsymbol{\beta}}(t) \in \mathbb{Q}(t)$ and $d_{\boldsymbol{\alpha},\boldsymbol{\beta}}(t) = 0$ unless $\boldsymbol{\beta} \succ \boldsymbol{\alpha}$ and $\boldsymbol{\beta} \not\sim \boldsymbol{\alpha}$.

(b) $Q^{\pm}_{\Lambda}(x;t)$ can be expressed as

$$Q^{\pm}_{\Lambda}(x;t) = \sum_{\beta \in Z^{0,0}} w_{\alpha,\beta}(t) s_{\beta}(x),$$

where $w_{\alpha,\beta}(t) \in \mathcal{A}$, and $w_{\alpha,\beta}(t) = 0$ unless $\beta \prec \alpha$ or $\beta \sim \alpha$. Moreover, $w_{\alpha,\beta} \in t\mathcal{A}$ if $\beta \neq \alpha$, and $w_{\alpha,\alpha} \in \mathcal{A}^*$.

Remark 3.6 We call $P^{\pm}_{\Lambda}$ the Hall–Littlewood function associated to $\Lambda \in Z^{r,s}_n$. It is likely that $P^{\pm}_{\Lambda}(x;t) \in \mathbb{Z}[x;t]$. In fact, in the classical case, i.e., the case where $e = 2, r = 2, s = 1$, the assertion follows from the known properties of Green functions.

The functions $P^{\pm}_{\Lambda}, Q^{\pm}_{\Lambda}$ have the stability property for the shift operation of symbols. Hence they can be regarded as functions with infinitely many variables. As a corollary to Theorem 3.5, we have

Corollary 3.7 *Let $\Omega(x,y;t)$ be as in Proposition 3.3. Then we have*

$$\Omega(x,y;t) = \sum_{\Lambda,\Lambda'} b_{\Lambda,\Lambda'}(t) P^{+}_{\Lambda}(x;t) P^{-}_{\Lambda'}(y;t). \tag{3.7.1}$$

$$\Omega(x,y;t) = \sum_{\Lambda} Q^{+}_{\Lambda}(x;t) P^{-}_{\Lambda}(y;t) = \sum_{\Lambda} P^{+}_{\Lambda}(x;t) Q^{-}_{\Lambda}(y;t). \tag{3.7.2}$$

where in (3.7.1), Λ, Λ' run over all the elements in $\bigcup_{n=1}^{\infty} Z^{r,s}_n$, and $b_{\Lambda,\Lambda'}(t) = 0$ unless $|\Lambda| = |\Lambda'|$ and $\Lambda \sim \Lambda'$. In (3.7.2), Λ runs over all the elements in $\bigcup_{n=1}^{\infty} Z^{r,s}_n$.

3.8 Let $\mathfrak{S}_{\mathbf{m}} = \mathfrak{S}_{m_0} \times \cdots \times \mathfrak{S}_{m_{e-1}}$ be a parabolic subgroup of $\mathfrak{S}_n$ of type $\mathbf{m}$. We denote by $\Xi_{\mathbf{m}} = \bigotimes_{k=0}^{e-1} \mathbb{Z}[x^{(k)}_1, \dots, x^{(k)}_{m_k}]^{\mathfrak{S}_{m_k}}$ the ring of symmetric polynomials (with respect to $\mathfrak{S}_{\mathbf{m}}$) with variables $x = (x^{(k)}_j)$. $\Xi_{\mathbf{m}}$ has a structure of a graded ring $\Xi_{\mathbf{m}} = \bigoplus_{i \ge 0} \Xi^i_{\mathbf{m}}$, where $\Xi^i_{\mathbf{m}}$ consists of homogeneous symmetric polynomials of degree i, together with the zero polynomial. We consider the inverse limit

$$\Xi^i = \varprojlim_{\mathbf{m}} \Xi^i_{\mathbf{m}}$$

with respect to homomorphisms $\rho_{\mathbf{m}',\mathbf{m}} : \Xi^i_{\mathbf{m}'} \to \Xi^i_{\mathbf{m}}$, where $\mathbf{m}' = (m'_0, \dots, m'_{e-1})$ with $m'_k = m_k + l$ for some integer $l \ge 0$, and $\rho_{\mathbf{m}',\mathbf{m}}$ is induced from the homomorphism $\bigoplus_k \mathbb{Z}[x^{(k)}_1, \dots, x^{(k)}_{m_k}] \to \bigoplus_k \mathbb{Z}[x^{(k)}_1, \dots, x^{(k)}_{m'_k}]$ given by sending $x^{(k)}_i$ to 0 for $i \ge m_k$, and leaving other $x^{(k)}_j$ unchanged. $\Xi = \bigoplus_{i \ge 0} \Xi^i$ is called the space of symmetric functions. Schur function $s_{\alpha}(x)$

with infinitely many variables $x_1^{(k)}, \dots$ is regarded as an element in Ξ^n with $n = |\boldsymbol{\alpha}|$, and the set $\{s_{\boldsymbol{\alpha}}(x)\}$ with $\boldsymbol{\alpha} \in Z_n^{0,0}$ forms a $\mathbb{Z}$–basis of Ξ^n. Put $\Xi^n_{\mathbb{Q}}[t] = \mathbb{Q}(t) \otimes_{\mathbb{Z}} \Xi^n$. It can be shown that the set $\{P^{\pm}_{\boldsymbol{\Lambda}} \mid \boldsymbol{\Lambda} \in Z_n^{r,s}\}$ (resp. the set $\{Q^{\pm}_{\boldsymbol{\Lambda}} \mid \boldsymbol{\Lambda} \in Z_n^{r,s}\}$, $\{q_{\boldsymbol{\alpha},\pm} \mid \boldsymbol{\alpha} \in Z_n^{0,0}\}$) forms a $\mathbb{Q}(t)$–basis of the space $\Xi^n_{\mathbb{Q}}[t]$. A similar property holds if one replaces Ξ^n by $\Xi^n_{\mathbf{m}}$

It follows from the above discussion that $\{s_{\boldsymbol{\alpha}}\}$ gives rise to a basis of the $\mathbb{Z}[t]$–module $\Xi[t] = \mathbb{Z}[t] \otimes_{\mathbb{Z}} \Xi$, hence so is $\{m_{\boldsymbol{\alpha}}\}$. Also, we see that $\{q_{\boldsymbol{\alpha},\pm}\}$, $\{Q^{\pm}_{\boldsymbol{\Lambda}}\}$, and $\{P^{\pm}_{\boldsymbol{\Lambda}}\}$ turn out to be bases of $\Xi_{\mathbb{Q}}[t]$. Moreover, $\{p_{\boldsymbol{\alpha}}\}$ gives a basis of $\mathbb{C}(t)$–space $\Xi_{\mathbb{C}}[t] = \mathbb{C}(t) \otimes_{\mathbb{Z}} \Xi$. We now define a scalar product on $\Xi_{\mathbb{Q}}[t]$ by the condition that

$$\langle q_{\boldsymbol{\alpha},+}(x;t), m_{\boldsymbol{\beta}}(x)\rangle = \delta_{\boldsymbol{\alpha},\boldsymbol{\beta}}, \tag{3.8.1}$$

and extend it to a sesquilinear form on $\Xi_{\mathbb{C}}[t]$. By using a similar argument as in [11] Chap. I, 4, it follows from Corollary 3.7 and Proposition 3.3 that we have

$$\begin{aligned} \langle m_{\boldsymbol{\alpha}}(x), q_{\boldsymbol{\beta},-}(x;t)\rangle &= \delta_{\boldsymbol{\alpha},\boldsymbol{\beta}}, \\ \langle P^{+}_{\boldsymbol{\Lambda}}(x;t), Q^{-}_{\boldsymbol{\Lambda}'}(x;t)\rangle &= \langle Q^{+}_{\boldsymbol{\Lambda}}(x;t), P^{-}_{\boldsymbol{\Lambda}'}(x;t)\rangle = \delta_{\boldsymbol{\Lambda},\boldsymbol{\Lambda}'}, \\ \langle p_{\boldsymbol{\alpha}}(x), p_{\boldsymbol{\beta}}(x)\rangle &= z_{\boldsymbol{\alpha}}(t)\delta_{\boldsymbol{\alpha},\boldsymbol{\beta}}. \end{aligned} \tag{3.8.2}$$

In the special case where $e = 2$, the function $q_{\boldsymbol{\alpha},+}$ (resp. $P^{+}_{\boldsymbol{\Lambda}}$, $Q^{+}_{\boldsymbol{\Lambda}}$) coincides with $q_{\boldsymbol{\alpha},-}$ (resp. $P^{-}_{\boldsymbol{\Lambda}}$, $Q^{-}_{\boldsymbol{\Lambda}}$). We also have $p_{\boldsymbol{\alpha}} = \bar{p}_{\boldsymbol{\alpha}}$ and $z_{\boldsymbol{\alpha}}(t) \in \mathbb{Q}(t)$. Hence in this case, the scalar product $\langle\, ,\, \rangle$ turns out to be symmetric.

The following result gives a characterization of $P^{\pm}_{\boldsymbol{\Lambda}}$ and $Q^{\pm}_{\boldsymbol{\Lambda}}$ in terms of the inner product.

Proposition 3.9 *The functions $P^{\pm}_{\boldsymbol{\Lambda}}$ are characterized by the following properties. (Then the function $Q^{\pm}_{\boldsymbol{\Lambda}}$ is obtained as the dual basis of $P^{\mp}_{\boldsymbol{\Lambda}}$ in the sense of (3.8.2)).*

(i) $P^{\pm}_{\boldsymbol{\Lambda}}(x;t)$ can be expressed in terms of $s_{\boldsymbol{\beta}}(x)$ as

$$P^{\pm}_{\boldsymbol{\Lambda}} = s_{\boldsymbol{\alpha}} + \sum_{\boldsymbol{\beta}} u^{\pm}_{\boldsymbol{\alpha},\boldsymbol{\beta}} s_{\boldsymbol{\beta}},$$

where $\boldsymbol{\Lambda} = \boldsymbol{\Lambda}(\boldsymbol{\alpha})$, and $u^{\pm}_{\boldsymbol{\alpha},\boldsymbol{\beta}} = 0$ unless $\boldsymbol{\beta} \prec \boldsymbol{\alpha}$ and $\boldsymbol{\beta} \not\sim \boldsymbol{\alpha}$.

(ii) $\langle P^{+}_{\boldsymbol{\Lambda}}, P^{-}_{\boldsymbol{\Lambda}'}\rangle = 0$ unless $\boldsymbol{\Lambda} \sim \boldsymbol{\Lambda}'$.

3.10 Let $K_{\pm}(t) = M(s, P^{\pm})$ be the transition matrix between Schur functions and Hall–Littlewood functions, i.e.,

$$s_{\boldsymbol{\beta}}(x) = \sum_{\boldsymbol{\Lambda}(\boldsymbol{\alpha})} K^{\pm}_{\boldsymbol{\beta},\boldsymbol{\alpha}}(t) P^{\pm}_{\boldsymbol{\Lambda}(\boldsymbol{\alpha})}(x;t).$$

We call $K^{\pm}_{\beta,\alpha}(t)$ Kostka functions associated to $Z_n^{r,s}$, in analogy to the case of $GL_n(\mathbf{F}_q)$. By Theorem 3.5, $K_{\pm}(t)$ is a block lower triangular matrix with identity diagonal blocks, with entries $K^{\pm}_{\alpha,\beta}(t)$ in $\mathbb{Q}(t)$. We put $\widetilde{K}^{\pm}_{\alpha,\beta}(t) = t^{a(\beta)}K^{\pm}_{\alpha,\beta}(t^{-1})$ and let $\widetilde{K}_{\pm}(t) = (\widetilde{K}^{\pm}_{\alpha,\beta}(t))$. Then the matrix $\widetilde{K}_{\pm}(t)$ satisfies a similar condition as P given in (2.4.2). Next we consider the matrix $D(t) = (b_{\Lambda,\Lambda'}(t))$, indexed by symbols in $Z_n^{r,s}$, where $b_{\Lambda,\Lambda'}(t)$ is as in Corollary 3.7. We write $D(t)^{-1} = (b'_{\alpha,\beta}(t))$, and define a matrix $\widetilde{\Lambda}(t)$ by $\widetilde{\Lambda}(t) = (d(t)t^{a(\beta)-a(\alpha)}b'_{\alpha,\beta}(t^{-1}))$, where $d(t) = t^{N^*-n}(t-1)^nP_W(t)$. Clearly, $\widetilde{\Lambda}(t)$ is a block diagonal matrix. Under this setting, we have the following theorem.

Theorem 3.11 *We have*

$$\widetilde{K}_-(t)\widetilde{\Lambda}(t)\,{}^t\widetilde{K}_+(t) = \Omega'.$$

In particular $P' = \widetilde{K}_-(t), P'' = \widetilde{K}_+(t)$ *and* $\Lambda' = \widetilde{\Lambda}(t)$ *gives a solution for the equation (2.5.2).*

Theorem 3.11 shows that the solution of (2.5.2), hence that of (2.4.2) can be described in terms of Kostka functions, just as in the case of Green polynomials of $GL_n(\mathbf{F}_q)$. In particular, the matrix P coincides with $\widetilde{K}_-(t)$.

Remarks 3.12 (i) As some examples show one can conjecture the following. $K^{\pm}_{\alpha,\beta}(t)$ is a polynomial in $\mathbb{Z}[t]$ with positive coefficients. Moreover, $\deg K^{\pm}_{\alpha,\beta}(t) \le a(\beta) - a(\alpha)$. This implies that $\widetilde{K}_{\alpha,\beta}(t)$ is also in $\mathbb{Z}[t]$. The conjecture also implies that $P^{\pm}_{\Lambda}(x;t)$ is a polynomial in $\mathbb{Z}[x;t]$.

(ii) From the discussion in 3.8, we see that $\{q_{\alpha,\pm}(x;t)\}$ and $\{m_{\alpha}(x)\}$, and $\{Q^{\pm}_{\Lambda}(x;t)\}$ and $\{P^{\mp}(x;t)\}$ are dual bases of each other. Hence we have

$$M(Q^{\mp}, q_{\mp}) = M(P^{\pm}, m)^* = (K_{\pm}(t)^{-1}K)^* = {}^tK_{\pm}(t)K^*, \tag{3.12.1}$$

where $K = M(s,m)$ is the Kostka matrix, and X^* denotes the transposed inverse of the matrix X. Since K^* is a matrix with entries in $\mathbb{Z}$, we see that $K_{\pm}(t)$ is a matrix with entries in $\mathbb{Z}[t]$ if and only if $M(Q^{\mp}, q_{\mp})$ is a matrix with entries in $\mathbb{Z}[t]$. It follows that $P^{\pm}_{\Lambda}(x;t)$ lie in $\mathbb{Z}[x;t]$ if an only if the coefficients $d_{\alpha,\beta}(t)$ in Theorem 3.5, (ii), (a) lie in $\mathbb{Z}[t]$ for any $\alpha, \beta \in Z_n^{0,0}$.

3.13. Concerning the above conjecture, we have some partial result in the case where $e = 2$. Now assume that $e = 2, r \ge 1$, and that $m_0 = m+1, m_1 = m$ for some $m > 0$. Let $\Lambda = (\Lambda_0, \Lambda_1) \in Z_n^{r,s}$ be a symbol. We write $\Lambda_0 : \nu_1 > \nu_3 > \cdots > \nu_{2m+1}$ and $\Lambda_1 : \nu_2 > \nu_4 > \cdots > \nu_{2m}$. We say that Λ is very special if $\nu_i \ge \nu_{i+1} + r$ for $i = 1, 2, \ldots$. Then we have the following.

Proposition 3.14 *Assume that $e = 2$ and that $\boldsymbol{\Lambda} = \boldsymbol{\Lambda}(\boldsymbol{\alpha}) \in Z_n^{r,s}$ is very special. Then the coefficient $d_{\boldsymbol{\alpha},\boldsymbol{\beta}}(t)$ of $q_{\boldsymbol{\beta},\pm}$ in $Q_{\boldsymbol{\Lambda}}^{\pm}$ is in $\mathbb{Z}[t]$. In particular, $K_{\boldsymbol{\beta},\boldsymbol{\alpha}}^{\pm}(t) \in \mathbb{Z}[t]$ for any $\boldsymbol{\beta} \in Z_n^{r,s}$. Moreover, we have $\deg K_{\boldsymbol{\beta},\boldsymbol{\alpha}}^{\pm} \le a(\boldsymbol{\alpha}) - a(\boldsymbol{\beta})$. Hence $\widetilde{K}_{\boldsymbol{\beta},\boldsymbol{\alpha}}^{\pm}(t) \in \mathbb{Z}[t]$.*

3.15 In the remainder of this section, we shall give explicit examples of Green functions associated to various $W = G(e,1,n)$. In each case, $Z_n^{0,0}$ are written as $\boldsymbol{\alpha}_1, \dots, \boldsymbol{\alpha}_h$, and the corresponding symbols in $Z_n^{r,s}$ are written as $\boldsymbol{\Lambda}_i = \boldsymbol{\Lambda}(\boldsymbol{\alpha}_i)$. In the matrix P of Green functions, the rows and columns are arranged along the order $\boldsymbol{\alpha}_1, \boldsymbol{\alpha}_2, \dots$, and the entries in the first column denote $\boldsymbol{\alpha}_i \in Z_n^{0,0}$. Throughout the examples, we assume that $r = 2$ and $s = 1$.

(A) $W = G(2,1,2)$. In this case W is the Weyl group of type B_2, and we give the table just for the sake of reference. $Z_n^{0,0}$ is given as

$$\boldsymbol{\alpha}_1 = (-;1^2), \boldsymbol{\alpha}_2 = (1^2;-), \boldsymbol{\alpha}_3 = (1;1), \boldsymbol{\alpha}_4 = (-;2), \boldsymbol{\alpha}_5 = (2;-).$$

The corresponding symbols and similarity classes are given as

$$\{\boldsymbol{\Lambda}_1 = (420;42)\}, \quad \{\boldsymbol{\Lambda}_2 = (31;1)\},$$
$$\{\boldsymbol{\Lambda}_3 = (30;2), \boldsymbol{\Lambda}_4 = (20;3)\}, \quad \{\boldsymbol{\Lambda}_5 = (2;-)\}.$$

The matrix P of Green functions is given in Table 1.

Table 1. $G(2,1,2)$

$(-;1^2)$	q^4				
$(1^2;-)$	q^2	q^3			
$(1;1)$	q^3+q	q	q		
$(-;2)$	q^2			q	
$(2;-)$	1	1	1		1

(B) $W = G(3,1,2)$. Then $Z_n^{0,0}$ is given as

$$\boldsymbol{\alpha}_1 = (-;1^2;-), \boldsymbol{\alpha}_2 = (-;-;1^2), \boldsymbol{\alpha}_3 = (1^2;-;-), \boldsymbol{\alpha}_4 = (-;1;1),$$
$$\boldsymbol{\alpha}_5 = (1;1;-), \boldsymbol{\alpha}_6 = (1;-;1), \boldsymbol{\alpha}_7 = (-;2;-), \boldsymbol{\alpha}_8 = (-;-;2), \boldsymbol{\alpha}_9 = (2;-;-).$$

The corresponding symbols and similarity classes are given as

$$\{\boldsymbol{\Lambda}_1 = (420;42;31), \boldsymbol{\Lambda}_2 = (420;31;42)\},$$
$$\{\boldsymbol{\Lambda}_3 = (31;1;1)\}, \quad \{\boldsymbol{\Lambda}_4 = (20;2;2)\}$$
$$\{\boldsymbol{\Lambda}_5 = (30;2;1), \boldsymbol{\Lambda}_6 = (30;1;2), \boldsymbol{\Lambda}_7 = (20;3;1) \boldsymbol{\Lambda}_8 = (20;1;3)\},$$
$$\{\boldsymbol{\Lambda}_9 = (2;-;-)\}.$$

The matrix P of Green functions is given in Table 2.

Table 2. $G(3,1,2)$

$(-;1^2;-)$	q^5								
$(-;-;1^2)$		q^5							
$(1^2;-;-)$	q^3	q^4	q^3						
$(-;1;1)$	q^3			q^3					
$(1;1;-)$	q^4+q		q	q	q				
$(1;-;1)$	q^2	q^3	q^2	q^2		q			
$(-;2;-)$	q^2		q^2				q		
$(-;-;2)$		q^2	q					q	
$(2;-;-)$	1		1	1	1				1

(C) $W = G(4,1,2)$. Then $Z_n^{0,0}$ is given as

$\boldsymbol{\alpha}_1 = (-;1^2;-;-), \boldsymbol{\alpha}_2 = (-;-;1^2;-;-), \boldsymbol{\alpha}_3 = (-;-;-;1^2),$
$\boldsymbol{\alpha}_4 = (1^2;-;-;-), \boldsymbol{\alpha}_5 = (-;1;1;-), \boldsymbol{\alpha}_6 = (-;1;-;1), \boldsymbol{\alpha}_7 = (-;-;1;1),$
$\boldsymbol{\alpha}_8 = (1;1;-;-), \boldsymbol{\alpha}_9 = (1;-;1;-), \boldsymbol{\alpha}_{10} = (-;2;-;-), \boldsymbol{\alpha}_{11} = (1;-;-;1),$
$\boldsymbol{\alpha}_{12} = (-;-;2;-), \boldsymbol{\alpha}_{13} = (-;-;-;2), \boldsymbol{\alpha}_{14} = (2;-;-;-).$

The corresponding symbols and similarity classes are given as

$\{\boldsymbol{\Lambda}_1 = (420;42;31;31), \boldsymbol{\Lambda}_2 = (420;31;42;31),\ \boldsymbol{\Lambda}_3 = (420;31;31;42)\},$
$\{\boldsymbol{\Lambda}_4 = (31;1;1;1)\}, \quad \{\boldsymbol{\Lambda}_5 = (20;2;2;1), \boldsymbol{\Lambda}_6 = (20;2;1;2), \boldsymbol{\Lambda}_7 = (20;1;2;2)\},$
$\{\boldsymbol{\Lambda}_8 = (30;2;1;1), \boldsymbol{\Lambda}_9 = (30;1;2;1), \boldsymbol{\Lambda}_{10} = (20;3;1;1),$
$\boldsymbol{\Lambda}_{11} = (30;1;1;2), \boldsymbol{\Lambda}_{12} = (20;1;3;1), \boldsymbol{\Lambda}_{13} = (20;1;1;3)\}$
$\{\boldsymbol{\Lambda}_{14} = (2;-;-;-)\}.$

The matrix P of Green functions is given in Table 3.

(D) $W = G(3,1,3)$. Then $Z_n^{0,0}$ is given as

$\boldsymbol{\alpha}_1 = (-;1^3;-), \boldsymbol{\alpha}_2 = (-;-;1^3), \boldsymbol{\alpha}_3 = (1^3;-;-), \boldsymbol{\alpha}_4 = (-;1^2;1),$
$\boldsymbol{\alpha}_5 = (-;1;1^2), \boldsymbol{\alpha}_6 = (1;1^2;-), \boldsymbol{\alpha}_7 = (-;21;-), \boldsymbol{\alpha}_8 = (1;-;1^2),$
$\boldsymbol{\alpha}_9 = (-;-;21), \boldsymbol{\alpha}_{10} = (1^2;1;-), \boldsymbol{\alpha}_{11} = (1^2;-;1), \boldsymbol{\alpha}_{12} = (21;-;-),$
$\boldsymbol{\alpha}_{13} = (1;1;1), \boldsymbol{\alpha}_{14} = (-;2;1), \boldsymbol{\alpha}_{15} = (-;1;2), \boldsymbol{\alpha}_{16} = (1;2;-),$
$\boldsymbol{\alpha}_{17} = (1;-;2), \boldsymbol{\alpha}_{18} = (2;1;-), \boldsymbol{\alpha}_{19} = (2;-;1), \boldsymbol{\alpha}_{20} = (-;3;-),$
$\boldsymbol{\alpha}_{21} = (-;-;3), \boldsymbol{\alpha}_{22} = (3;-;-).$

Table 3. $G(4,1,2)$

$(-;1^2;-;-)$	q^6													
$(-;-;1^2;-)$		q^6												
$(-;-;-;1^2)$			q^6											
$(1^2;-;-;-)$	q^4			q^4										
$(-;1;1;-)$	q^3	q^5			q^3									
$(-;1;-;1)$	q^4		q^4			q^3								
$(-;-;1;1)$		q^3	q^5				q^3							
$(1;1;-;-)$	q^5+q			q	q			q						
$(1;-;1;-)$	q^2	q^4		q^2	q^2				q					
$(-;2;-;-)$	q^2				q^2					q				
$(1;-;-;1)$	q^3		q^3	q^3		q^2					q			
$(-;-;2;-)$		q^2					q^2					q		
$(-;-;-;2)$			q^2			q							q	
$(2;-;-;-)$	1			1	1			1						1

The corresponding symbols and families are given as

$\{\boldsymbol{\Lambda}_1 = (6420;642;531), \boldsymbol{\Lambda}_2 = (6420;531;642)\}$,
$\{\boldsymbol{\Lambda}_3 = (531;31;31)\}$, $\quad \{\boldsymbol{\Lambda}_4 = (420;42;41), \boldsymbol{\Lambda}_5 = (420;41;42)\}$,
$\{\boldsymbol{\Lambda}_6 = (520;42;31), \boldsymbol{\Lambda}_7 = (420;52;31), \boldsymbol{\Lambda}_8 = (520;31;42), \boldsymbol{\Lambda}_9 = (420;31;52)\}$,
$\{\boldsymbol{\Lambda}_{10} = (31;2;1), \boldsymbol{\Lambda}_{11} = (31;1;2)\}$, $\qquad \{\boldsymbol{\Lambda}_{12} = (41;1;1)\}$,
$\{\boldsymbol{\Lambda}_{13} = (30;2;2), \boldsymbol{\Lambda}_{14} = (20;3;2), \boldsymbol{\Lambda}_{15} = (20;2;3)\}$,
$\{\boldsymbol{\Lambda}_{16} = (30;3;1), \boldsymbol{\Lambda}_{17} = (30;1;3)\}$,
$\{\boldsymbol{\Lambda}_{18} = 40;2;1), \boldsymbol{\Lambda}_{19} = (40;1;2), \boldsymbol{\Lambda}_{20} = (20;4;1), \boldsymbol{\Lambda}_{21} = (20;1;4)\}$,
$\{\boldsymbol{\Lambda}_{22} = (3;-;-)\}$.

The matrix P of Green functions is given in Table 4, and 5. The columns in Table 4 correspond to $\boldsymbol{\alpha}_1, \dots, \boldsymbol{\alpha}_9$, and those in Table 5 correspond to $\boldsymbol{\alpha}_{10}, \dots, \boldsymbol{\alpha}_{22}$.

Table 4. $G(3,1,3)$, the former part ($\boldsymbol{\alpha}_1 \sim \boldsymbol{\alpha}_9$)

$(-;1^3;-)$	q^{12}								
$(-;-;1^3)$		q^{12}							
$(1^3;-;-)$	q^9		q^9						
$(-;1^2;1)$	$q^{10}+q^7$	q^{10}		q^7					
$(-;1;1^2)$	q^8	$q^{11}+q^8$			q^7				
$(1;1^2;-)$	$q^{11}+q^8+q^5$		q^5	q^5		q^5			
$(-;21;-)$	q^9+q^6			q^6			q^5		
$(1;-;1^2)$	q^7	$q^{10}+q^7$	q^7		q^6			q^5	
$(-;-;21)$		q^9+q^6			q^5				q^5
$(1^2;1;-)$	$q^{10}+q^7+q^4$		q^7+q^4	q^4		q^4			
$(1^2;-;1)$	q^8+q^5	q^8	q^8+q^5	q^5				q^3	
$(21;-;-)$	q^6+q^3		q^6+q^3	q^3		q^3			
$(1;1;1)$	$q^9+2q^6+q^3$	q^9+q^6	q^6+q^3	q^6+q^3	q^5	q^3		q^4	
$(-;2;1)$	q^7+q^4	q^7		q^4	q^6		q^3		
$(-;1;2)$	q^5	q^8+q^5		q^5	q^4				q^4
$(1;2;-)$	$q^8+q^5+q^2$		q^2	q^5+q^2		q^2	q^4		
$(1;-;2)$	q^4	q^7+q^4	q^4	q^4	q^3			q^2	q^3
$(2;1;-)$	q^7+q^4+q		q^4+q	q^4+q		q^4+q			
$(2;-;1)$	q^5+q^2	q^5	q^5+q^2	q^2	q^4	q^2		q^3	
$(-;3;-)$	q^3			q^3			q^2		
$(-;-;3)$		q^3			q^2				q^2
$(3;-;-)$	1		1	1		1			

Table 5. $G(3,1,3)$, the latter part ($\boldsymbol{\alpha}_{10} \sim \boldsymbol{\alpha}_{22}$)

$(-;1^3;-)$													
$(-;-;1^3)$													
$(1^3;-;-)$													
$(-;1^2;1)$													
$(-;1;1^2)$													
$(1;1^2;-)$													
$(-;21;-)$													
$(1;-;1^2)$													
$(-;-;21)$													
$(1^2;1;-)$	q^4												
$(1^2;-;1)$		q^4											
$(21;-;-)$	q^3		q^3										
$(1;1;1)$	q^3			q^3									
$(-;2;1)$					q^3								
$(-;1;2)$						q^3							
$(1;2;-)$	q^2			q^2			q^2						
$(1;-;2)$		q^3				q^2		q^2					
$(2;1;-)$	q		q	q			q		q				
$(2;-;1)$	q^2		q^2	q^2						q			
$(-;3;-)$					q^2						q		
$(-;-;3)$						q		q				q	
$(3;-;-)$	1		1	1			1		1				1

References

1. P. Deligne and G. Lusztig; Representations of reductive groups over finite fields, Ann. of Math. 103 (1976), 103–161.
2. J.A. Green; The characters of the finite general linear groups, Trans. Amer. Math. Soc. 80 (1955), 402–447.
3. M. Geck and G. Malle; On special pieces in the unipotent variety, Experimental Math. **8** (1999).
4. G. Lusztig; Irreducible representations of finite classical groups, Invent. Math. **43** (1977), 125–175
5. G. Lusztig; Intersection cohomology complexes on a reductive group, Invent. Math. **75** (1984), 205–272.
6. G. Lusztig; Character sheaves V, Advances in Math. **61** (1986), 103–155.
7. G. Lusztig; On the character values of finite Chevalley groups at unipotent elements, J. Algebra, **104** (1986), 146–194.
8. G. Lusztig; Green functions and character sheaves, Annals of Mathematics, 131 (1990), 355–408.
9. G. Lusztig; Notes on unipotent classes, Asian J. Math. **1** (1997), 194–207.
10. G. Lusztig and N. Spaltenstein; On the generalized Springer correspondence for classical groups, in "Advanced Studies in Pure Math." – Algebraic groups and related topics –, **6**, 1985, pp. 289–316.
11. I.G. Macdonald; Symmetric functions and Hall Polynomials, second edition. Clarendon Press. Oxford 1995.
12. G. Malle; Unipotent Grade imprimitiver komplexer Spiegelungsgruppen, J. Algebra, **177** (1995), 768–826.
13. T. Shoji; On the Green polynomials of classical groups. Inventiones Math., **74** (1983), 239 - 267.
14. T. Shoji; Character sheaves and almost characters of reductive groups, II, Adv. in Math. **111** (1995), 314 - 354.
15. T. Shoji; Green functions associated to complex reflection groups. Preprint.

SIMPLIFICATION OF THERMODYNAMIC BETHE-ANSATZ EQUATIONS

MINORU TAKAHASHI

Institute for Solid State Physics, University of Tokyo
Kashiwanoha 5-1-5, Kashiwa, Chiba, 277-8581 Japan
E-mail: mtaka@issp.u-tokyo.ac.jp

Thermodynamic Bethe ansatz equations for XXZ model at $|\Delta| \geq 1$ is simplified to an integral equation which has one unknown function. This equation is analytically continued to $|\Delta| < 1$.

1 Introduction

Thermodynamic Bethe ansatz equations for exactly solvable one-dimensional systems have many unknown functions.[1] About the XXZ model at $|\Delta| \geq 1$, Gaudin-Takahashi equation contains infinite unknown functions.[2,3] Here I can simplify this set of equations to an integral equation which contains only one unknown function. For Hamiltonian

$$\mathcal{H}(J, \Delta, h) = -J \sum_{l=1}^{N} S_l^x S_{l+1}^x + S_l^y S_{l+1}^y + \Delta(S_l^z S_{l+1}^z - \frac{1}{4}) - 2h \sum_{l=1} S_l^z,$$
$$h \geq 0, \tag{1}$$

thermodynamic Bethe ansatz equation at temperature T is

$$\begin{aligned}
&\ln \eta_1(x) = \frac{2\pi J \sinh \phi}{T\phi} \mathbf{s}(x) + \mathbf{s} * \ln(1 + \eta_2(x)), \\
&\ln \eta_j(x) = \mathbf{s} * \ln(1 + \eta_{j-1}(x))(1 + \eta_{j+1}(x)), \quad j = 2, 3, ..., \\
&\lim_{l \to \infty} \frac{\ln \eta_l}{l} = \frac{2h}{T}.
\end{aligned} \tag{2}$$

Here we put

$$\Delta = \cosh \phi, \; Q \equiv \pi/\phi, \; \mathbf{s}(x) = \frac{1}{4} \sum_{n=-\infty}^{\infty} \mathrm{sech}\Big(\frac{\pi(x - 2nQ)}{2}\Big),$$

$$\mathbf{s} * f(x) \equiv \int_{-Q}^{Q} \mathbf{s}(x - y) f(y) \mathrm{d}y. \tag{3}$$

The free energy per site is

$$f = \frac{2\pi J \sinh \phi}{\phi} \int_{-Q}^{Q} \mathbf{a}_1(x) \mathbf{s}(x) \mathrm{d}x - T \int_{-Q}^{Q} \mathbf{s}(x) \ln(1 + \eta_1(x)) \mathrm{d}x,$$

$$\mathbf{a}_1(x) \equiv \frac{\phi \sinh\phi/(2\pi)}{\cosh\phi - \cos(\phi x)}. \tag{4}$$

On the contrary new equation is

$$u(x) = 2\cosh(\frac{h}{T}) + \oint_C \frac{\phi}{2}\Big(\cot\frac{\phi}{2}[x-y-2i]\exp[-\frac{2\pi J\sinh\phi}{T\phi}\mathbf{a}_1(y+i)]$$
$$+\cot\frac{\phi}{2}[x-y+2i]\exp[-\frac{2\pi J\sinh\phi}{T\phi}\mathbf{a}_1(y-i)]\Big)\frac{1}{u(y)}\frac{dy}{2\pi i}, \tag{5}$$

and free energy is given by

$$f = -T\ln u(0). \tag{6}$$

Contour C is an arbitrary closed loop counterclockwize around 0. $2nQ, n \neq 0$ and $\pm 2i + 2nQ$ should be outside of this loop. This loop should not contain zeros of $u(y)$. It is expected that $u(y)$ has not zero in region $|\Im y| \leq 1$. This equation can be calculated even at imaginary ϕ, namely, $|\Delta| < 1$ case. This equation converges numerically at least $T/J > 0.07$. The results coincide with those of older equations. Then this equation unifies Gaudin-Takahashi equation for $\Delta \geq 1$ and Takahashi-Suzuki equation for $|\Delta| < 1$.[4]

2 Derivation

We should note that if

$$g(x) = \mathbf{s} * h(x), \tag{7}$$

it stands

$$g(x+i) + g(x-i) = h(x). \tag{8}$$

The Fourier transform of (7) is

$$\tilde{g}(\omega) = \frac{1}{e^{\omega}+e^{-\omega}}\tilde{h}(\omega), \quad \omega = \frac{\pi}{Q}n.$$

Then we have

$$(e^{\omega}+e^{-\omega})\tilde{g}(\omega) = \tilde{h}(\omega),$$

and obtain (8).

Then set of equations (2) is rewritten as

$$\eta_1(x+i)\eta_1(x-i) = \exp\Big[\frac{2\pi J\sinh\phi}{T\phi}\sum_n \delta(x-2nQ)\Big](1+\eta_2(x)),$$
$$\eta_j(x+i)\eta_j(x-i) = (1+\eta_{j-1}(x))(1+\eta_{j+1}(x)), \quad j = 2,3,...,$$
$$\lim_{l\to\infty}\frac{\ln\eta_l}{l} = \frac{2h}{T}. \tag{9}$$

At $J = 0$ this set of equation becomes a difference equation and we have analytical solution,

$$\eta_j = \left(\frac{\sinh(j+1)h/T}{\sinh h/T}\right)^2 - 1.$$

We can expand perturbatinally as power series of J/T. Very surprisingly $\eta_j(x)$ has singularity only at $x = \pm ji$, $\pm(j+2)i + 2nQ$. So we assume that $\eta_j(x)$ is univalent on the complex plane of x and $1 + \eta_j(x)$ is factorized as follows:

$$1 + \eta_j(x) = A_j(x - ji)\overline{A_j}(x + ji)B_j(x - (j+2)i)\overline{B_j}(x + (j+2)i). \quad (10)$$

Here functions $A_j(x)$, $B_j(x)$ are periodic with periodicity $2Q$ and have singularity only at $x = 2nQ$ and

$$\overline{A_j}(x) \equiv \overline{A_j(\overline{x})}, \quad \overline{B_j}(x) \equiv \overline{B_j(\overline{x})}.$$

One should note that $\overline{A_j(x)}$ is not an analytic function of x but $\overline{A_j}(x)$ is analytic. Then second equation of (9) becomes

$$\begin{aligned}
&\eta_j(x+i)\eta_j(x-i) = \\
&A_{j-1}(x-(j-1)i)\overline{A_{j-1}}(x+(j-1)i)B_{j-1}(x-(j+1)i)\overline{B_{j-1}}(x+(j+1)i) \\
&\times A_{j+1}(x-(j+1)i)\overline{A_{j+1}}(x+(j+1)i)B_{j+1}(x-(j+3)i)\overline{B_{j+1}}(x+(j+3)i).
\end{aligned} \quad (11)$$

If we assume that $\eta_j(x)$ is factorized as

$$\eta_j(x) = X(x - ji)Y(x + ji)Z(x - (j+2)i)W(x + (j+2)i),$$

$\eta_j(x+i)\eta_j(x-i)$ is

$$\begin{aligned}
&X(x-(j-1)i)Y(x+(j-1)i)X(x-(j+1)i)Y(x+(j+1)i) \\
&Z(x-(j+1)i)W(x+(j+1)i)Z(x-(j+3)i)W(x+(j+3)i).
\end{aligned}$$

Comparing the singularity at $x = \pm(j-1)i$ and $x = \pm(j+3)i$ we have $X \sim A_{j-1}$, $Y \sim \overline{A_{j-1}}$, $Z \sim B_{j+1}$ and $W \sim \overline{B_{j+1}}$. Then $\eta_j(x)$ should be factorized as:

$$\eta_j(x) = A_{j-1}(x - ji)\overline{A_{j-1}}(x + ji)B_{j+1}(x - (j+2)i)\overline{B_{j+1}}(x + (j+2)i). \quad (12)$$

Considering the singularity at $x = \pm(j+1)i$ we have

$$\frac{A_{j-1}(x)}{B_{j-1}(x)} = \frac{A_{j+1}(x)}{B_{j+1}(x)}. \quad (13)$$

Using the first equation of (9) and noting that

$$\delta(x) = \frac{1}{2\pi}\left\{\frac{-i}{x - i\epsilon} + \frac{i}{x + i\epsilon}\right\},$$

we have

$$A_0(x) = \exp(\frac{J\sinh\phi}{T\phi}\sum_n \frac{-i}{x-2nQ-i\epsilon}) = \exp\Big(\frac{J\sinh\phi}{2Ti}\cot\frac{\phi}{2}(x-i\epsilon)\Big),$$
$$B_0(x) = 1. \tag{14}$$

From (13) we have

$$\frac{A_{2j}(x)}{B_{2j}(x)} = \frac{A_0(x)}{B_0(x)} = A_0(x), \quad \frac{A_{2j+1}(x)}{B_{2j+1}(x)} = \frac{A_1(x)}{B_1(x)} = A'_0(x). \tag{15}$$

One can show that $A_0(x) = A'_0(x)$. Consider $\eta_{2j+1}^{-1}(x+(2j+1)i)$. This quantity approaches 0 in the limit of infinite j. Of course near the singularity at $2i, 0, -(4j+2)i$ and $-(4j+4)i$ the deviation from 0 becomes big. Nevertheless the region on the complex plane where $|\eta_{2j+1}^{-1}(x+(2j+1)i)| > \epsilon$ is expected to be narrower as j goes to infinity. So we have

$$1 = \lim_{j\to\infty}\frac{1+\eta_{2j+1}(x+(2j+1)i)}{\eta_{2j+1}(x+(2j+1)i)} = \lim_{j\to\infty}\frac{A'_0(x)}{A_0(x)}\frac{\overline{A'_0}(x+(4j+2)i)}{\overline{A_0}(x+(4j+2)i)}$$
$$\times\Big[\frac{B_{2j+1}(x)\overline{B_{2j+1}}(x+(4j+2)i)}{B_{2j}(x)\overline{B_{2j}}(x+(4j+2)i)}\cdot\frac{B_{2j+1}(x-2i)\overline{B_{2j+1}}(x+(4j+4)i)}{B_{2j+2}(x-2i)\overline{B_{2j+2}}(x+(4j+4)i)}\Big]. \tag{16}$$

The first fraction in the big bracket should go to $e^{h/T}$ and the second should go to $e^{-h/T}$. The bracket in (16) becomes 1 in the limit of infinite j. Then we have

$$\frac{A'_0(x)}{A_0(x)} = \frac{\overline{A_0}(i\infty)}{\overline{A'_0}(i\infty)} = \alpha.$$

As $\overline{A_0}(i\infty)/\overline{A'_0}(i\infty) = \overline{A_0(-i\infty)/A'_0(-i\infty)} = 1/\overline{\alpha}$, we have $|\alpha| = 1$. If we choose the phase factor α is 1, the ratio of $A_j(x)$ and $B_j(x)$ is always $A_0(x)$. From (10) and (12) we have

$$A_{j-1}(x-ji)\overline{A_{j-1}}(x+ji)B_{j+1}(x-(j+2)i)\overline{B_{j+1}}(x+(j+2)i)+1$$
$$= A_j(x-ji)\overline{A_j}(x+ji)B_j(x-(j+2)i)\overline{B_j}(x+(j+2)i). \tag{17}$$

At $j=1$ we have

$$B_1(x-i)\overline{B_1}(x+i)B_1(x-3i)\overline{B_1}(x+3i)$$
$$= \frac{1}{A_0(x-i)\overline{A_0}(x+i)} + B_2(x-3i)\overline{B_2}(x+3i). \tag{18}$$

B_1, B_2 are unknown functions. But this equation and condition

$$\lim_{x\to i\infty} B_1(x-i)\overline{B_1}(x+i)B_1(x-3i)\overline{B_1}(x+3i) = (2\cosh h/T)^2, \tag{19}$$

are sufficient to determine these functions. Put

$$u(x) = B_1(x-2i)\overline{B_1}(x+2i). \tag{20}$$

Equation (18) is written as

$$u(x+i) = \frac{1}{A_0(x-i)\overline{A_0}(x+i)u(x-i)} + \frac{B_2(x-3i)\overline{B_2}(x+3i)}{u(x-i)}. \tag{21}$$

The l.h.s. has singularity at $i, -3i$. The first term of r.h.s. has at $i, -i, 3i$. The second term of r.h.s has at $3i, -3i, -i$. Assume that $u(x)$ is expanded as follows

$$u(x) = 2\cosh(\frac{h}{T}) + \sum_{j=1}^{\infty}\sum_{n}\frac{c_j}{(x-2nQ-2i)^j} + \sum_{j=1}^{\infty}\sum_{n}\frac{\overline{c_j}}{(x-2nQ+2i)^j}. \tag{22}$$

Consider the contour integral around $x = i$. Coefficients c_j is determined by

$$c_j = \oint \frac{(x-i)^{j-1}}{A_0(x-i)\overline{A_0}(x+i)u(x-i)}\frac{\mathrm{d}x}{2\pi i} = \oint \frac{y^{j-1}}{A_0(y)\overline{A_0}(y+2i)u(y)}\frac{\mathrm{d}y}{2\pi i}. \tag{23}$$

The first sum of r.h.s. of (22) is

$$\begin{aligned}
&\sum_{j=1}^{\infty}\oint\sum_{n}\frac{\exp[-\frac{2\pi J\sinh\phi}{T\phi}\mathbf{a}_1(y+i)]}{(x-2nQ-2i)^j}\frac{y^{j-1}}{u(y)}\frac{\mathrm{d}y}{2\pi i} \\
&= \oint\sum_{n}\frac{\exp[-\frac{2\pi J\sinh\phi}{T\phi}\mathbf{a}_1(y+i)]}{x-y-2nQ-2i}\frac{1}{u(y)}\frac{\mathrm{d}y}{2\pi i} \\
&= \oint\frac{\phi}{2}\cot\frac{\phi}{2}(x-y-2i)\exp[-\frac{2\pi J\sinh\phi}{T\phi}\mathbf{a}_1(y+i)]\frac{1}{u(y)}\frac{\mathrm{d}y}{2\pi i}.
\end{aligned} \tag{24}$$

The second sum is calculated in similar way. Thus we get (5). From equation (10) we have

$$1 + \eta_1(x) = \exp\Big(\frac{2\pi J\sinh\phi}{T\phi}\mathbf{a}_1(x)\Big)u(x+i)u(x-i). \tag{25}$$

Substituting this into eq.(4) we get eq.(6)

$$f = -T\int_{-Q}^{Q}\mathbf{s}(x)[\ln u(x+i) + \ln u(x-i)]\mathrm{d}x = -T\ln u(0). \tag{26}$$

3 Analytical solutions

Ising limit
In this limit we put $\Delta \to \infty, J = J_z/\Delta$. As ϕ goes to ∞, $2Q$ becomes 0. Then function $\frac{2\pi J \sinh\phi}{T\phi}\mathbf{a}_1(y)$ is 0 at $|\Im y| > 1$ and J_z/T at $|\Im y| < 1$. Function $u(y)$ is $2\cosh h/T$ at $|\Im y| > 2$ and also a constant $u(0)$ at $|\Im y| < 2$. From eq.(5) we have

$$u(0) = 2\cosh(h/T) + \frac{1 - \exp(-J_z/T)}{u(0)}. \tag{27}$$

So we have $u(0) = \cosh(h/T) + \sqrt{\sinh^2(h/T) + \exp(-J_z/T)}$ and known free energy of the Ising model.

Kuniba, Sakai and Suzuki [5] showed that Takahashi-Suzuki equation for $|\Delta| < 1$, $h = 0$ can be derived from the quantum transfer matrix and its fusion hierarchy matrices. We are confident to derive eqs.(5, 6) from the quantum transfer matrix and its fusion hierarchies. Details will be published elsewhere.

Acknowledgments

This research was supported in part by Grants-in-Aid for the Scientific Research (B) No. 11440103 from the Ministry of Education, Science and Culture, Japan.

References

1. Takahashi, M. (1999) Thermodynamics of One-Dimensional Solvable Models, Cambridge University Press.
2. Takahashi, M. (1971) *Prog. Theor. Phys.* **46**, 401.
3. Gaudin, M. (1971) *Phys. Rev. Lett.* **26**, 1301.
4. Takahashi, M. and Suzuki, M. (1972) *Prog. Theor. Phys.* **46**, 2187.
5. Kuniba, A. , Sakai, K. and Suzuki, J. (1998) *Nucl. Phys.* **B 525**, 597.

A BIRATIONAL REPRESENTATION OF WEYL GROUP, COMBINATORIAL R-MATRIX AND DISCRETE TODA EQUATION

YASUHIKO YAMADA
Department of Mathematics, Kobe University
Rokko, Kobe 657-8501, Japan
E-mail: yamaday@math.kobe-u.ac.jp

We study certain birational representation of symmetric group arising from inverse ultra discretization of the combinatorial R-matrix. This representation is related with the Weyl group action on the crystal bases and Lusztig's bijection on the canonical bases. We apply the result to the discrete Toda equation.

1 Introduction

Recently, a relation between the box and ball systems [27,28,30] and the theory of crystal bases [14,16] was discovered. [11,9,4] This remarkable relation has been further studied and extended in the papers. [10,7,6,8] In these studies, the combinatorial R-matrix of the quantum algebra [16,14] plays the fundamental role. An explicit piecewise linear formula of the combinatorial R-matrix was obtained for the case of symmetric tensors of $U_q(\widehat{sl_n})$ by Hatayama *et al.* [6]

In this paper, we study the inverse ultra discretization [29] of the formula [6] and give a direct proof of their braid relations. As a result, we obtain a birational representation of the symmetric group. We construct conserved quantities (invariants) of the birational mappings which play essential role in our proof.

In Sec. 2, we present the main results and their proofs. Then we discuss the relation of our representation with

- the combinatorial R-matrix, [24,6] and Weyl group action, [17]
- Lusztig's bijection on parametrizations of the canonical bases, [20,21,22]
- (ultra) discrete Toda equations, [12,13]

in Sec. 3, 4 and 5 respectively. In the Appendix we give some determinant formulas used in the main text.

2 A birational map R

Let N, L be positive integers and $K = \mathbf{C}(x)$ be the field of rational functions on NL variables x_i^a $(i = 1, \cdots, N,\ a = 1, \cdots, L)$. We extend the index i of x_i^a for $i \in \mathbf{Z}$ by the condition $x_{i+N}^a = x_i^a$ and define a cyclic shift operator $\pi : K \to K$ by $\pi(x_i^a) = x_{i+1}^a$.

Definition 2.1 *Define algebra automorphisms $R_a : K \to K$ $(a = 1, \cdots, L-1)$ as follows:*

$$\begin{aligned} R_a(x_i^a) &= x_i^{a+1} \frac{Q_{i-1}(x^a, x^{a+1})}{Q_i(x^a, x^{a+1})}, \\ R_a(x_i^{a+1}) &= x_i^a \frac{Q_i(x^a, x^{a+1})}{Q_{i-1}(x^a, x^{a+1})}, \\ R_a(x_i^b) &= x_i^b, \quad (b \neq a, a+1), \end{aligned} \tag{1}$$

where $Q_j(x, y) = Q_j(x_1, \cdots, x_N, y_1, \cdots, y_N)$ is given by

$$Q_j(x, y) = \sum_{k=1}^{N} \left(\prod_{i=1}^{k-1} x_{i+j} \prod_{i=k+1}^{N} y_{i+j} \right). \tag{2}$$

We also define operators L_i $(i \in \mathbf{Z})$ by

$$L_i = (\pi + x_i^1)(\pi + x_i^2) \cdots (\pi + x_i^L), \tag{3}$$

where x_i^a are interpreted as multiplication operators.

The following theorem is our main result of this paper.

Theorem 2.2 *We have the following relations:*

$$R_a \pi = \pi R_a, \quad (a = 1, \cdots, L-1), \tag{4}$$

$$R_a(L_i) = L_i, \quad (a = 1, \cdots, L-1, i = 1, \cdots, N), \tag{5}$$

$$R_a^2 = 1, \quad (a = 1, \cdots, L-1), \tag{6}$$

$$R_a R_{a+1} R_a = R_{a+1} R_a R_{a+1}, \quad (a = 1, \cdots, L-2). \tag{7}$$

In words, the automorphisms R_a $(a = 1, \cdots, L-1)$ generate a birational representation of the symmetric group S_L on K. The operator L_i, or its coefficients $I_{k,i}$, defined by

$$L_i = \sum_{k=0}^{L} I_{k,i} \pi^{L-k}, \tag{8}$$

are invariants of the S_L action. The relation between R_a and L_i is an analog of that between screening operators and W-currents. [5] Explicit form of $I_{k,i}$ is given by

$$\begin{aligned} I_{1,i} &= \sum_{1\le a\le L} x^a_{i+(a-1)}, \\ I_{2,i} &= \sum_{1\le a<b\le L} \left(x^a_{i+(a-1)} x^b_{i+(a-1)+(b-2)} \right), \\ I_{3,i} &= \sum_{1\le a<b<c\le L} \left(x^a_{i+(a-1)} x^b_{i+(a-1)+(b-2)} x^c_{i+(a-1)+(b-2)+(c-3)} \right), \\ &\cdots \\ I_{L,i} &= \prod_{a=1}^{L} x^a_i, \end{aligned} \tag{9}$$

[*Proof of the Theorem 2.2*].

The relation (4) is obvious from the explicit formula.

Proof of Eq. (5). Since R_a acts non-trivially only on the two factors $(\pi + x^a_i)(\pi + x^{a+1}_i)$ in L_i, it is enough to show in case of $L = 2$. So we prove that

$$x'_i y'_i = x_i y_i, \quad x'_i + y'_{i+1} = x_i + y_{i+1}, \tag{10}$$

where

$$x'_i = y_i \frac{Q_{i-1}(x,y)}{Q_i(x,y)}, \quad y'_i = x_i \frac{Q_i(x,y)}{Q_{i-1}(x,y)}. \tag{11}$$

The first equation is trivial. The second one is due to the equation

$$x'_i = x_i - \frac{\delta}{Q_i(x,y)}, \quad y'_i = y_i + \frac{\delta}{Q_{i-1}(x,y)}, \tag{12}$$

which follows from

$$x_i Q_i(x,y) - y_i Q_{i-1}(x,y) = \prod_{i=1}^{N} x_i - \prod_{i=1}^{N} y_i \equiv \delta. \tag{13}$$

Proof of Eq. (6). It is also enough to show $RR = 1$ in case of $L = 2$ as above. Using the Lemma 5.3 in Appendix, we see that the polynomial $Q_j(x,y)$ can be expressed as a determinant of the following $(N-1) \times (N-1)$ matrix,

$$\begin{pmatrix} x_{j+1} + y_{j+2} & 1 & & \\ x_{j+2}y_{j+2} & x_{j+2} + y_{j+3} & 1 & \\ & \ddots & \ddots & 1 \\ & & x_{j+N-1}y_{j+N-1} & x_{j+N-1} + y_{j+N} \end{pmatrix}. \tag{14}$$

Then **we have** $Q_j(x', y') = Q_j(x, y)$ and hence

$$\begin{aligned} RR(x_i) &= R(x_i') = R\left(y_i \frac{Q_{i-1}(x,y)}{Q_i(x,y)}\right) \\ &= y_i' \frac{Q_{i-1}(x',y')}{Q_i(x',y')} = x_i \frac{Q_i(x,y)}{Q_{i-1}(x,y)} \frac{Q_{i-1}(x',y')}{Q_i(x',y')} = x_i, \end{aligned} \tag{15}$$

$RR(y_i) = y_i$ is similar.

Proof of Eq. (7). It is enough to show

$$R_1R_2R_1 = R_2R_1R_2 \tag{16}$$

on $3N$ variables (x_i, y_i, z_i). Consider

$$R_2R_1(x_j) = R_2\left(y_j \frac{Q_{j-1}(x,y)}{Q_j(x,y)}\right) = y_j' \frac{Q_{j-1}(x,y')}{Q_j(x,y')} = z_j \frac{Q_{j-1}(y,z)}{Q_j(y,z)} \frac{Q_{j-1}(x,y')}{Q_j(x,y')}, \tag{17}$$

where

$$y_j' = R_2(y_j) = z_j \frac{Q_{j-1}(y,z)}{Q_j(y,z)}. \tag{18}$$

We have

$$\begin{aligned} &Q_{1,j}(x,y,z) \equiv Q_j(y,z)Q_j(x,y') \\ &= Q_j(y,z) \sum_{k=1}^{N} \left(\prod_{i=1}^{k-1} x_{i+j} \prod_{i=k+1}^{N} y_{i+j}' \right) \\ &= Q_j(y,z) \sum_{k=1}^{N} \left(\prod_{i=1}^{k-1} x_{i+j} \prod_{i=k+1}^{N} \left(z_{i+j} \frac{Q_{i+j-1}(y,z)}{Q_{i+j}(y,z)} \right) \right) \\ &= \sum_{k=1}^{N} \left(\prod_{i=1}^{k-1} x_{i+j} \prod_{i=k+1}^{N} z_{i+j} \right) Q_{k+j}(y,z) \\ &= \sum_{k=1}^{N} \left(\prod_{i=1}^{k-1} x_{i+j} \prod_{i=k+1}^{N} z_{i+j} \right) \sum_{l=1}^{N} \left(\prod_{i=1}^{l-1} y_{i+j+k} \prod_{i=l+1}^{N} z_{i+j+k} \right) \\ &= \sum_{k,l=1}^{N} \left(\prod_{i=1}^{k-1} x_{i+j} \prod_{i=k+1}^{k+l-1} y_{i+j} \prod_{i=k+N+1}^{2N} z_{i+j} \prod_{i=k+l+1}^{k+N} z_{i+j} \right) \\ &= \sum_{k,l=1}^{N} \left(\prod_{i=1}^{k-1} x_{i+j} \prod_{i=k+1}^{k+l-1} y_{i+j} \right) \prod_{i=k+l+1}^{2N} z_{i+j}. \end{aligned} \tag{19}$$

Again, by Lemma 5.3 in Appendix, the quantity

$$\sum_{\substack{1\le k,l\le N\\ k+l=m}} \left(\prod_{i=1}^{k-1} x_{i+j} \prod_{i=k+1}^{m-1} y_{i+j} \right), \tag{20}$$

can be expressed as an $(N-1)\times(N-1)$ minor determinant of the following $(N-1)\times(N+1)$ matrix:

$$\begin{pmatrix} x_{j+1}y_{j+1} & x_{j+1}+y_{j+2} & 1 & & \\ & x_{j+2}y_{j+2} & x_{j+2}+y_{j+3} & 1 & \\ & & \ddots & \ddots & \\ & & x_{j+N-1}y_{j+N-1} & x_{j+N-1}+y_{j+N} & 1 \end{pmatrix}, \tag{21}$$

where two columns, 1-st and m-th (for $2 \le m \le N+1$) or $m-N$-th and $N+1$-st (for $N+1 \le m \le 2N$), are deleted. So we see that the expression

$$R_2R_1(x_j) = z_j \frac{Q_{1,j-1}(x,y,z)}{Q_{1,j}(x,y,z)}, \tag{22}$$

is invariant under the action of R_1. From this and obvious fact $R_2(x_j) = x_j$ we have

$$R_2R_1R_2(x_j) = R_2R_1(x_j) = R_1R_2R_1(x_j). \tag{23}$$

In a similar way, we have

$$R_1R_2R_1(z_j) = x_j \frac{Q_{2,j}(x,y,z)}{Q_{2,j-1}(x,y,z)} = R_2R_1R_2(z_j), \tag{24}$$

where

$$Q_{2,j}(x,y,z) = \sum_{k,l=1}^{N} \prod_{i=1}^{k+l-2} x_{i+j} \left(\prod_{i=k+l}^{k+N-1} y_{i+j} \prod_{i=k+N+1}^{2N} z_{i+j} \right). \tag{25}$$

Finally, $R_1R_2R_1(y_j) = R_2R_1R_2(y_j)$ follows from

$$R_1R_2R_1(x_jy_jz_j) = x_jy_jz_j = R_2R_1R_2(x_jy_jz_j). \tag{26}$$

The proof of the Theorem 2.2 is completed. ∎

Remark 2.3 *We have*

$$R_1R_2R_1(x_i) = z_i \frac{Q_{1,i-1}}{Q_{1,i}},$$
$$R_1R_2R_1(y_i) = y_i \frac{Q_{1,i}Q_{2,i-1}}{Q_{1,i-1}Q_{2,i}},$$

$$R_1R_2R_1(z_i) = x_i\frac{Q_{2,i}}{Q_{2,i-1}}. \tag{27}$$

It is interesting to note the similarity with the dressed vacuum form of analytic Bethe ansatz. [5]

The relation (10) can be written in matrix form as

$$X(x)X(y) = X(x')X(y'), \tag{28}$$

where

$$X(x) = \begin{pmatrix} x_1 & 1 & & & \\ & x_2 & 1 & & \\ & & \ddots & \ddots & \\ & & & x_{N-1} & 1 \\ z & & & & x_N \end{pmatrix}. \tag{29}$$

By similar computation as in the Proof of Eq. (7), we have the following formula

$$R_pR_{p-1}\cdots R_1(x_j^1) = x_j^{p+1}\frac{P_{j-1}(x^1,\cdots,x^{p+1})}{P_j(x^1,\cdots,x^{p+1})}, \tag{30}$$

where

$$P_j(x^1,\cdots,x^{p+1}) = \sum_{m=p}^{pN} c_{j,m}(x^1,\cdots,x^p)\prod_{i=m+1}^{pN} x_{i+j}^{p+1}. \tag{31}$$

The coefficient $c_{j,m}(x^1,\cdots,x^p)$ is given explicitly as

$$\sum_{\substack{1\le k_1,\cdots,k_p\le N\\ k_1+\cdots+k_p=m}}\prod_{a=1}^{p}\left(\prod_{i_a=k_1+\cdots+k_{a-1}+1}^{k_1+\cdots+k_a-1} x_{i_a+j}^a\right). \tag{32}$$

The coefficient $c_{j,m}(x^1,\cdots,x^p)$ has a simple expression in terms of minor determinants of the matrix $X(x^1)\cdots X(x^p)$. For $N\times N$ matrix X, let $C_b^a(X)$ $(1\le a,b\le N)$ be the following minor determinant:

$$C_b^a(X) = \det\left(X_{i,j}\right)_{\substack{1\le i(\ne a)\le N\\ 1\le j(\ne b)\le N}}. \tag{33}$$

Note that

$$C_b^a(X(x)) = \prod_{k=a+1}^{b+N-1} x_k, \tag{34}$$

where $x_{k+N} = x_k$, and a factor $(x_1 \cdots x_N)$ should be replaced with $(-1)^N z$ when $b > a$. Then we have

$$C_b^a \left(X(x^1) \cdots X(x^p) \right) = \sum_{k=0}^{p} (-1)^{Nk} z^k c_{a,(p-k)N+b-a}(x^1, \cdots, x^p). \quad (35)$$

Remark 2.4 *The representation of symmetric group S_L can be extended to the (extended) affine Weyl group of type $A_{L-1}^{(1)}$ by including the automorphism $p : K \to K$ such that*

$$p(x_i^a) = x_i^{a+1}, (a = 1, \cdots, L-1), \quad p(x_i^L) = x_i^1. \quad (36)$$

Then the translations such as $T_1 = pR_{L-1} \cdots R_1$ etc. define a system of commuting rational maps that can be viewed as a version of generalized discrete Painlevé equation. [25] The conserved quantities of this system are generated by the characteristic polynomials of the matrix

$$X(x^1)X(x^2) \cdots X(x^L). \quad (37)$$

It is remarkable that this system can be ultra discretized.

3 Combinatorial R-matrix

We have obtained a representation of symmetric group S_L as birational mapping R_a on LN variables x_i^a ($a = 1, \cdots, L$, $i = 1, \cdots, N$). Note that all the coefficients of the rational functions R_a and invariants $I_{k,i}$ are nonnegative integers. This property is natural since this representation is expected to be an inverse ultra discretization of its combinatorial counterpart. [6] Here, we consider the ultra discrete analog of our representation obtained by the standard procedure, [29]

$$x + y \to \min(x, y), \quad xy \to x + y. \quad (38)$$

For objects in the previous section such as x_i^a, R_a, Q_i, we denote their ultra discretized analogs by the same symbols. For instance, in case of $L = 2$ we have

$$R(x_i) = y_i + Q_{i-1} - Q_i, \quad R(y_i) = x_i + Q_i - Q_{i-1}, \quad (39)$$

where

$$Q_j = \min_{1 \le k \le N} \left(\sum_{i=1}^{k-1} x_{i+j} + \sum_{i=k+1}^{N} y_{i+j} \right). \quad (40)$$

Then, the ultra discretized maps R_a $(a = 1, \cdots, L-1)$ define bijections on the set

$$X = \{x_i^a \in \mathbf{Z}_{\geq 0} \mid a = 1, \cdots, L,\ i = 1, \cdots, N\}, \tag{41}$$

and satisfy the relation

$$\begin{aligned} &R_a^2 = 1, \quad (a = 1, \cdots, L) \\ &R_a R_{a+1} R_a = R_{a+1} R_a R_{a+1}, \quad (a = 1, \cdots, L-1). \end{aligned} \tag{42}$$

One can identify the set X with the crystal bases of L-fold tensor product of complete symmetric representations of $U_q(\widehat{sl_N})$

$$B = B_{l_1} \otimes \cdots B_{l_L}, \tag{43}$$

where x_i^a is interpreted as the i-th weight of a-th element $b_a \in B_{l_a}$. $(l_a = x_1^a + \cdots + x_N^a)$. On the bases B, we have a bijection $\tilde{R}_a$ that acts on the $B_a \otimes B_{a+1}$ as the isomorphism of crystals. [24]

Proposition 3.1 *Under the identification of X and B, we have*

$$R_a = \tilde{R}_a, \quad (a = 1, \cdots, L-1). \tag{44}$$

This fact follows by direct comparison of the expression Eq. (39) for R and the expression for $\tilde{R}$ in (the proof of) Proposition 4.1 of the Ref. [6]

As a ultra discrete analog of Eq. (9), we obtain the following conserved quantities for the action of R_a

$$\begin{aligned} I_{1,i} &= \min_{1 \leq a \leq L} x_{i+(a-1)}^a, \\ I_{2,i} &= \min_{1 \leq a < b \leq L} \left(x_{i+(a-1)}^a + x_{i+(a-1)+(b-2)}^b \right), \\ I_{3,i} &= \min_{1 \leq a < b < c \leq L} \left(x_{i+(a-1)}^a + x_{i+(a-1)+(b-2)}^b + x_{i+(a-1)+(b-2)+(c-3)}^c \right), \\ &\cdots \\ I_{L,i} &= \sum_{a=1}^{L} x_i^a. \end{aligned} \tag{45}$$

The last one is nothing but the total weight.

There is a remarkable relation between L and N. By replacing the role of the indices i and a in the formula for R_a in Eq. (39), we define bijections s_i $(i = 1, \cdots, N-1)$.

Proposition 3.2 *The bijection s_i is nothing but the Weyl group action on the crystal bases*

$$B = B_{l_1} \otimes \cdots B_{l_L}. \tag{46}$$

This fact[a] follows from the coincidence of the signature rule for the Weyl group action [17] and the rule for the combinatorial R. [24]

As a consequence of this Proposition, we see that the actions R_a and s_i are mutually commute. This commutativity seems also hold in the inverse ultra discretized version, namely

Conjecture 3.3 *Define birational automorphism s_i by the same formula for R_a in Eq. (1) replacing the indices a's and i's. Then the birational actions R_a and s_i mutually commute.*

Remark 3.4 *A similar combinatorial representation of symmetric group has been also obtained by A. N. Kirillov.* [18] *His construction is based on Bender-Knuth transformation and Schützenberger involution.* [2,3,19]

4 Lusztig's bijection $R_{\mathbf{i}}^{\mathbf{i}'}$

Let $\mathcal{X}$ be the set of all sequences $\mathbf{i} = (i_1, \cdots, i_\nu)$ such that $s_{i_1} s_{i_2} \cdots s_{i_\nu}$ is a reduced element[b] in a symmetric group S_n. For $\mathbf{i}, \mathbf{i}' \in \mathcal{X}$, Lusztig defined a bijection $R_{\mathbf{i}}^{\mathbf{i}'} : \mathbf{N}^\nu \to \mathbf{N}^\nu$ as follows, [20,21]

(a) If $\mathbf{i} = (\cdots, i, j, i, \cdots)$ and $\mathbf{i}' = (\cdots, j, i, j, \cdots)$ with $|i-j| = 1$, then

$$R_{\mathbf{i}}^{\mathbf{i}'}(\cdots, a, b, c, \cdots) = (\cdots, b + c - \min(a,c), \min(a,c), a + b - \min(a,c), \cdots). \tag{47}$$

(b) If $\mathbf{i} = (\cdots, i, j, \cdots)$ and $\mathbf{i}' = (\cdots, j, i, \cdots)$ with $|i-j| > 1$, then

$$R_{\mathbf{i}}^{\mathbf{i}'}(\cdots, a, b, \cdots) = (\cdots, b, a, \cdots). \tag{48}$$

For any two reduced decompositions $\mathbf{i}'$ and $\mathbf{i}$, there is a sequence $\mathbf{i}_1, \mathbf{i}_2, \cdots, \mathbf{i}_p$ such that $\mathbf{i}_1 = \mathbf{i}$, $\mathbf{i}_p = \mathbf{i}'$ and any consecutive pair satisfy the conditions (a) or (b), then

$$R_{\mathbf{i}}^{\mathbf{i}'} = R_{\mathbf{i}_{p-1}}^{\mathbf{i}_p} \cdots R_{\mathbf{i}_2}^{\mathbf{i}_3} R_{\mathbf{i}_1}^{\mathbf{i}_2}. \tag{49}$$

Note that $R_{\mathbf{i}}^{\mathbf{i}'}$ is independent of the sequence $\mathbf{i}_1, \mathbf{i}_2, \cdots, \mathbf{i}_p$. This fact plays essential role in our construction.

The action $R_{\mathbf{i}}^{\mathbf{i}'}$ is a ultra discretization of the following $\bar{R}_{\mathbf{i}}^{\mathbf{i}'}$:

(a) If $\mathbf{i} = (\cdots, i, j, i, \cdots)$ and $\mathbf{i}' = (\cdots, j, i, j, \cdots)$ with $|i-j| = 1$, then

$$\bar{R}_{\mathbf{i}}^{\mathbf{i}'}(\cdots, a, b, c, \cdots) = (\cdots, \frac{bc}{a+c}, a + c, \frac{ab}{a+c}, \cdots). \tag{50}$$

---

[a] I would like to thank M. Okado for pointing this fact to me.

[b] Lusztig considered the case of longest element.

(b) If $\mathbf{i} = (\cdots, i, j, \cdots)$ and $\mathbf{i}' = (\cdots, j, i, \cdots)$ with $|i-j| > 1$, then

$$\bar{R}_{\mathbf{i}}^{\mathbf{i}'}(\cdots, a, b, \cdots) = (\cdots, b, a, \cdots). \tag{51}$$

One can interpret this in matrix form as follows. [22] Let $E_i(x)$ $(i = 1, \cdots, n-1)$ be a matrix

$$E_i(x) = 1 + xE_{i,i+1} = \exp(xE_{i,i+1}), \tag{52}$$

where E_{ij} is the matrix unit. Then we have
(a) for $|i-j| = 1$,

$$E_i(a)E_j(b)E_i(c) = E_j(\frac{bc}{a+c})E_i(a+c)E_j(\frac{ab}{a+c}), \tag{53}$$

(b) for $|i-j| > 1$,

$$E_i(a)E_j(b) = E_j(b)E_i(a). \tag{54}$$

That is, for two sequences $\mathbf{i}$ and $\mathbf{i}'$, we have

$$E_{i_1}(x_1)\cdots E_{i_k}(x_\nu) = E_{i'_1}(x'_1)\cdots E_{i'_k}(x'_\nu), \tag{55}$$

if the parameters are related as

$$\bar{R}_{\mathbf{i}}^{\mathbf{i}'}(x_1, \cdots, x_\nu) = (x'_1, \cdots, x'_\nu). \tag{56}$$

Explicit formulas for $R_{\mathbf{i}}^{\mathbf{i}'}$, $\bar{R}_{\mathbf{i}}^{\mathbf{i}'}$ and their application for the totally positive matrices were obtained. [1]

For $a \leq b$ and $x = (x_a, \cdots, x_b)$, define

$$Z_{a,b}(x) = E_b(x_b)\cdots E_{a+1}(x_{a+1})E_a(x_a) = 1 + \sum_{i=a}^{b} x_i E_{i,i+1}. \tag{57}$$

Proposition 4.1 *We have an exchange algebra*

$$Z_{a,b-1}(y)Z_{a,b}(x) = Z_{a,b}(x')Z_{a+1,b}(y'), \tag{58}$$

where

$$x'_i = \frac{P_i}{P_{i-1}}, \quad y'_i = x_i y_{i-1}\frac{P_{i-2}}{P_{i-1}}, \quad P_j = \sum_{k=a}^{b}\left(\prod_{i=a}^{k-1} x_i \prod_{i=k}^{b} y_i\right). \tag{59}$$

These parameters x_i, y_i, x'_i, and y'_i satisfy the following relations:

$$x'_i y'_{i+1} = x_{i+1}y_i, \quad x'_i + y'_i = x_i + y_i, \tag{60}$$

for $i = a, \cdots, b$ $(y'_a = y_b = 0)$. This map $(x, y) \to (x', y')$ is a finite (i.e. non periodic) analog of our R operator in section 2.

5 Discrete Toda equation

A relation between the birational and/or combinatorial R and non-autonomous discrete KP equation was discovered by Hatayama *et al.* [6] Here, we consider the relation between the map R and the discrete Toda equation. [12]

$$\begin{aligned} I_i^{t+1} V_i^{t+1} &= I_{i+1}^t V_i^t, \\ I_i^{t+1} + V_{i-1}^{t+1} &= I_i^t + V_i^t. \end{aligned} \tag{61}$$

Here $i \in \mathbf{Z}$ and $t \in \mathbf{Z}$ are discrete coordinates of space and time and I_i^t and V_i^t are the dependent variables. We will consider the case with periodic boundary conditions $I_{i+N}^t = I_i^t$ and $V_{i+N}^t = V_i^t$. If we put

$$x_i = I_{i+1}^t, \quad y_i = V_i^t, \quad x_i' = V_i^{t+1}, \quad y_i' = I_i^{t+1}, \tag{62}$$

we have the relation Eq. (10). Hence, the time evolution of the periodic discrete Toda system is essentially given by the map R. Explicitly

$$I_i^{t+1} = I_{i+1}^t \frac{G_{i+1}^t}{G_i^t}, \quad V_i^{t+1} = V_i^t \frac{G_i^t}{G_{i+1}^t}, \tag{63}$$

$$G_j^t = \sum_{k=1}^{N} \left(\prod_{i=1}^{k-1} I_{i+j}^t \prod_{j=k}^{N-1} V_{i+j}^t \right). \tag{64}$$

This system can be ultra discretized as follows:

$$I_i^{t+1} = I_{i+1}^t + G_{i+1}^t - G_i^t, \quad V_i^{t+1} = V_i^t + G_i^t - G_{i+1}^t, \tag{65}$$

$$G_j^t = \min_{1 \le k \le N} \left(\sum_{i=1}^{k-1} I_{i+j}^t + \sum_{j=k}^{N-1} V_{i+j}^t \right). \tag{66}$$

It is known that the system Eq. (61) can be written as, [12,13]

$$L^{t+1} R^{t+1} = R^t L^t, \tag{67}$$

where

$$L^t = \begin{pmatrix} 1 & & & & V_N^t \\ V_1^t & 1 & & & \\ & V_2^t & 1 & & \\ & & \ddots & \ddots & \\ & & & V_{N-1}^t & 1 \end{pmatrix}, \quad R^t = \begin{pmatrix} I_1^t & 1 & & & \\ & I_2^t & 1 & & \\ & & \ddots & \ddots & \\ & & & I_{N-1}^t & 1 \\ 1 & & & & I_N^t \end{pmatrix}. \tag{68}$$

Proposition 5.1 *The polynomials* $P_k = P_k(I^t, V^t)$ *defined by*

$$\sum_{k=0}^{N} P_k z^{N-k} = \det(R^t L^t + z1), \tag{69}$$

are conserved quantities of the discrete Toda equation Eq. (61).

One may choose $\mathrm{tr}\,(R(t)L(t))^k$ as another complete set of conserved quantities. [12] The choice here is convenient for ultra discretization.

Using the determinant formulas in Appendix, the explicit form of the conserved quantities P_k can be written as follows:

$$\begin{aligned} P_k &= \sum_{m=0}^{k} \left(\sum_{\{i\},\{j\}} I^t_{i_1} \cdots I^t_{i_{k-m}} V^t_{j_1} \cdots V^t_{j_m} \right), \quad (k = 0, \cdots, N-1), \\ P_N &= (1 + I^t_1 \cdots I^t_{i_N})(1 + V^t_1 \cdots V^t_N), \end{aligned} \tag{70}$$

where the sum $\sum_{\{i\},\{j\}}$ is taken over all the indices $1 \le i_1 < \cdots < i_{k-m} \le N$ and $1 \le j_1 < \cdots < j_m \le N$ such that $i_a \neq j_b, j_b + 1$ for all a, b. The ultra discrete analog of these conserved quantities coincide with the result by Nagai. [23]

Remark 5.2 *A non-commutative analog of discrete Toda equation has been studied by Kashaev-Reshetkhin,* [15] *where the 2-diagonal matrices like* $R(t), L(t)$ *appear as the "minimal" representation of* $U_q(\hat{sl}_N)$.

Acknowledgments

The author would like to thank K. Hasegawa, G. Hatayama, R. M. Kashaev, K. Kajiwara, A. Kirillov, A. Kuniba, A. Lascoux, M. Noumi, M. Okado, T. Takagi and T. Tokihiro for valuable discussions and comments.

Appendix

We give determinant formulas used in the text. Let $X = (x^i_j)_{1\le i,j\le N}$ be a $N \times N$ matrix. For index sets $I = (1 \le i_1 < \cdots < i_m \le N)$ and $J = (1 \le j_1 < \cdots < j_m \le N)$ we define its $m \times m$ minor determinant $\xi^I_J(X)$ as

$$\xi^I_J(X) = \det(x^{i_a}_{j_b})_{1\le a,b\le m}. \tag{71}$$

The basic formula is

$$\xi^I_J(XY) = \sum_K \xi^I_K(X) \xi^K_J(Y), \tag{72}$$

where X, Y are $N \times N$ matrix and $K = (1 \leq k_1 < \cdots < k_m \leq N)$.

Another useful formula is

$$\det(X + 1z) = \sum_L z^{N-m} \xi_L^L(X), \tag{73}$$

where $L = (1 \leq l_1 < \cdots < l_m \leq N)$ and $0 \leq m \leq N$.

Lemma 5.3 *If the matrix Y is of '2-diagonal' form*

$$Y = \begin{pmatrix} y_1^1 & y_2^1 & & & \\ & y_2^2 & y_3^2 & & \\ & & \ddots & \ddots & \\ & & & y_{N-1}^{N-1} & y_N^{N-1} \\ & & & & y_N^N \end{pmatrix}, \tag{74}$$

the formula Eq. (72) can be written as

$$\xi_J^I(XY) = \sum_K \xi_K^I(X) y_{j_1}^{k_1} \cdots y_{j_m}^{k_m}. \tag{75}$$

This simple formula is a key of our explicit computation. In particular, any minor determinant of a matrix which is a product of 2-diagonal matrices can be expressed as sum of monomials with nonnegative coefficients.

The most fundamental object in soliton theory, the τ function, is difficult to handle in ultra discrete formalism, since it is a determinant (or Pfaffian) with alternating signs and in the ultra discretization, terms with negative coefficient are problematic. As we have seen in the main text, the above treatment of the determinant is useful to overcome this difficulty.

References

1. A. Berenstein, S. Fomin and A. Zelevinsky, *Parametrizations of canonical bases and totally positive matrices*, Adv. Math. **122**(1996) 49–149.
2. A. Berenstein and A. N. Kirillov, *Domino tableaux, Schützenberger involution, and the symmetric group action* q-alg/9709010.
3. A. Berenstein and A.N. Kirillov, *Groups generated by involutions, Gelfand-Tsetlin patterns and combinatorics of Young tableaux* Algebra i Analiz, **7** (1995) 92–152; translation in St. Petersburg Math. J. **7** (1996) 77–127.
4. K. Fukuda, M. Okado and Y. Yamada, *Energy functions in box ball systems*, Int. Mod. Phys. **A15** (2000) 1379–1392.

5. E. Frenkel and N. Reshetikhin, *The q-characters of representations of quantum affine algebras and deformed W-algebras*, Cont. Math. **248** (1999) 163–206.
6. G. Hatayama, K. Hikami, R. Inoue, A. Kuniba, T. Takagi and T. Tokihiro, *The* $A_M^{(1)}$ *automata related to crystals of symmetric tensors*, q-alg/9912209.
7. G. Hatayama, A. Kuniba M. Okado and T. Takagi, *Combinatorial R matrices for a family of crystals :* $C_n^{(1)}$ *and* $A_{2n-1}^{(2)}$ *cases*, math/9909068.
8. G. Hatayama, A. Kuniba M. Okado, T. Takagi and Y. Yamada, *Scattering rules in soliton cellular automata associated with crystal bases*, q-alg/0007175.
9. G. Hatayama, A. Kuniba and T. Takagi, *Soliton cellular automata associated with crystal bases*, solv-int/9907020.
10. G. Hatayama, A. Kuniba and T. Takagi, *Factorization of combinatorial R matrices and associated cellular automata*, math/0003161.
11. K. Hikami, R. Inoue and Y. Komori, *Crystallization of the Bogoyavlensky lattice*, J. Phys. Soc. Jpn. 68 (1999) 2234–2240.
12. R. Hirota and S. Tsujimoto, *Conserved quantities of a class of nonlinear difference-difference equations*, J. Phys. Soc. Japan **64** (1995), 3125–3127.
13. R. Hirota, S. Tsujimoto and T. Imai, *Difference scheme of soliton equations* in *Future directions of nonlinear dynamics in physics and biological systems*, ed. P. L. Christiansen, *et al*, (Plenum, New York, 1993) p.7.
14. S-J. Kang, M. Kashiwara, K. C. Misra, T. Miwa, T. Nakashima and A. Nakayashiki, *Affine crystals and vertex models*, Int. J. Mod. Phys. A **7** (suppl. 1A), 449-484 (1992).
15. R. M. Kashaev and N. Reshetikhin, *Affine Toda field theory as a 3-dimensional integrable system* Commun. Math. Phys. **188** (1997) 251–266.
16. M. Kashiwara, *On crystal bases of the q-analogue of universal enveloping algebras*, Duke Math. J. **63**, 465-516 (1991).
17. M. Kashiwara, *Crystal bases of modified quantized enveloping algebra*, Duke Math. J. **73**, 383-413 (1994).
18. A. N. Kirillov, *Discrete Hirota equations, Schützenberger involution and rational representations of extended affine symmetric group*, Talk presented at the workshop at Okinawa (Dec. 1998) organized by M. Okado *et al.*
19. A. Lascoux, M.-P. Schützenberger, *Le monoide plaxique* in *Noncommutative structures in algebra and geometric combinatorics*, (Naples, 1978), 129–156, Quad. "Ricerca Sci.", 109, CNR, Rome, 1981.

20. G. Lusztig, *Canonical bases arising from quantized enveloping algebras*, J. Amer. Math. Soc. **3** (1990) 447–498.
21. G. Lusztig, *Canonical bases arising from quantized enveloping algebras II*, Progr. Theor. Phys. Suppl. **102** (1990) 175–201.
22. G. Lusztig, *Total positivity in reductive groups*, in *Lie theory and geometry: In honor of Bertram Kostant*, Progress in Mathematics, Vol.123, Birkhäuser, Basel, 1994.
23. A. Nagai, Talk presented at the meeting of Japan Math. Soc. (Hiroshima, 1999).
24. A. Nakayashiki and Y. Yamada, *Kostka polynomials and energy functions in solvable lattice models*, Selecta Mathematica, New Ser. **3**, 547-599 (1997).
25. M. Noumi and Y. Yamada, *Affine Weyl groups, discrete dynamical systems and Painlevé equations*, Comm. Math. Phys. **199** (1998) 281–295.
26. M. Noumi and Y. Yamada, *Birational Weyl group actions arising from nilpotent Poisson algebras*, preprint (2000).
27. D. Takahashi, *On some soliton systems defined by using boxes and balls*, Proceedings of the International Symposium on Nonlinear Theory and Its Applications (NOLTA '93), (1993) 555–558.
28. D. Takahashi and J. Satsuma, *A soliton cellular automaton*, J. Phys. Soc. Jpn. **59** (1990) 3514–3519.
29. T. Tokihiro, D. Takahashi, J. Matsukidaira and J. Satsuma, *From soliton equations to integrable cellular automata through a limiting procedure*, Phys. Rev. Lett. **76** (1996) 3247–3250.
30. T. Tokihiro, A. Nagai and J. Satsuma, *Proof of solitonical nature of box and ball systems by means of inverse ultra-discretization*, preprint (1999).